Scientist of empire

Sir Roderick Impey Murchison, Bart, KCB (1792–1871)

Scientist of empire

Sir Roderick Murchison, scientific exploration and Victorian imperialism

ROBERT A. STAFFORD

Department of History, La Trobe University

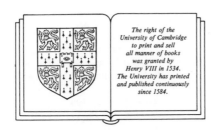

The right of the
University of Cambridge
to print and sell
all manner of books
was granted by
Henry VIII in 1534.
The University has printed
and published continuously
since 1584.

CAMBRIDGE UNIVERSITY PRESS

CAMBRIDGE

NEW YORK   PORT CHESTER

MELBOURNE   SYDNEY

Published by the Press Syndicate of the University of Cambridge
The Pitt Building, Trumpington Street, Cambridge CB2 1RP
40 West 20th Street, New York, NY 10011, USA
10 Stamford Road, Oakleigh, Melbourne 3166, Australia

First published 1989

Printed in Great Britain at the University Press, Cambridge

British Library cataloguing in publication data
Stafford, Robert A.
Scientist of empire: Sir Roderick Murchison, scientific exploration and
Victorian imperialism
1. Commonwealth countries. Geological features
I. Title
551'.09171'241081

National Library of Australia cataloguing in publication data
Stafford, Robert A.
Scientist of empire: Sir Roderick Murchison, scientific exploration and
Victorian imperialism.
Bibliography.
Includes index.
ISBN 0 521 33537 X.
1. Murchison, Sir Roderick Impey, 1792–1871.
2. Geologists – Great Britain – Biography.
I. Title.
550'.92'4

Library of Congress cataloguing in publication data
Stafford, Robert A.
Scientist of empire: Sir Roderick Murchison, scientific exploration and
Victorian imperialism/Robert A. Stafford.
p.   cm.
Bibliography.
Includes index.
ISBN 0-521-33537-X
1. Murchison, Roderick Impey, Sir, 1792–1871.
2. Geologists – Great Britain – Biography.
I. Title.
QE22.M8S82   1989
550'.92'4 – dc 19   88-34134 CIP
[B]

ISBN 0 521 33537 X

CE

141124

For Sally

CONTENTS

ILLUSTRATIONS

PREFACE

Many people and institutions have contributed to the production of this book. Special thanks are due my parents, Carol and James Stafford, for without their confidence and generous financial support this work could not have been accomplished. Aldon Bell of the University of Washington first inspired my interest in British imperial history and encouraged me to pursue post-graduate study in this field at the University of Oxford. Under the detailed supervision of Colin Newbury and the general direction of Ronald Robinson, this work took shape in the stimulating atmosphere provided by Balliol College and Oxford's venerable Imperial and Commonwealth History Seminar. A post-doctoral research fellowship in the Department of History and Philosophy of Science at the University of Melbourne allowed me to refine the text in the company of colleagues whose viewpoints helped me appreciate the meaning of imperial science from the peripheral as well as the metropolitan end of the equation. The opportunity to live in the beautiful Dandenong Ranges outside Melbourne has also given me the pleasure of revising my work on the King of Siluria's overseas activities in an area where memories of the gold-rush he claimed to have predicted are still very much alive, and where forests of tree ferns reminiscent of the flora of the upper Palaeozoic era repose, appropriately, on Devonian rocks surrounded by good Silurian strata (see plate 1). The award of a research fellowship in the Department of History at La Trobe University has permitted me to continue my research regarding imperial science.

For criticism of the text in its various incarnations I wish to thank Colin Newbury, Ronald Robinson, and Jack Morrell at the English end, and Roderick Home, Homer Le Grand, and Neil Archbold in Australia. Invaluable advice and aid of various kinds have also been provided during the research and writing of this book by Andrew Goudie, Margaret Gowing, Paul McCartney, James Secord, and John Thackray. I wish to express my particular thanks to Christine Kelly, Archivist, and David Wileman, Librarian, for assistance in locating documents at the Royal Geographical Society. Allen Howell, Conservationist, Mrs E. R. Nutt, former Librarian, and John Thackray, Honorary Archivist, were equally courteous and helpful at the Geological Society of London. The staff of the British Geological Survey Library and Archive, which was housed in the Geological Museum, London, when I used it, also deserve thanks. Ann Pottage of the Graphics Design Studio, Centre for the Study of Higher Education at the University of Melbourne, provided crucial assistance in preparing the illustrations.

For permission to use and quote from unpublished manuscripts in their care or ownership, I am grateful to the American Philosophical Society; British Association for the Advancement of Science; British Geological Survey Archives; British Library; Bodleian Library, Oxford; the Syndics of Cambridge University Library; the Earl of Clarendon, owner of the Clarendon Papers deposited in the Bodleian Library, Oxford; the Hon. David Lytton Cobbold, owner of the Lytton Papers deposited in the Hertfordshire Record Office; Edinburgh University Library; Geological Society of London; Hocken Library, Dunedin; Imperial College, London (Archives); India Office Library and Records; McGill University Library, Montreal; Mitchell and Dixon Libraries, State Library of New South Wales, Sydney; John Murray, Publisher, London; National Archives Depot, Pietermaritzburg, South Africa; National Archives of Zimbabwe; National Library of Scotland; National Museum of New Zealand, Wellington; National Museum of Wales; Public Record Office, Kew; the Trustees, Rhodes House, Oxford; Royal Botanic Gardens, Kew; Royal Commission on Historical Manuscripts, London; Royal Commonwealth Society Library, London; Royal Engineers Museum, Chatham; Royal Geographical Society; Royal Society of London; St Andrews University Library; Charles J. Sawyer, Bookseller, London; Scott Polar Research Institute, Cambridge; Staffordshire Record Office, Stafford; the Masters and Fellows of Trinity College, Cambridge; Alexander Turnbull Library, Wellington; University of Cape Town Library; the Library of University College, London; W. J. Kennedy, Curator of the Geological Collections, University Museum, Oxford; University of the Witwatersrand Library; Beinecke Rare Book and Manuscript Library, Yale University.

For permission to use small portions of text previously published in other essays (Stafford, 1988a, 1988b), I wish to thank, respectively, the Press Syndicate of the University of Cambridge, and Taylor and Francis Ltd.

Finally, and most especially, I owe my wife Sally an immense debt for her affection, patience, and incisive editorial criticism throughout an endeavour which has spanned six years and taken us to three continents.

Introduction

In the formal and informal empires of Victorian Britain, natural scientists played a primary role in reconnoitring natural resources, stimulating their exploitation, and advising government policy makers on related issues. Geology and botany led the other sciences in support received and research accomplished because of the immediate economic usefulness of their results. In the second quarter of the nineteenth century the Geological Survey of Great Britain was established and the Royal Botanic Gardens at Kew transformed into a state institution, to direct official activity in these disciplines and function as centralised information repositories for the empire. Geography found quasi-official expression in the Royal Geographical Society. The initiative for founding these institutions came from London scientists. Because of the nation's expanding overseas interests and the precedent for official patronage provided by the Royal Society, metropolitan savants realised the opportunities for research abroad which an alliance with government could provide. They were able to realise their ambitions because of their skill at organised lobbying, for by this period Britain's governing elite had been imbued with enthusiasm for the goals of science and belief in its power to promote economic progress. Still, in an era of retrenchment and reform, public science had to perform to earn its keep. It did so by providing the nation with information of utilitarian value as well as cultural capital worthy of Britain's status as a great imperial power.[1]

If imperialism is defined according to the formula proposed by David Fieldhouse – 'the deliberate act or advocacy of extending or maintaining a state's direct or indirect political control over any other inhabited territory'[2] – a study of Sir Roderick Murchison's career offers important insights into several facets of this phenomenon in mid-nineteenth-century Britain. If the connotation of the term is widened to include the exportation of metropolitan culture to the periphery, Murchison assumes an even larger historical significance. The interlocking relationships between his high office in several scientific societies, his official position as Director-General of the Geological Survey, and his vast social and political influence made him a crucial connection between the decentralised structure of British science and the imperial government. From this pivotal position, previously occupied by Sir John Banks and less authoritatively by his successor Sir John Barrow, second Secretary of the Admiralty,[3] Murchison manoeuvred to institutionalise natural science as an integral component of both imperial administration and foreign policy. By promoting exploration, resource reconnaissance, commercial expansion, and imperial development and security, he stimulated systematic

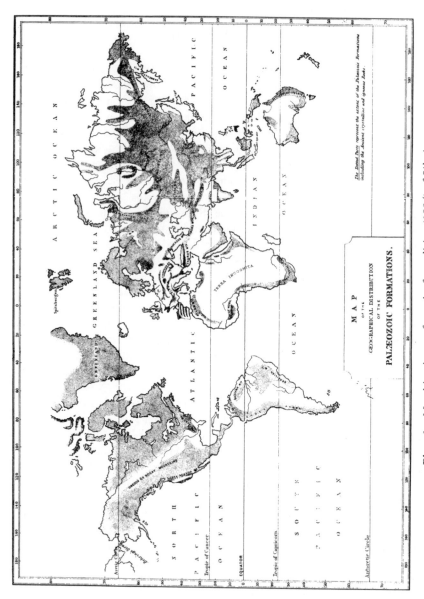

Plate 1 Murchison's map from the first edition (1854) of *Siluria*

exploitation of the empire and the entire periphery. By mediating the resultant flow of information reaching the home audience, he also played a significant role in defining metropolitan perceptions of this zone.

The standard biography of Murchison is now over a century old. As a work Murchison commissioned his protégé Archibald Geikie to compose, it hardly represents a critical treatment.[4] The results of extensive research regarding specific aspects of Murchison's geological work have recently appeared,[5] and analysis has begun of the interplay between science and empire in his career.[6] In what follows I develop these themes and present a comprehensive assessment of Murchison's imperial activities throughout their chronological and geographical range. While addressing the general subject of science and empire, my account also provides partial answers to more specific questions.

I endeavour to clarify the relationship between the growing demand of Britain's governing and commercial elites for accurate data about the extra-European periphery, and the simultaneous search for increased status and authority by scientists willing to provide such information. I seek to demonstrate how colonial data and career opportunities influenced the development of British scientific disciplines, and thus how imperialism formed part of the social matrix in which nineteenth-century British science was embedded. I attempt to elucidate how British scientists exported their disciplines to the gigantic laboratory of the colonies, where they throve as resource intelligence services and motors of economic development. This study contributes to an understanding of when, where, and why key sectors of colonial infrastructure such as railways, mines, and port facilities were installed. It illuminates how the official mind of imperialism procured the facts upon which it based its decisions, how its priorities could be manipulated by interest groups or even individuals, and how it often proceeded to its goals through unofficial channels such as the metropolitan scientific societies. Finally, it helps explicate the way in which public reaction to imperial adventures could be orchestrated through the medium of organisations such as the Royal Geographical Society and how the debate among British savants about the appropriate function and style of science in the imperial context played itself out on this institutional stage.

The first chapter is biographical, emphasising those factors in Murchison's background, personality, and career which predisposed him to a keen interest in and personal identification with the British empire. In the succeeding six chapters I present regional studies of Murchison's efforts to elaborate a symbiotic relationship between science and the forces of imperial development and national expansion. In the conclusion, I evaluate the evidence set forth according to the hypotheses advanced in Chapter 1 about Murchison's significance and the larger issues of science, empire, and public opinion illustrated by his career.

I

The King of Siluria

Where, where was Roderick then!
One blast upon his bugle horn
Were worth a thousand men!
Walter Scott, *The Lady of the Lake*

Roderick Murchison's heritage rendered him as much a child of empire as a son of Scotland. An old family of Highland landowners, the Murchisons had suffered as a result of their loyalty to the Stuart cause during the rebellions of 1715 and 1745. Roderick's father Kenneth joined the diaspora of Scots into the British empire resulting from the breakdown of the Highland clan system. He rescued his family from decline by amassing a fortune as a surgeon in the service of the East India Company between 1769 and 1786. In the lurid style set by the nabobs of the day, he also managed to kill a man in a duel, father three illegitimate children, and augment his lucrative salary with a reward granted by a native prince for curing his daughter's ailments. When Kenneth retired to Scotland, he purchased the estate of Tarradale in Easter Ross, married, and fathered two sons – Roderick Impey, born in 1792, and Kenneth, two years his junior. The elder Kenneth died in 1796 and his widow soon married Colonel Robert Murray MacGregor, who had been a close friend in India.

As Kenneth Murchison's eldest legitimate son, Roderick inherited a joint interest in empire and applied science, for Kenneth's medical training had been the foundation of his success in India. Roderick's second name expressed this link by honouring Sir Elijah Impey, first Chief Justice of the Supreme Court of Bengal and later the boy's guardian. The career of Roderick's brother Kenneth (1794–1854) would exemplify this imperial heritage even more directly, for he served the East India Company as Governor of Singapore and Resident at Penang in the Straits Settlements.[1]

In 1799 Roderick began his formal education at the Durham Grammar School. Performing poorly in the classroom but excelling at dangerous escapades, he was soon inspired to a military career by his uncle, Lieutenant-General Sir Alexander Mackenzie. In 1805 Murchison entered the Military College of Great Marlow in Buckinghamshire. Here he learned topographical appraisal and military draughtsmanship, exercises which were to contribute significantly to his future success as a geologist. At the same time, Murchison demonstrated a marked deficiency in mathematics, a trait which would also influence his scientific career. In 1807 Murchison was gazetted an Ensign in the 36th Infantry Regiment, of which his uncle was Lieutenant-Colonel. The following year he joined an expeditionary force sent to Spain to support the rising against Napoleon, landing in Portugal at the side of Sir Arthur Wellesley, the future Duke of Wellington. He carried the colours of the 36th in the battle of Vimieira, fought in numerous skirmishes, and, promoted Lieutenant, took part in the retreat to Corunna.[2]

After six months in the field Murchison returned to England with his regiment: he had seen the last of active soldiering. In 1809 he was posted to Sicily as aide-de-camp to his uncle, but Mackenzie's failing health soon brought a return to garrison duty near London. After a few months General Mackenzie was appointed to the military command of northern Ireland, and Murchison joined his staff again at Armagh. The post, however, was a backwater. As the Lieutenant's hopes of rapid advancement faded, he threw himself into riding, hunting, and walking, punctuating these exertions with drinking bouts and visits to London to parade as a dandy. He also purchased a captaincy and ran heavily into debt, borrowing against the value of his Ross-shire patrimony. In 1812 he made an experiment in self-improvement by joining the Royal Institution and attending a series of lectures by the chemist Sir Humphry Davy. The following year Murchison came of age. A visit to Tarradale, during which he supported the local Tory candidate in a parliamentary election, convinced him that he must keep his debt-ridden estate in order to retain some position in his native Highlands. With the signing of the Treaty of Paris in 1814, Murchison's military career seemed finished. He retired on half pay and continued his gay life in London. At the end of that year he visited Paris, and when Napoleon returned from Elba, was forced to beat a hasty retreat to Calais. He exchanged into a cavalry regiment in the hope of participating in the impending campaign, but was once more disappointed.[3]

Murchison then met and married Charlotte Hugonin, the only child of a wealthy general. He resigned his commission and, while considering an improbable career as a clergyman, decided to embark on a tour of the Continent with his intelligent and cultured wife. In part, Charlotte had engineered this foreign excursion to wean Murchison from his idle habits, but the ultimate reason for their departure was a deterioration in finances brought about by a virtual cessation of rents from Tarradale and Murchison's unwillingness to reduce his spending. The couple spent the years 1816 to 1818 touring the scenery and galleries of France, Switzerland, and Italy. Roderick performed great feats of walking in the Alps and began to take an interest in geological formations. Yet the soldier was still apparent in his observations on fortifications, military tactics, and the potentially threatening growth of the United States Navy. Militaristic overtones aside, this interlude developed Murchison's appreciation for landscape features, accurate observation, and steady intellectual labour.[4]

When the Murchisons returned home, Roderick sold the failing estate of Tarradale for £27,000. The income from the investment of this sum and Charlotte's dowry allowed the couple to set up housekeeping in County

Durham, where Murchison fell back on his old habits and became an enthusiastic fox-hunter. He spent five years at this strenuous life and, as at Ross-shire, his friends were predominantly Tory landowners. In the general election of 1820, he endured electioneering violence while canvassing for the unsuccessful government candidate, but his political views remained staunchly conservative.[5] Still living beyond his means, Murchison borrowed from his father-in-law. His financial problems were compounded in 1822 when he moved to Leicestershire and enlarged his establishment in order to gratify further his passion for fox-hunting. Murchison also began speculating in foreign funds on the advice of aristocratic friends. He obtained an annual income of around £2,000, but his rate of expenditure forced him to draw upon his capital. Nor were these investments secure. Murchison lost heavily on a notorious Peruvian mining loan and forfeited some £6,000 hazarded in Spanish government funds in the hope of capitalising upon the revolution of 1820.[6]

In 1823, when a drastic change in his way of life became unavoidable, a new vocation which might combine a love of field sports with somewhat reduced circumstances presented itself to Murchison. Shooting partridges with Sir Humphry Davy, he was inspired to consider moving to London and applying himself to science. Charlotte concurred with this advice as a promising means of developing his husband's nascent intellectual ambitions by means of the inquiries in natural history which she herself found rewarding. The immediate spur to change, however, was again financial, and after living for a year with Charlotte's parents, the couple rented a house in London befitting their battered but still considerable income. At the age of thirty-two, Roderick was free to throw himself with characteristic vigour into the study of science.[7] He began by returning to the Royal Institution. Science had become integral to the cultural life of the capital, and the Royal Institution, as the chief venue for purveying the new knowledge to the landed classes and the rising bourgeoisie, was the logical point of departure for Murchison's initiative at re-education. He attended lectures on chemistry and related subjects by Professor William Brande. In keeping with the utilitarian bent of the Institution, these lectures dwelt largely on matters of practical concern to agriculturalists, manufacturers, and merchants.[8] Brande's comments on geology soon captured Murchison's attention. In 1825 Davy and Brande secured his re-election as a member of the Royal Institution, and in 1828 he became one of its managers, an office held repeatedly in later years.[9] The precedents this organisation established for deploying entrepreneurial science would inspire Murchison's own attempts to win public patronage for science.

In 1825 Davy also secured Murchison's admission as a Fellow of the Geological Society of London, and the aspirant began attending its celebrated debates. The Society's leaders soon convinced Murchison that geology, then the most dynamic and fashionable of the sciences, offered a gentlemanly avenue for achieving intellectual distinction. A link was also established in his mind between geology and exploration when he attended a lecture to indoctrinate the naval officers about to accompany John Franklin's second expedition in search of a North-west Passage.[10] In the same year Murchison had his first field lesson from William Buckland, Reader in Geology at Oxford University. The experience of watching Buckland point out from a commanding height the processes which had transformed the features of the Thames Valley not only reawakened his interest in topography, but impressed him with a Napoleonic image of the geologist.[11]

Thus inspired, Murchison began independent fieldwork, and soon submitted his first paper to the Geological Society. This memoir described the stratigraphy, or chronological sequence and character, of the sedimentary rocks around his father-in-law's Hampshire estate.[12] In 1826 Murchison became an Honorary Secretary of the Geological Society, while his wealth, zeal, and friendship with Davy also secured his election as a Fellow of the Royal Society. Advised by Buckland to work out the stratigraphic age of the Brora coalfield in Sutherlandshire, Scotland, he established these seams to be Oolitic (Jurassic) strata rather than an isolated outcropping of the older Carboniferous-age coal formations found to the south. At the outset of his geological career, Murchison thus accomplished a stratigraphic determination of considerable economic significance, for the Jurassic coals were not then considered worthy of exploitation because of the superior steam-producing characteristics of Carboniferous fuel. Furthermore, he had made his pronouncement solely on fossil evidence, for the lithology, or rock types involved, suggested the alternative conclusion.[13] These two characteristics − reliance on palaeontological dating and interest in economic geology − would distinguish his subsequent work.

During 1827 Murchison returned to Scotland in the company of Adam Sedgwick, Professor of Geology at Cambridge University, and in 1828 he toured the volcanic district of Auvergne in France with Charles Lyell.[14] Murchison gloried in the physical exercise of fieldwork. He conducted the French tour like a forced march, striding over the countryside at such a relentless pace that he walked the legs off the younger Lyell.[15] At this time Murchison came under the temporary influence of Lyell's theory of gradual geological change, but subsequent work undertaken with the catastrophist Sedgwick convinced him to alter his views of geological dynamics in favour of greater intensity of forces in the past.[16]

In 1828 Murchison was elected to a two-year term as Foreign Secretary of the Geological Society. As the recipient in this capacity of a stream of reports and specimens forwarded by Britons abroad, he acquired direct knowledge of untrodden fields of research in other continents. The Society had long recommended the prosecution of geological inquiries overseas, and Murchison now appreciated the usefulness of this far-flung network of observers providing grist for the mill of metropolitan science. He also established close social and working relationships with all of the great contemporary European geologists

which helped him 'broaden his hold upon the general scientific activity of his
time'.[17]

Murchison's initial burst of scientific energy resulted in his election to the
presidency of the Geological Society in 1831. The same year he turned his
attention to the problem of establishing a conformable, or unbroken,
stratigraphic sequence downward from the previously classified Secondary
(Mesozoic) series in the west of England into the undifferentiated 'Transition'
or 'Grauwacke' rocks underlying them in Wales. Murchison did not at first
envision this as a major research programme, but rather as a minor exercise to
clear up unresolved boundary problems.[18] Yet in a situation frequently
paralleled in imperial history, what began as a frontier skirmish soon
developed into a full-scale campaign of pacification and annexation, for the
'invasion of Grauwacke'[19] led Murchison to the stratigraphic key necessary to
relate the ancient and highly distorted strata of Wales and other outlying areas
in the west of Britain to the rocks of the rest of England.

As a result of this important discovery, which in Murchison's retrospective
accounts would be portrayed as the outcome of a deliberate search rather than a
stroke of unanticipated fortune, the ambitious geologist focused his attention
on the older rocks. For six more seasons he returned to Wales to work out the
succession of what came to be termed the Palaeozoic rocks – that is, the ancient
fossil-bearing sedimentary strata sandwiched between the overlying Old Red
Sandstone (Devonian) formations and the underlying basement of igneous
rocks. While Murchison hammered at the rocks in South Wales and the Welsh
Borders, his companion Sedgwick toiled amidst the even older and more
tortuous strata of North Wales. Their labours were initially complementary,
but they eventually fell out over differing interpretations of the two stratigraphic divisions defined – Murchison's Silurian System and Sedgwick's
underlying Cambrian System. What had been a famous friendship and a
fruitful scientific partnership was to devolve into one of the bitterest
controversies in the history of science.[20] Murchison's work in Wales was
greatly aided by the availability of Ordnance Survey maps which permitted the
tracing of rock strata in the field. This experience convinced Murchison that
accurate topographical maps were not only of fundamental economic and
administrative value, but that they were crucial to a whole range of scientific
endeavours.

For fifteen years, Murchison laboured in the field of Palaeozoic stratigraphy,
building his scientific reputation with an impressive record of research and a
relentless campaign of self-promotion. Murchison was an excellent observer
with almost incredible physical stamina, a voracious capacity for correlating
detail, and an 'eye for country' developed by his military and hunting
background. His scientific style, based on the conviction that fossils were
superior to lithological characteristics as indices of age and thus as criteria for
long-range stratigraphic correlation, contrasted with that of such other
geologists as Sedgwick who relied on the mineralogical features and geometry

of rock masses to interpret their chronological relationships.[21] Murchison also sought to explain past geological events according to current or actual processes whenever the evidence permitted, but like many contemporaries he rejected Lyellian uniformitarianism. While admitting the efficacy of agents of gradual change, Murchison had no hesitation in invoking catastrophic outbursts of these same physical forces in former ages to account for the more dramatic transformations which had occurred in the earth's crust and the record of life. An ardent believer in organic progression, Murchison, like many other progressionists, rejected Darwin's theory of evolution.[22]

Murchison believed, as did the founders of the Geological Society, that the inductive accumulation of data as a preliminary to theory construction represented a superior methodology to the postulation of theories from which derivative phenomena might be deduced.[23] This approach was expressive of the new style of descriptive science propagated by Murchison's hero Alexander von Humboldt. The great Prussian polymath had given a huge impetus to the collection, graphic display, and correlation of precise physical data of all sorts.[24] The new science of geology, especially, was developing its own 'visual language' to articulate the wide range of three-dimensional phenomena with which it dealt.[25] Murchison also followed Humboldt in forming the axiom that 'the geologist is but the physical geographer of former periods', and his own work combined these two spatial disciplines.[26] As a promoter of geographical exploration he would attempt to extend the horizontal axis of earth science, while as a geologist he plumbed the vertical axis, seeking explanations for the changes which the features of the globe had undergone over time.

Murchison rose rapidly to prominence in the world of London science. He was elected to a second term as President of the Geological Society in 1842, and served as a constant member of its executive Council between 1826 and 1869. He also sat on the Council of the Royal Society for many years and served as its Vice-President in 1849. In the course of his career he became a Trustee of the British Museum and of the Royal College of Surgeons' Hunterian Museum, the recipient of honorary degrees from four British universities, a fellow of numerous foreign academies, and a founding member of the Athenaeum Club. Murchison was truly 'a man born to fill chairs',[27] and his institutional affiliations, like his widening network of scientific and social contacts, served to increase his influence in public affairs. Aside from the Geological Society, however, he associated himself in particular with two scientific organisations.

In 1830 Murchison became the youngest member of the small band of enthusiasts led by Sir John Barrow which founded the London Geographical Society in order to further scientific exploration overseas. Similar societies had been established in Paris in 1821 and Berlin in 1828. The Geographical Society was one of several learned societies formed in London during this era to organise specific fields of knowledge while providing rational amusement for the educated classes. It grew out of a nucleus of interest represented by the

African Association, established by Sir Joseph Banks in 1788 to promote exploration in Africa and other distant regions, and the Raleigh Club, which had been formed by a secession of serious explorers from the Travellers Club in 1826.[28] Murchison had joined the Raleigh in 1829, and on his advice the new geographical entity was organised on the model of the Geological Society.[29] The founders of the Royal Geographical Society (hereafter referred to as the RGS), as it was soon styled, included politicians, diplomats, military officers, scientists, linguists, and antiquaries. Many geologists joined the RGS, and its early membership also comprised cartographers, surveyors, instrument makers, engineers, enthnographers, orientalists, and merchants, as well as many fashionable dilettantes.[30] At its inception, the geographers made clear the relationship of mutual service which they hoped would prevail between their organisation and their country's imperial establishment, stating that the RGS would prove of the utmost use 'to the welfare of a maritime nation like Great Britain, with its numerous and extensive foreign possessions'.[31] In this same spirit they elected as their first President Viscount Goderich, Secretary of State for War and the Colonies in Grey's Reform Ministry. Murchison joined the Society's Council in 1831, served as Vice-President in 1836, and was elected President for the first time in 1843–44, but his time was essentially occupied with geology until the 1850s.

In 1931 Murchison also helped found the British Association for the Advancement of Science, which sought to organise research, lobby for government funding, and promote intercourse between scientists throughout the empire. Again, the leaders of the Geological and Geographical Societies were prominent among the initiators of this public resource. Realising its potential for mobilising a broader base of national support for science than the socially exclusive and London-focused Royal Society, Murchison became one of the Association's staunchest supporters. He served continuously as a Trustee until 1870, as General Secretary from 1836 to 1845, as President in 1846, and as a sectional President or Vice-President on many occasions. During the decade of his secretaryship Murchison largely directed the activities of the British Association. By actively developing its relationship with government, promoting social display at its annual meetings in provincial cities, and recruiting powerful peers as presidents, he increased the political leverage of science while enhancing his own status.[32]

Murchison's sustained programme of research culminated in 1839 in the publication of *The Silurian System*.[33] This *magnum opus* earned its author worldwide renown for establishing from the apparently chaotic Palaeozoic succession of the Welsh Borders the oldest fossil-bearing stratigraphic classification then defined. Having been warned that 'No man meets twice in his life with a terra incognita; therefore you must take full & real possession of it',[34] Murchison took this advice with a vengeance. He named his system after the Silures, a British tribe indigenous to the region where he had described the Silurian formations and noteworthy for resisting the Roman invasion. He looked upon the restoration of the ancient name as an act of patriotism, and gloried in identifying himself with Caractacus, the Silures' war leader. While considering Siluria a distinct period in the history of life bounded by a system of characteristic fossils, Murchison simultaneously viewed his creation as a geographical territory, a personal 'kingdom' or 'empire' which extended wherever these organic remains might be found.[35]

At the same time, Murchison realised that his system had important economic implications which might increase the public authority and prestige of geology and its practitioners. Since Silurian fossils were almost entirely marine invertebrates which predated the appearance of terrestrial vegetation, recognition of them would 'reliably indicate a baseline beneath which it was pointless to search for coal'.[36] Mining experience and geological research also suggested that valuable metal ores were most frequently associated with such ancient strata. Murchison reigned jealously for the rest of his life as the world authority on Silurian rocks, supporting propagators of his doctrines and arguing with detractors who sought to limit the extent of his writ either through geological time or across geographical space. His long-running controversy with Adam Sedgwick over the Silurian-Cambrian stratigraphic boundary[37] is merely the most celebrated of innumerable skirmishes, waged like those of the British army on the imperial frontiers, to defend the borders and glory of an ever-expanding realm.

During 1839 and 1840, working in conjunction with Sedgwick, Murchison established the Devonian System to distinguish the Palaeozoic strata which lay between the Silurian and Carboniferous rocks. Again, the term commemorated the region where the system's stratigraphy had first been deciphered, and once more Murchison carried his campaign to the Continent, where he discovered Devonian fossils and convinced influential geologists of the new system's validity.[38] Acceptance of such nomenclature abroad aided the export of British theories about the structure of the earth and the history of life, so that the establishment of the British stratigraphic sequence as an international standard contributed to Britain's hegemony in the wider intellectual sphere. In his active proselytising, Murchison clearly emerges as an agent of cultural imperialism. His role in this regard is most evident in his habitual resort to metaphors of conquest to describe the expansion of his domain,[39] but his biographer also draws a telling parallel between Murchison's scientific strategy and that of the founder of a commercial enterprise promoting his overseas interests.[40] The next phase in Murchison's geological career well illustrates these themes.

In 1840 Murchison learned of the existence of vast stretches of undisturbed and fossiliferous ancient rocks in the Russian Baltic provinces. He hoped by investigating these strata to confirm the stratigraphic sequence established among the highly contorted Palaeozoic districts of Western Europe. Murchison seized upon an invitation to visit this geologically little known area with Baron Alexander von Meyendorf, a Russian official planning a fact-finding tour.[41] He invited the French palaeontologist Edouard de Verneuil to accompany him, and after securing introductions from Baron Brunnow, the Russian Ambassador in London, the two set off for St Petersburg. Here they rendezvoused with Meyendorf as well as welcoming the naturalist Count Alexander von Keyserling to their party, and Murchison explained his goals to Count Cancrin, the Minister of Finance whose responsibilities included mineral development. Cancrin introduced Murchison to General Konstantin Tchevkin, head of the Imperial School of Mines, who detailed the mineralogist Lieutenant Nikola Koksharov to accompany the British geologist as his aide-de-camp.

The party travelled north from the capital, but the scientists soon parted from Meyendorf and pressed on to the White Sea. They searched for tell-tale fossils and vertical exposures of strata, noting the strike (direction of the outcropping edges) and dip (direction and angle of inclination) of the latter. Topographical observation also enabled Murchison to seize upon the salient features of the country's physical structure, so that in mapping its stratigraphy he was employing the methods of both the geographer and the geologist. In the style set by Humboldt, Murchison also collected ethnological and natural history specimens, noted local commercial patterns, and observed the state of the peasantry. The vastness of the Russian empire fascinated him, its potential for economic development fired his enthusiasm for action, and the spectacle of a seemingly well-ordered society in which all classes knew their function appealed to his conservatism.

Near Lake Ladoga Murchison discovered a sequence of strata whose fossils linked the rocks of the Devonian System with those of the underlying Silurian. Farther north he found a similar passage between the Devonian formations and those of the overlying Carboniferous System. After confirming this succession on a southern loop through Moscow, Murchison and Verneuil rushed back to Britain to publicise results which proved the validity of the Devonian System, positively established fossils as the surest indices for stratigraphic dating, and extended Siluria to the shores of the Arctic Ocean.[42] These were remarkable accomplishments for a summer's fieldwork traversing 4,500 miles, and Murchison's triumph was completed by his display of the first accurate geological map of Russia. The Silurian conqueror carried all before him: unlike Napoleon, he returned victorious from his invasion of Russia.

Murchison at this time considered an American campaign in order to establish his Palaeozoic classifications in the New World, but he and Verneuil had made preliminary arrangements to extend their Russian researches east to the Ural Mountains. This fieldwork was primarily intended to annex new territories to Murchison's scientific empire, but it also had important implications for Russian industrial development. Murchison's awareness that Russia seemed geologically unsuited to large-scale coal production suggested that he could enhance his standing with Cancrin and Tsar Nicholas I by determining the stratigraphy of known deposits as a guide to efficient exploration. Providing information on the ores of the Urals might likewise bestow great benefits on Russia's primitive mining industry. At this time Murchison published a popular account of his first Russian tour which praised Nicholas as a model ruler, dwelt on the historic alliance between Britain and Russia, and reproved 'the conquering thundering Palmerston' for his aggressive policy regarding Russia and the Eastern Question. Murchison advocated that British ministers moderate criticism of the pace of Russian reform and encourage Russia to adopt free trade. He believed that coal, 'the motor and the metre of all commercial nations', constituted a critical element in the economic and political equation between the two countries. 'Without it', he warned, 'no modern people can become great, either in manufactures or in the *naval art of war.*' While Britain and her colonies were well endowed with this 'mighty agent', Russia possessed only limited proven deposits located far from major population centres and oceanic trade routes. Russia was already importing most of her coal from Britain, Murchison pointed out, and he argued that she should continue to do so in order to cement a co-operative relationship between the paramount maritime and continental powers.[43]

Murchison's publicity exercise had its desired effect. In 1841 Cancrin invited him and Verneuil to return to continue their Russian research with the aid of Keyserling and Koksharov and to report on the ore and coal reserves of the districts visited. The two geologists travelled once more to St Petersburg, where they met the Tsar and laid their route according to both scientific and economic priorities. It was another programme of breathtaking scope for a single summer, but Murchison was eager to conquer the remainder of European Russia in a single lightning campaign. In this martial mood, Murchison led his 'geological division' towards Moscow, and from there sped eastward by carriage and six-horse team. Ever alert to comparisons between the British and Russian imperial systems, Murchison remarked upon witnessing a string of prisoners departing for Siberia: 'Thank God! in England we have the sea for our high-road to banishment, for such scenes are very harassing.'[44] In the copper-mining province of Perm, the geologists observed vast deposits of red sandstone which would soon lead Murchison to designate his third Palaeozoic system, the Permian, separating Carboniferous strata from younger Triassic rocks at the base of the Secondary (Mesozoic) series.

The party then crossed the Urals to the gold-mining centre of Ekaterinburg, where Murchison marvelled at the efficacy of Russian mining enterprise as a colonising force.[45] Here, as at Zlatoust to the south, he studied the occurrence of gold ores and alluvia, for he was developing a theory to supersede the general postulations of Humboldt and directly link the origin and distribution of the precious metal with rocks of Silurian age.[46] Hampered by a lack of reliable topographical maps, the geologists constructed a new map of the Ural range based on their own discoveries and Russian data.[47] For Murchison, however, the hazards of travel in virtual wilderness proved as exciting as scientific labour. At Troitsk, a caravan terminus which marked the party's furthest penetration eastward, Murchison gained a palpable sense of the mysterious khanates of Central Asia. He also encountered the nomadic Bashkirs, who represented the limit of Russian authority over the peoples of the steppe: beyond lay the unconquered Kirghis tribes. He viewed the Russian advance as a dynamic campaign of conquest and consolidation akin to the European occupation of other peripheral regions. 'The Russ is certainly the best land colonist in the world', he noted. 'By sea we flatter ourselves that we are.'[48] As the Russians pushed inexorably south, Murchison predicted that the vanquished Kirghis and Caucasians, like the Bashkirs and Cossacks before them, would be converted into frontier guards in a process analogous to the expansion of imperial Rome.

From Troitsk, the scientists proceeded south to Orsk and then westward to Orenberg. Here they were entertained by the provincial governor, a military hero who received Murchison as a British general. Orenberg too was a terminus for southern caravan routes, and Murchison, eager to serve British interests, interrogated tribesmen in the bazaar about the dangerous political situation in Afghanistan.[49] The geologists concluded that the Ural range comprised a central ridge of igneous rocks flanked by successive deposits of Silurian, Devonian, and Carboniferous strata. On the western side, this Palaeozoic series was topped by the formations of the newly defined Permian System. True to his catastrophist proclivities, Murchison also decided that the chain marked an immense crustal fracture through which had periodically poured the igneous rocks responsible for uplifting and metamorphosing the sedimentary strata, as well as impregnating them with metallic ores. Murchison's final traverse of the Urals near Orenberg cemented his romantic attachment to the scene of his most adventurous geological exploit. He relished the danger of potential Kirghis attacks and took an officer's pride in the panache of his Bashkir escort. As a fellow scientist remarked with regard to the Russian tour, Murchison had 'reforged the sword of the old Peninsular soldier into the hammer of the geologist'.[50]

Murchison and Verneuil then turned westward towards the Donetz coalfield where they confirmed that the seams were of Carboniferous age and consequently of enormous economic significance. They also pointed out where the government should explore for further deposits and recommended improvements in mining methods.[51] In St Petersburg, Murchison and Verneuil presented their official report to Cancrin, and the Tsar awarded them each a minor decoration. It was also hinted that Nicholas might patronise the book the geologists were planning, and Murchison was granted permission to explore the Siberian Altai range, whose few known fossils and burgeoning gold production suggested that another major development of Silurian strata lay far to the east in the Tsar's dominions.

Murchison then returned to England to present his findings to the Geological Society, of which he had been elected President for the second time. Predictably, his Anniversary Address of 1842 advertised his personal contributions to Palaeozoic stratigraphy as well as the spread of Silurian nomenclature around the globe.[52] Taken together, the two Russian campaigns constituted a stunning success. Murchison and his companions had confirmed the European Palaeozoic succession as far east as the Ural Mountains, defined the Permian System, and contributed substantially to knowledge of the structure of mountains, the occurrence of gold, and the former extension of the Caspian Sea. Murchison's ego expanded to match his Russian conquests: in almost conscious emulation of the Russian autocrat, he henceforth became more authoritarian and intolerant in nature.[53] His adulation of Tsar Nicholas and his thirst for honours provoked widespread ridicule,[54] but the Russian exploits greatly enhanced Murchison's reputation. His financial position also improved as a result of his wife's inheritance, and the grand mansion in which he established himself in London's Belgrave Square was transformed into a leading intellectual salon. The fashionable soirées which he and Charlotte organised served to increase his own influence as well as the social standing of scientists in general.[55]

With the aid of Verneuil and Keyserling, Murchison began composing a new book to enshrine the complete results of the Russian tours. Further excursions to Scandinavia and Central Europe were required to accomplish this goal. While President of the Royal Geographical Society, Murchison at this time also published his new map of the Urals.[56] During 1843 he received the gift of a monumental stone vase from the Tsar. He used this additional distinction to pressure the British government to permit him to wear his Russian order and to adopt a more liberal policy of rewarding scientific achievement.[57] Murchison had already requested a baronetcy from Sir Robert Peel on the grounds of his geological accomplishments and Tory political convictions, and he kept his name before the Prime Minister by providing information on Russian gold production which proved useful in bringing in the Bank Charter Act of 1844.[58] He simultaneously applied pressure on Lord Aberdeen, Foreign Secretary and a fellow Scot, and William Gladstone, President of the Board of Trade and an old family friend.[59] These efforts failed, and Murchison's political allegiance swung away from the Tories. He complained of Gladstone as an 'upstart minister' who ignored all merit 'save aristocratical & parliamentary interest', and sneered at Peel as a man whose concept of science rose no higher than 'turnips, drains, ploughs, & etc.'[60] Murchison's lust for increased prestige was somewhat assuaged, however, by attending the Tsar at Buckingham Palace.[61]

In 1845 *The Geology of Russia in Europe and the Ural Mountains* was published on a lavish scale befitting its imperial patron.[62] Reflecting Murchison's joint preoccupation with geology and geography, it was as much a narrative of exploration and travel as a scientific treatise on stratigraphy, orogeny, faunal extinction, and the formation and distribution of metallic ores. Murchison dedicated the book to Nicholas I, and he and Verneuil returned to St Petersburg to present a copy to the Tsar. Murchison's plans for Siberian research were now abandoned, however, because the Russian naturalist Pëtre Chikhachëv had already revealed the presence of Devonian and other Palaeozoic rocks in the tracts he had hoped to visit.[63] Murchison received the Tsar's thanks for advising his son Konstantin on the founding of an Imperial Geographical Society modelled on the RGS[64] and promised to continue to combat anti-Russian sentiment in England.

He was also awarded the Grand Cross of St Stanislaus, elected to the Imperial Academy of Sciences, and urged to accept an appointment as 'Inspector of Imperial Geological Explorations'. This was enough for Murchison to win permission to wear his Russian orders in Britain.[65] Murchison also discussed with his hosts the Russian campaign to conquer the Caucasus. He agreed that Schamil Beg, the resistance leader of the Caucasian tribes, required to be 'tranquillized'. But in comparing him with Abd-el-Kader, the Algerian chieftain who was proving an obstacle to French imperial ambitions in north Africa, Murchison considered that a wiser policy than straightforward assault might be to grant the Caucasians semi-autonomous status and incorporate them into the Russian army in the manner which had proved successful with the tribesmen of the steppes.[66]

Murchison found upon his return to London that Peel had awarded him a knighthood. Embittered by the tardiness of this honour, however, he had deserted the Tory cause and gone over to the Whigs.[67] Despite the manner in which Murchison had intrigued for the reward, his promotion was seen as an official recognition of the value of science in general. He entertained Grand Duke Konstantin during a Russian naval visit to Britain,[68] and enjoyed the additional honour of presiding over the Southampton meeting of the British Association in 1846. 'Here for the first time', he boasted, 'I was duly canonized'.[69] He surrounded himself with a galaxy of scientific talent and secured the attendance of Palmerston, Prince Albert, and the Duke of Wellington. In his presidential address Murchison dwelt upon the necessity of the government's maintaining its scientific patronage in order to promote progress in the practical arts and sustain Britain's cultural prestige. Murchison's speech was replete with expressions of patriotism, but it also struck an international note echoing his recent experiences on the Continent. The Siberian explorer Alexander von Middendorf, recently awarded an RGS gold medal at Murchison's urging, became the President's lion, visiting Palmerston's country house with him after the meeting.[70]

For the rest of his life Murchison continued to champion Russia as Britain's natural ally. He hosted many Russian scientists, mining officials, and military engineers who visited England and he provided introductions for Britons travelling in Russia. He also retained close ties with the Imperial Geographical Society, working to co-ordinate its labours with those of the RGS. Publicly and privately, Murchison thus acted as a conduit for cultural and technical exchange which influenced the diplomatic and commercial equations between the two nations.[71]

When the accolades for the Russian work died down, Murchison seemed to have 'no new kingdoms to conquer'.[72] Whereas the fieldwork which estalished the Silurian System had been accomplished in the Welsh Borders, then a frontier of English geological science, Murchison's sweeping Russian researches had been performed in another transition zone where the overlapping frontiers of conquest, colonisation, and scientific classification marked the very edge of Europe. Russia represented Murchison's ultimate quest as a knight of the hammer. Roaming over scientifically unexplored territory with his subordinates, he had been free to assert his ego on a grand scale. At Murchison's age it would have been difficult to find another such convenient *terra incognita*. Yet other empires did await conquest, and generals might win campaigns by superior strategy in the staff room rather than brilliance in the field.

In the late 1840s, however, Murchison contented himself with geological excursions to Italy and the Alps. His response to political unrest witnessed in Florence and Rome during 1847–48 was characteristically reactionary, and news of the insurrections in France, Austria, and Germany likewise awakened deep-seated fears of Jacobinism, social chaos, and dispossession in the recently created knight.[73] He turned in dismay 'from the dark vista of communism and destruction with which the political horizon is shrouded to Nature and Nature's works'.[74] But because of the environmental determinism which informed his science, the study of geology forced him to confront again Italy's intertwined social, political, and economic problems. In contrast to Britain, the natural features and resources of Italy seemed to preclude the prosperity upon which he believed independence and stability rested. Murchison's solution was authoritarian: 'A succession of good despots would be a like blessing to this country as to wretched Ireland.'[75]

If the Italians were victimised by their physical geography, so too were the equally volatile French. The Swiss, on the other hand, were praised as valiant and sensible mountaineers akin to Highlanders, Silures, and Caucasians.[76] Murchison's despair at the spectre of revolution was coloured not only by his class allegiance and battlefield experience as an opponent of Napoleon, but by his childlessness. Having renounced a military career and turned to science as an alternative field of endeavour in which his conquests might be permanent and his empire everlasting, the perpetuation of his name – if only through scientific nomenclature – could only be achieved by the cultural free trade between nations that wars and revolutions disrupted. In consequence, he felt his intellectual as well as his real property threatened by such upheavals, for his social rank depended as much upon his scientific accomplishments as his inherited wealth.[77] Bellicose as he was by nature and training, Murchison paradoxically found himself often driven into adopting an anti-war stance.[78]

Returning to England in 1848, Murchison lapsed into a fit of depression arising from the trauma of the Italian insurrections and the unwonted lack of direction in his life. Partial compensation came the following year with his election as Vice-President of the Royal Society and the award of the Society's Copley Medal for his work on Palaeozoic stratigraphy. At the meeting of the British Association, Murchison also enjoyed a mock enthronement as the 'King of Siluria' amid a welter of patriotic speeches linking the territorial implications of Murchison's stratigraphy to Britain's industrial might.[79] It was therefore fitting that in 1851 Murchison began writing a third book to pull together the results of other geologists in applying the Silurian classification to foreign rocks. He kept several other irons in the fire as well. In 1853, with the death of his uncle General Mackenzie, he inherited a considerable additional fortune. He was urged to use this new wealth in the cause of science by canvassing for the presidency of the Royal Institution.[80] Murchison had a different agenda, however. Just as he had used his enthronement as King of Siluria to seek a peerage through the Earl of Clarendon, an influential member of Russell's Whig government, he now wrote to Aberdeen, the leader of a Whig–Peelite coalition, requesting that Mackenzie's baronetcy be conferred upon him.[81]

Murchison was again unsuccessful, but other issues seemed to promise opportunities for fresh laurels. In 1853 he took a leading role in approaching Aberdeen's government on behalf of the Royal Society with a request that the metropolitan scientific societies be housed in a central edifice testifying to the national utility and authority of science.[82] Murchison had campaigned unsuccessfully during his first term as President of the RGS for the government to provide quarters for the geographers, and as President once more he now petitioned both Aberdeen and the Queen to grant the Geographical Society rooms in the proposed building on the grounds of 'its connexion with the Foreign, Colonial, and other offices of Her Majesty's Government, and from the useful nature of its publications in developing new marts of commerce and industry'.[83] The result was an annual subsidy rather than a place in the new national science headquarters, but the episode demonstrates Murchison's growing partiality for the RGS as the most perfect embodiment of the alliance between science and the national interest which he sought to consolidate.

During the same period Murchison was involved in a more direct move to secure an increased public voice for science. In 1853 he gave his support to a conservative scheme for parliamentary reform which called for direct representation of 'the educated intelligence of the country'. The Earl of Harrowby, a friend of science who tabled this petition, had in mind scientists, doctors, lawyers, and other professional groups including retired military officers, but he also argued for the election of MPs from colonial constituencies.[84] The scheme fell through because of Tory intransigence and the Crimean War, but while it was under consideration Murchison played a prominent part in advising Harrowby's circle and securing publicity for the proposal.[85] Given the flattening trajectory of Murchison's career, it is quite probable that he hoped to enter Parliament himself on behalf of London science had the reform succeeded. In the light of the simultaneous flowering of his interest in overseas exploration and imperial development, it is also significant that Murchison here aligned himself with the conservative interests anxious to bind the colonies to the mother country.

With the publication in 1854 of *Siluria*, a semi-popular volume that synthesised new Palaeozoic research around the world and setting forth information concerning the occurrence of coal and gold, Murchison's reputation as the premier practical geologist of his day was secured.[86] In Russia he had worked out what he believed to be a second economic aspect of his Silurian System as important as its role in providing a baseline for coal exploration – that is, that exploitable deposits of gold could only be found where Silurian strata occurred. In 1844 he had predicted the discovery of gold in Australia on the basis of this theory. When the first Antipodean rush began in 1851, he was accorded credit for the forecast and thereafter acknowledged as the leading expert on the world's gold supply.[87] Historians of science have for the most part interpreted British geology as having developed in the first half of the nineteenth century almost exclusively within a gentlemanly tradition of ornamental learning divorced from economic reality.[88] Yet Murchison maintained ties with the mining industry throughout his scientific career, both as a source of data and because he felt it to be a public duty of geologists to protect investors from wasteful expenditure fostered by sheer ignorance or the misrepresentations of unscrupulous speculators. He continued to visit mines and smelters until the last years of his life, kept in touch with mine owners with overseas interests, and conned the *Mining Journal* for news of mineral discoveries, variations in the occurrence of ores, and British investments around the world. He was also well connected with the profession of engineering and consulted by its practitioners.[89]

Truly, Murchison 'virtually personified geology to the general public'.[90] In 1860 he considered accepting an offer to enter Parliament as a member for the University of London in order to represent the interests of science. Murchison's age, other duties, and desire to maintain a reputation as the '*slave of no party*' despite his conservative-cum-Whiggish political leanings, prevented his standing for this safe Liberal seat.[91] His Palaeozoic research played a crucial role in establishing the international prestige of British geology, and he viewed the spread of his systems' nomenclature across the world's geological maps as a scientific form of imperial expansion. 'Just as our goods are patterns for the world', he told Sir Robert Peel, 'so may our geological types be recognized in the remotest parts of the world. This is my deepest geological aspiration.'[92] He never ceased to enjoy being addressed as the 'King of Siluria', having his dominion over Palaeozoic geology compared to the autocracy of the Tsar, and describing his geological excursions as 'raids', 'campaigns', 'invasions', or 'conquests'.[93] The tendencies towards militarism and imperialism consistently displayed in Murchison's earlier career finally found full expression in his simultaneous leadership of the Royal Geographical Society and the Geological Survey of Great Britain.

As his stamina for field work declined by the early 1850s and the larger

Plate 2   The Gentleman Geologist

questions of European stratigraphy were answered, Murchison realised that the evolving science of geology offered diminishing returns for his overweening ambitions. He had remarked upon his return from Russia in 1840 that 'nothing short of Continental masses will now suit my palate', and he complained that his subsequent Italian research was 'fragmentary work, and unworthy of me'. By 1853 he was forced to acknowledge that his recent inquiries in Britain had been 'pottering work, but necessary in these pottering times, the chief work being over'. Murchison also felt threatened by the changing style of science. Having heard a paper on organic chemistry in 1850, he admitted that it 'astonished me, & proved to me (an old pupil of Brande & Faraday) that I was incapable of understanding the elements and phraseology of the science as it is now carried on'.[94] In geology itself, the development of new subdisciplines was demonstrating the value of chemistry, physics, mathematics, and laboratory techniques to an improved understanding of even the structural relationships which the stratigraphers of Murchison's generation had largely passed over in their eagerness to map formation boundaries.[95]

The accelerating pace of British activity abroad, however, promised new opportunities for dramatic geological and geographical exploits using the same scientific methods that had produced Murchison's early successes. Perceiving that a niche had opened for an institution capable of organising and evaluating the flood of travellers' information then deluging Britain from overseas – an institution capable, moreover, of gathering accurate intelligence on potential markets and sources of raw materials by means of scientific expeditions orchestrated as patriotic adventures – he decided to commandeer the Royal Geographical Society as the vehicle for a final drive for increased personal status.

The Society's revival from a period of mismanagement began in 1849 under the presidency of Admiral William H. Smyth, and it owed a good deal to the general economic recovery from the 'Hungry Forties'. But the driving force of Murchison's personality, promotional expertise, and political and social connections really established the RGS as a key national institution.[96] Murchison was elected to his second term as President in 1851: the previous year he had engineered the alteration of the structure of the British Association to recognise geography's status as an independent science. Section C – 'Geology and Physical Geography' – became simply 'Geology', while Section E was established as 'Geography and Ethnology'. Murchison had manoeuvred for such a realignment since his presidency of the Association in 1846, and when the RGS organised the new section he became its first President in 1851. He had been subsectional Vice-President for Geography six times before the reorganisation: he was to serve as the new section's President on seven occasions. In this capacity he rapidly transformed it into the most popular feature of the Association's annual meetings.[97]

The coincidence of this reorganisation with Murchison's resumption of command at the RGS symbolised the shift of emphasis in his career from

European geology to overseas exploration and imperial development. As he later admitted, 1850 marked the year in which he devoted himself to advancing geology 'in what I consider its enlarged sense . . . the sister science of geography'.[98] Murchison was re-elected to the RGS presidency for a third term in 1856, and he held the office for nine consecutive years after 1862. He presided for a total of sixteen years, usually managing the Society in intervening periods for figurehead peers. By making the RGS at once more rigorous and more popular, Murchison increased its usefulness to government, its attractiveness to scientists, and its *éclat* as the definitive institution for that section of the elite bent on overseas expansion. While many scientists welcomed the reconstituted RGS as a source of accurate data and, to a lesser extent, a forum for discourse between specialists of various disciplines, several key British savants remained aloof. Though all were Fellows, Charles Lyell, Charles Darwin, Joseph Hooker, Thomas Huxley, and Alfred Wallace shunned the Society's meetings because they believed Murchison's promotional efforts were devaluing science.[99]

In the larger cultural perspective, however, Murchison was eminently successful. For most mid-Victorians, Murchison *was* the RGS, just as the Society's indomitable explorers personified imperial Britain. His lengthy anniversary addresses on the progress of geography became major events of the London season. At RGS meetings the display of outlandish curios, maps of hitherto-unknown territories, dramatic reports from explorers in the field, and returning geographical heroes attracted wide publicity and hundreds of new members. With the help of influential Council members and Assistant Secretary Norton Shaw, Murchison rapidly made the Society what it had always aspired to be – Britain's quasi-official directorate of exploration. He succeeded in securing an annual government subsidy of £500 for the maintenance of a public map room in 1854, a Royal Charter in 1859, a freehold building in 1870, ongoing accessions to the library and map collection from British, colonial, and foreign sources, and, by the time he died, a membership of more than 2,300 Fellows – by far the largest of any scientific society in London. By skilfully lobbying government ministers and organising public appeals he raised the further funds necessary to finance a breathtaking series of expeditions.

During Murchison's reign the style 'F.R.G.S.' became an eagerly sought distinction, while the RGS gold medal emerged as one of the most coveted decorations in the world. *Hints to Travellers*, first issued by the Society in 1854, soon achieved international fame as a field manual for scientific exploration.[100] The RGS headquarters constituted a metropolitan repository for data of all kinds about the undeveloped world, as well as an important venue for the public discussion of topics concerning overseas expansion. The Geographical Club, formed by Murchison as an exclusive dining clique in 1854, permitted RGS officers to meet explorers on terms of intimacy and market their organisation to visiting politicians and makers of public

Plate 3   RGS headquarters, 15 Whitehall Place, 1854–1870

opinion.[101] Geographical soirées at Murchison's Belgrave Square mansion attended by scientists, politicians, explorers, editors, diplomats, officers, and colonial governors added to the fame of the Society and helped intensify the ambience of expansionist enthusiasm in which his power-brokerage flourished. Murchison was also President of the Hakluyt Society from its foundation in 1846 until his death. By publishing historical accounts of early geographical discoveries, this affiliate of the RGS emphasised the causal links between exploration and global power while simultaneously serving the needs of scholarship.[102] The Hakluyt Society also functioned as an historical intelligence bureau, delving into British and foreign archives for 'lost' information concerning trade routes, mines, and native products discovered by ancient civilisations and early European navigators.[103] Murchison likewise corresponded with classicists and Biblical scholars to pinpoint the gold-mining sites of the ancient world.[104]

In 1855 Murchison succeeded Sir Henry De la Beche as Director-General of the Geological Survey of the United Kingdom. The first national geological survey in the world, this institution had been established in 1835 through the initiative of De la Beche coupled with the support of the Geological Society. Its purpose was to map the nation's rock formations and provide an authoritative evaluation of mineral wealth to guide its efficient exploitation. Trained like Murchison at the Military College of Great Marlow, De la Beche had pursued the career of a gentleman geologist until the failing income from his Jamaican sugar estate forced him to seek an alternative income as a government scientist.[105] He had gradually expanded his Survey to encompass the whole of Great Britain and Ireland and to include laboratories, a Mining Record Office, and a Museum of Practical Geology. As a result of the enthusiasm for scientific development of British industry sparked by the Great Exhibition, a Government School of Mines was also added to the Survey establishment in 1851. Prince Albert pointed out that the School was 'intended to direct the researches of science, and to apply their results to the development of the immense mineral riches granted by the bounty of Providence to our isles and their numerous colonial dependencies'.[106]

De la Beche promoted his institution's usefulness by emphasising economic geology rather than the theoretical researches which absorbed the interest of his distinguished scientific staff.[107] Before his death he was also called upon to recommend geologists to conduct several colonial surveys. In a role analogous to that played by the Royal Botanic Gardens, the British Survey served as an official co-ordinating agency, reference bureau, and training academy for the colonial surveys, gradually supplanting the Geological Society as the empire's central repository of geological information and expertise. Its palaeontologists classified fossils to help colonial geologists date unfamiliar strata, while chemists and metallurgists reported upon the commercial value of colonial ore and coal specimens. De la Beche was the advisor to the Colonial Office on mineral leases of Crown lands, and government functionaries throughout the world forwarded specimens to his Museum. The British Survey and the School of Mines not only provided the personnel for the colonial surveys, but courses were offered to naval, military, and East India Company officers at half price, and naval explorers departing on voyages of discovery received special geological instructions.

Between 1845 and 1850 De la Beche also conducted an analysis of global coal resources for the Admiralty.[108] Faced with expanding commitments abroad during the transition from sail to stream, the Royal Navy was anxious to locate sources of Carboniferous-age coal along imperial shipping lanes. Murchison thus inherited a national institution dedicated to the systematic mapping of the strata and economic minerals of Britain and Ireland, the development of imperial mineral resources, and the gathering of geological intelligence on a global scale.[109]

Murchison was recommended to the post of Director-General by the professors of the School of Mines, leading geologists, Palmerston and at least two members of his cabinet, and Prince Albert.[110] The appointment seemed unlikely in some respects for a scientist of Murchison's age and status: 'I fancy he would have thought it *infra dig*', commented one palaeontologist.[111] Twenty years earlier, when science remained almost exclusively the preserve of gentlemanly devotees, De la Beche had endured sneers of jobbery for accepting government employment. Ironically, such criticisms had even come from Murchison, who, in defending the validity of the Devonian System from De la Beche's conflicting opinion, had attacked his competence to the point of threatening the Survey's survival.[112] During the 1840s, however, Murchison had come round to supporting the Survey. He had advocated the founding of a national mining school as early as 1838, and in 1851 he attended the formal opening of the School of Mines and Museum of Practical Geology with other scientific and political celebrities, many of whom were associated with imperial affairs.[113] He would later describe the new building housing these joint facilities as 'the first palace ever raised from the ground in Britain, which is entirely devoted to the advancement of science'.[114]

By the 1850s, therefore, the post of Director-General had become attractive to Murchison for several reasons. Throughout De la Beche's tenure he had urged the founding of colonial surveys and the extension of geological inquiry into unexplored overseas regions, so that the appointment now presented an opportunity to link his scientific expertise and promotional abilities in patriotic service to the British empire. As one congratulator put it: 'You have now done enough in the field with your hammer, & a new & vast field is opened to you whereby you will be able to advance Geology during the rest of your life.'[115] The Devonian dispute had also demonstrated the advantage which the commander of the 'trigonometrical forces' of the Survey enjoyed in enforcing his own scientific authority, and Murchison realised, as one of Sedgwick's supporters put it, that the appointment would give him 'another line of defence to fight for Silurian extensions' in the ongoing frontier dispute with

Cambria.[116] In practice Murchison's power to impose his views on the Survey proved quite limited,[117] but the director-generalship provided the perfect means to push for overseas additions to his Silurian domain.

By mid-century, the writing was on the wall for geology's gentlemanly devotees, as Murchison's behaviour during the Devonian controversy and his career anxiety following the Russian campaign attest.[118] He had striven to publish *The Silurian System* before his Welsh research was overtaken by the advancing Survey, and nearly all of his subsequent work had been accomplished in areas beyond the pale of systematic geological surveying. Just as Britain, faced with severe competition later in the century, would turn increasingly to the empire for secure markets, sources of supply, and ideological reassurance of its continuing status as the premier power, the ageing Murchison now found opportunities for continuation of familiar stratigraphic conquests on a scale commensurate with his ego. Murchison's personal needs and interests, as much as changes in the economic climate, thus explain why the Survey would focus its attentions on colonial mapping and mineral development much more obviously under his direction than it had under De la Beche.

There was the further danger that a non-geologist might take over the Survey, and Murchison wished to preserve one of his discipline's principal institutions from decline or dismemberment. Geological friends urged him to accept the director-generalship in order to prevent its going to a man whom the debilitated De la Beche had suggested as his successor. De la Beche had grown anxious that his heir apparent, Andrew Ramsay, Local Director for Great Britain, lacked the stature to prevent the break-up of the Survey in an imminent bureaucratic reorganisation. In consequence, and despite his distaste for the incompetence and favouritism personified by the aristocracy and its hangers-on,[119] he had turned his consideration to John Francis Campbell, a connection of the politically powerful Duke of Argyll who had an amateur enthusiasm for geology. Campbell's views were unscientific to the point of being bizarre, however, and the British geological community, acting in defence of science in general, was unwilling to tolerate the imposition of such a placeman upon the Survey.[120] Murchison was therefore unanimously drafted, and he promptly accepted the post as '*the only man* who could unite all parties & work the concern efficiently'.[121]

When Palmerston announced Murchison's appointment to a cheering House of Commons, the Survey palaeontologist John Salter remarked in relief that 'we shall get his influence in high quarters & his energetic will to help us carry out what is wanted'.[122] The geographers, meanwhile, worried unnecessarily that Murchison's new duties might 'lessen his attentions' at the RGS.[123] To the honours-hungry Murchison, the post also represented an official distinction with which to cap his career and a personal favour to Prince Albert, who in the aftermath of the Great Exhibition wished to expand the School of Mines into a national centre of scientific and technical education. Furthermore, Murchison was a personal friend of the Prime Minister and his Foreign Secretary Lord Clarendon, both of whom had used the Survey in the past to generate strategic intelligence about overseas territories. His control of the RGS, his international prestige, and his unrivalled knowledge of foreign mineral resources were politically useful, and the Australian prediction had won him wide respect as a scientific oracle capable of conjuring treasure from barren colonial ground. As Murchison realised, he was in fact uniquely qualified to direct Britain's geological and exploratory enterprises.

By providing control over mineral reconnaissance abroad and privileged access to official data-collecting agencies, the Survey post perfectly complemented Murchison's role at the RGS. From these interlocking power bases he was able to dominate British exploration for the remainder of his life, largely defining the goals, methods, and personnel of nearly every expedition to leave the home islands during the spate of exploration that characterised Britain's mid-century zenith as a great power. Here was a task of epic scope worthy of his talents, and his performance would bring the increased rank he had long coveted. In 1863 he was promoted K.C.B., and in 1866 created a baronet. Despite his protestations to the contrary, Murchison had no intention of renouncing the glory of new triumphs abroad for the relatively insipid distinction of presiding over an expansion of education at home. Preferring that the School of Mines remain a purely geological academy, and fearing the gradual subordination of his own authority to increasing levels of bureaucracy, he therefore obstructed the implementation of Albert's plans until his own death, fulfilling his ambitions at the cost of letting Britain fall behind her industrial rivals in the field of scientific education. [124]

During Murchison's lifetime the Survey had been subordinated to the Ordnance Department (1839–45), the Department of Woods and Forests (1845–54), and the Department of Science and Art under, successively, the Board of Trade (1854–56) and the Education Committee of the Privy Council (1857–1919). Yet he ensured that it remained a virtually autonomous department dedicated to field research and mapping, and that his own role as Director-General continued to be that of commanding a regiment of geologists spread across the empire as an auxiliary corps to his RGS explorers. Successive governments acquiesced in Murchison's intransigence because of his unassailable personal prestige, the coincidence of his tenure of office with a period of high prosperity that tended to reinforce the commitment to *laissez-faire*, and the fact that the scientific data gathered by his agents constituted vital intelligence in Britain's ongoing struggle for new resources, new markets, flourishing colonies, and secure frontiers.

After 1855 Murchison's energies were divided between the RGS and the Survey, but both organisations served the purpose of investigating the resource endowment of British colonies and foreign territories in order to promote imperial economic development and extend scientific knowledge of the physical world. These activities meshed so nicely that the field activities of the

North America
1. Murchison Cape
2. Murchison Sound
3. Murchison Promontory
4. Murchison River
5. Murchison Cape
6. Mount Murchison and
   Murchison Glacier
7. Murchison Island

Africa
8. Murchison Falls
9. Murchison Stream
10. Murchison Bay
11. Murchison Cataract
12. Murchison Mountains

Figure 1   The British empire at Murchison's death, depicting topographical features named in his honour. Numerous minor British possessions and several political features bearing Murchison's name have been omitted for clarity.

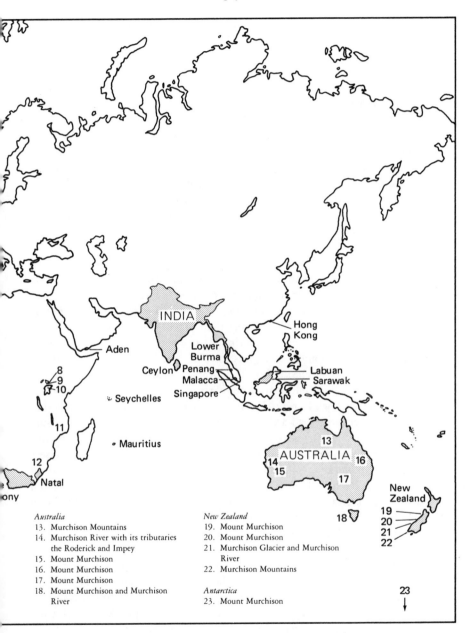

Australia
13. Murchison Mountains
14. Murchison River with its tributaries
    the Roderick and Impey
15. Mount Murchison
16. Mount Murchison
17. Mount Murchison
18. Mount Murchison and Murchison
    River

New Zealand
19. Mount Murchison
20. Mount Murchison
21. Murchison Glacier and Murchison
    River
22. Murchison Mountains

Antarctica
23. Mount Murchison

geologist and the explorer were often indistinguishable. Murchison commented in *Siluria* that governments should justly appreciate 'the value of the
application of geological science as a prelude to the settlement of a new
country', and he expressed his opinion to the British Association that gold was
'the most seductive of baits to entice the traveller'.[125] Warington Smyth,
Professor of Mining and Mineralogy at the School of Mines and the son of
former RGS President Admiral William Smyth, remarked similarly in 1864
that 'the events of the last few years have abundantly proven that the discovery
of mineral wealth is the most powerful incentive to the exploration and
settlement of distant lands'.[126] In his dual role Murchison encouraged the
topographical mapping of unexplored continents, the hydrographical charting
of coasts and rivers, and the detailed geological mapping of British colonies.
He urged the strengthening of naval defences, the development of imperial
trade routes, and the construction of colonial roads, railways, harbours, and
telegraphs.

As Director-General, Murchison oversaw the ongoing geological field work
in Great Britain and Ireland, but his interests focused overseas where scientific
mapping and the discovery of exploitable minerals were concomitant objectives. He believed the function of geological surveys was 'to prevent the public
from being misled by false data' rather than to advertise or validate ventures
underway,[127] and although he continued to advise his aristocratic friends
about mining development and investments,[128] he was careful to avoid any
personal conflicts of interest in his official capacity. Murchison was convinced
that reliable geological information could stimulate imperial economic development, and that while private investors were responsible for interpreting
facts, the government had a clear duty to provide them. His high-pressure
tactics in advising the government to install geologists in every British
dependency had immediate impact: fifty per cent more official overseas
geological explorations were undertaken during his tenure as Director-General
than in De la Beche's. Unlike his predecessor, who had waited upon random
discoveries to create interest in colonial geological surveying, Murchison
promoted it as an indispensable aspect of economic development. In colonies
unwilling to fund permanent surveys, he encouraged the labours of amateur
geologists.

Murchison also accomplished much more than De la Beche in reconnoitring
mineral opportunities in the larger sphere of British influence beyond the
imperial frontiers. Undertaken in conjunction with RGS, military, boundary
delimitation, or hydrographic expeditions, these surveys provided intelligence
about the resources of strategic or commercially significant territories and gave
Murchison research opportunities in regions he would not otherwise have been
able to explore. Wherever they penetrated, RGS explorers collected specimens
for the Museum of Practical Geology, and many received special geological
instructions before their departure. Murchison also remained vigilant for
opportunities to investigate mineral resources in undeveloped nations and

foreign colonies. He perused the results of foreign geological surveys and reports received from British geologists employed abroad for additional information. His laboratories continued to analyse foreign coal samples for the Admiralty and to conduct chemical and metallurgical analyses for the Foreign, Colonial, and India Offices. The outstanding graduates of the Royal School of Mines, as it was known after 1862, were consistently appointed to staff colonial geological surveys or accompany exploring expeditions. Many other graduates, as well as occasional students who did not take degrees, made their careers in the empire or other remote parts of the world as mining engineers, exploration geologists, chemists, metallurgists, and assayers. [129]

Murchison likewise strove as Director-General to further the policy of scientific instruction for military officers. In 1857 he arranged with Clarendon that such training be offered to diplomats as well. [130] When a panic seized the nation in 1866 that its coal might soon be exhausted, Murchison organised a detailed evaluation of Britain's deposits and testified to a Royal Commission about domestic, colonial, and foreign reserves. [131] At Clarendon's request, he also instructed the diplomatic and consular corps about more efficient methods of collecting information regarding foreign mineral assets. [132] All of these agents formed a global network of observers who collected raw information which Murchison's metropolitan organisations processed into data for the advancement of British science and the maintenance of Britain's world supremacy as a commercial, industrial, and military power.

As a major public figure, Murchison also served the cause of empire in ways which fell beyond the purview of his institutional capacities. In the 1860s, for example, he played a key role in defending Governor Edward Eyre against public persecution for the brutal suppression of a black uprising in the colony of Jamaica. With other leading conservatives, imperialists, and concerned intellectuals, Murchison organised the Eyre Defence Committee in order to combat Eyre's prosecution for murder by the coalition of liberals, radicals, and philanthropists which formed the Jamaica Committee. Eyre had won the gold medal of the RGS for his Australian explorations in the 1830s and had subsequently helped Murchison maintain funding for the Geological Survey of the West Indies. [133] Murchison believed Eyre had 'saved Jamaica & put down what could have been as bad as the India Mutiny', and that his hanging of the insurrection's leader, George William Gordon – 'the Nana Sahib of Jamaica' – was not only justified but commendable. [134] He wrote to *The Times* praising Eyre for 'preventing a great possession of the British Crown from merging into a black Republic like that of Hayti' and repeatedly, though unsuccessfully, petitioned the Colonial Office to reinstate the acquitted governor. [135]

In the era of *laissez-faire*, science had to demonstrate its usefulness in order to justify its mounting demands for public subsidy. Perforce, the Victorian scientists most successful at winning funds for large-scale research projects usually clothed their disciplinary goals in the trappings of utility. None was more adept at this tactic than the arch-patriot and consummate promoter Murchison. His success with the government sprang from his ability to identify the aspirations of natural scientists with the practical and emotional needs of the nation. By catalysing existing trends in the Geological Society, the British Association, the RGS, the Geological Survey, and, to a lesser extent, the Royal Society, Murchison was able to match the corporate goals of these institutions with his own ambitions and the shifting topography of public interest in expansion – an interest which he himself helped to create, to shape, and to actualise.

2

The Antipodes

Britain's Antipodean colonies held Murchison's attention more consistently than any other overseas region. Only in the last decade of his life did Africa take precedence in his interest, though even then he continued to promote the systematic exploration of Australia and New Zealand because they had greater immediate potential for scientific development. His imperial patriotism here found its widest vent in promoting emigration, new settlements, pastoral, agricultural, and mining development, steam and telegraph communication, and colonial defence. Australia, especially, held primacy of place in the imperial vision of the most powerful metropolitan savant since Sir Joseph Banks to champion Antipodean colonisation.

Metropolitan geologists sought correlations between the strata and fossils of the Antipodes and those found elsewhere in order to explain the region's unique physical features and apparently anomalous flora and fauna. Such research had economic implications, for the discovery of coal and ore deposits affected the pace of colonial development. Since the time of Cook and Banks, naval expeditions to the region had conducted geological investigations. From 1803 to 1812 a mineralogist had been maintained on the civil establishment of New South Wales, and by the 1820s collections made during inland explorations were finding their way to the Geological Society.[1] Murchison watched the progress of the Antipodes with proprietary concern after the 1840s, for his prediction of the discovery of Australian gold enhanced his scientific reputation with the lustre of a prophecy of colonial prosperity. He believed that the role of the scientist as a pioneer of economic development was being enacted with miraculous success in the southern colonies, and he promoted further research to press the advantages won and stimulate a similar approach in other British colonies and spheres of interest.

Australia

During his first term as President of the Geological Society in 1832–33, Murchison encouraged research to augment the Australian material received in the previous decade.[2] He was soon rewarded with specimens collected in the new colony of Western Australia by Captain James Mangles, R.N., a co-founder of the RGS.[3] In 1838 Murchison provided the emigrating clergyman–geologist William B. Clarke with introductions to his cousin John Murchison and two key members of the small scientific community in New South Wales, Phillip Parker King and William S. MacLeay.[4] King, a celebrated naval surveyor, had met Murchison during an earlier sojourn in

London. As a member of the Legislative Council of New South Wales and resident commissioner of the powerful Australian Agricultural Company, he had been well placed to help John Murchison, a former army officer, establish himself as a pastoralist in the growing colony.[5] MacLeay was the son of the entomologist Alexander McLeay [sic], an influential official in New South Wales. A prominent zoologist himself, the younger MacLeay had contributed to *The Silurian System*, and after Clarke delivered to him the first copy of this work to arrive in Australia, he began providing Murchison with specimens to aid the intercontinental expansion of his Palaeozoic domain.[6]

Murchison promoted the metropolitan reputations of King, MacLeay, and Clarke in return for their proselytising of his geological views. In common with other London savants, however, he attempted to maintain the authority of metropolitan science by means of influence over publications and the distribution of honours and through collaboration with emigrant scientists. According to this scheme, the ultimate arbitration of data would remain the preserve of the scientific elite in the imperial capital, while colonial researchers, like their provincial counterparts, were consigned to the subordinate role of field agents.[7]

Another early Australian contact of Murchison's was Sir John Franklin, FRS, FGS, FRGS, Lieutenant-Governor of Van Diemen's Land from 1837 to 1843. Franklin had shared his first lessons in geology with Murchison, proposed him for the Raleigh Club, and often met him socially before going out to administer the colony.[8] Franklin played a decisive role in creating a scientific community in Tasmania, believing that research could directly aid development. At the same time, metropolitan scientists were showing increasing interest in the island as a relict environment harbouring living and fossil organisms which might explain what were seen in European eyes as the anomalies of Antipodean biology.[9] In 1838 Franklin introduced Murchison to the Czech naturalist John Lhotsky, who had explored southern New South Wales and examined Tasmania's coal formations.[10] From Lhotsky's specimens, as well as other fossils collected by Charles Darwin and French explorers – the latter brought to his attention by Verneuil, his companion in Russia – Murchison judged that Palaeozoic strata might prove common in Tasmania. When he returned from his second Russian tour in 1842, he therefore asked Franklin for further fossils. Explaining that he had now extended Siluria into Asia, he pleaded: 'I must try to do the same through my allies in the distant colonies and as you are one of my earliest playfellows in geology, I count upon your aid.'[11] In reciprocation, Murchison observed from the chair of the Geological Society that Franklin was 'rendering Van Diemen's Land a school of natural knowledge'.[12]

Franklin immediately enlisted the services of Count Paul Edmund de Strzelecki, a Polish geologist exploring south-eastern Australia.[13] Together with Joseph Beete Jukes, geologist on the H.M.S. *Fly* hydrographic expedition,[14] and armed with a copy of Murchison's *Silurian System*, Strzelecki discovered an excellent series of Palaeozoic strata in New South Wales and Tasmania, though without any identifiable Silurian fossils. Murchison probably received news of this find in 1843 through Sedgwick, Jukes' former professor.[15] In 1844, following Strzelecki's return to Britain with an introduction to the King of Siluria provided by Franklin, Murchison examined his specimens as well as the manuscript of the geological map and book on the physical geography of New South Wales and Tasmania for which he was awarded an RGS gold medal in 1845.[16] Murchison would later be instrumental in securing Strzelecki's election to both the Royal Society and the RGS.

Franklin, recalled from his governorship for political reasons, also presented Murchison with a collection of Tasmanian geological specimens in 1845.[17] Murchison joined other London savants at this time in defending Franklin from the opprobrium brought about by his recall, and he attended a meeting of the Raleigh Club to see the explorer off on the ill-fated Arctic expedition by which he hoped to vindicate his name.[18] By the 1840s Murchison thus realised not only that Australian researchers could supply the metropolis with invaluable new data, but that science could enhance its local position and larger imperial opportunities by serving colonial governments in the cause of economic development.

As the founder of the Silurian System, Murchison developed a special interest in the formation and distribution of gold. His theories about the metal derived from various hypotheses regarding geological dynamics, including the French geologist Elie de Beaumont's theory of mountain building, which postulated that entire systems of mountain chains had each been formed by specific catastrophic events.[19] According to Elie de Beaumont, chains created simultaneously should display parallel compass directions, both in the strike of their strata and in geographical trend. Adding his own observations in the Urals and those of Humboldt in South America and Siberia to Elie de Beaumont's concept, Murchison theorised that mountain ranges formed of Silurian strata had been heaved up throughout the world during the Permian period and later infused with gold by quartz veins. Gold would therefore occur in mountains exhibiting these features, he maintained, and would prove especially abundant at the junctions where transverse chains cut these ranges.

Murchison believed the major meridional chains were the result of the Permian orogeny, and the principal gold-producing ranges did happen to be north–south chains displaying Silurian strata metamorphosed by igneous intrusions.[20] By the 1860s, after the gold-rushes confirmed and somewhat modified his opinions, Murchison had come to assume that any auriferous metamorphosed rocks could be dated as Silurian. Gold discoveries, like the discernment of diagnostic fossils, thus represented extensions of his system.[21]

Murchison had returned from the Ural mining districts full of enthusiasm about gold. In his first presidential address to the RGS in 1844, he compared the geology of Australia's Great Dividing Range as set forth by Strzelecki with that of the Urals, remarking that no gold had yet been found in the Australian chain.[22] In 1846, having learned of the discovery of gold in New South Wales and South Australia, Murchison reiterated his initially vague inference in stronger terms and sought confirmation of this forecast by urging unemployed Cornish tin miners to emigrate to Australia and search for gold.[23] During the next two years he received samples of the metal from William Tipple Smith, a prospector in New South Wales inspired by his prediction, as well as evidence suggesting its occurrence in the copper-rich meridional ranges of South Australia.[24] Murchison then advised Earl Grey, Secretary of State for the Colonies, to commission a geological survey of New South Wales in 1848. If the results proved encouraging, he hoped the government might either commence state gold-mining with convict labour or clarify the laws on mineral rights to encourage private exploitation.[25] Though Clarke, Strzelecki, King, and Jukes had also campaigned for such a survey, Grey did nothing, fearing as he later admitted to Murchison 'that the discovery of gold would be very embarrassing to a wool-growing colony'.[26] The colonial authorities remained unresponsive until 1849, when South Australia's copper boom and further reports of gold from New South Wales convinced them to follow the geologists' advice.

Requested to recommend a geological surveyor, De la Beche consulted King about Australian conditions before choosing Samuel Stutchbury, a museum curator experienced as a coal viewer who had previously visited the colony. De la Beche stipulated that Stutchbury conduct his research according to the methods of the British Survey and that he collect specimens of economic minerals for the Museum of Practical Geology. In return, the Director-General promised to help Stutchbury date strata and evaluate ores in the interests of mining development.[27] De la Beche also hinted to the Colonial Office that a geologist be appointed in Western Australia to survey its rich deposits of copper and lead. The Western Australian government had briefly employed the German geologist Ferdinand von Sommer, and the availability of his expertise, as well as that of the colonial land surveyors, probably dissuaded the authorities from following De la Beche's suggestion.[28] The example of Stutchbury's employment, coupled with increasing copper production and discoveries of other ores, also prompted South Australia to request a geological surveyor in 1851. Though the South Australian Company had employed another German geologist, Johann Menge, during the period of initial settlement, the colony's Legislative Council now refused to meet the contract terms insisted upon by De la Beche. The Director-General then fell back upon a recommendation that the civil engineer Benjamin Babbage, son of the mathematician Charles Babbage, be appointed to conduct a temporary survey.[29]

In New South Wales, Stutchbury's efforts were resented by W. B. Clarke – who had coveted such an appointment himself – and frowned upon by the colonists as impractical. They were upstaged in 1851, however, by the decisive gold-strike of Edward Hargraves, a prospector recently returned from

California.[30] Murchison boasted for the rest of his life that his 'scientific' prediction of 1844, as realised by W. T. Smith in 1847–48, had anticipated Hargraves' find by seven years, though Lhotsky, Strzelecki, Clarke, and others had reported gold discoveries even earlier. While the first of these discoverers had simply been ignored, Strzelecki and Clarke had been enjoined to silence by a governor fearful of social disruption during the transition from penal colony to free society. Despite Murchison's less than disinterested efforts on his behalf, the unfortunate Smith died without receiving due credit for the discovery which formed the linchpin of Murchison's claim to scientific clairvoyance. As a result of the informal censorship of Strzelecki and Clarke and his own privileged access to the organs of opinion in the imperial capital, Murchison nevertheless managed to neutralise Strzelecki's claim by a gentleman's agreement,[31] stave off Clarke's bid for full metropolitan recognition, reduce Hargraves' officially rewarded claim to the luck of a mere pick-and- shovel prospector, and reserve for himself the lion's share of scientific credit for what was in reality a fortuitous prognostication.[32] The King of Siluria even persuaded the authors of several popular accounts of the gold discoveries to adopt his version of events,[33] and the Australian newspapers had 'commanded in . . . thundering peals . . . "Fall down and worship the unimpeachable science of Sir R. Murchison."'[34]

Murchison's struggle with Clarke involved theory as well as priority. Clarke maintained that gold occurred in more varied conditions than were permitted by the 'golden constants' which Murchison insisted governed the precious metal's distribution. Further, Clarke believed that auriferous mountain ranges occurred along parallel and equidistant meridians.[35] Following the Hargraves strike, he was commissioned to reconnoitre the gold districts of New South Wales while Stutchbury continued his general survey. Both appointments terminated in 1853, but Clarke confirmed Strzelecki's evidence of an important development of Murchison's Silurian System in the colony.[36] Murchison and Clarke then forged an uneasy truce based on recognition of their respective metropolitan and colonial claims to priority for the gold prediction and jointly attacked Hargraves' discovery as unscientific. Murchison also sponsored Clarke's election to the RGS, and the latter subsequently helped the Society promote Australian exploration.[37] At the same time, however, Murchison arranged Hargraves' election to the Geological Society, a ploy which won his allegiance in the periodically flaring controversy with Clarke.[38] Murchison thus used his position in the imperial capital to play off his colonial rivals in order to maintain his own ascendancy as the premier authority on Australian gold.

The dispute with Clarke, like the controversy with Sedgwick, reflects the King of Siluria's extreme jealousy in regard to scientific credit, as well as his habit of attempting to alter the historical record for the purpose of self-aggrandisement. Like the Devonian conflict with De la Beche, it also exemplifies how clashes of career interest could provoke bitter rivalries between gentlemanly scientists and those defending professional or semi-professional reputations.[39] While the issue of priority for the gold prediction represented to Murchison a triumphant vindication of his prescience as a geological sage, for the impecunious Clarke it meant maintaining the scientific credibility in which consisted his only hope for further government employment or reward.[40] Further, the Murchison–Clarke dispute illustrates the tendency of the London savants to quell bids by colonial scientists for equal status as threats to the domination of the metropolis. As always, Murchison showed himself eager to carry the battle to the wider world of public opinion beyond the scientific arena where he could frequently carry his point by sheer weight of influence. The dispute with Hargraves, on the other hand, ranged the geologists Clarke and Murchison against a prospector whose procedural methods reeked of charlatanism. Winning this struggle also involved commandeering public opinion, and the subordinate cultural status of colony *vis-à-vis* metropolis was demonstrated by Hargraves largely triumphing over Clarke in New South Wales as the harbinger of golden prosperity, while Murchison received wide credit in Britain for initiating this profound change in imperial affairs.

As the City of London resounded with the news of successive gold-strikes during the 1840s and 1850s, alarm mounted in conservative circles that the gold standard might be undermined and the value of government bonds destroyed. In 1854 Sir Henry Holland, the Queen's physician, cautioned Murchison that he was 'moving upwards of £25,000 in Consols, which I consider too much of mere *anxiety* in these days of Australian gold'.[41] To allay these fears and advance his own reputation by demonstrating the utility of geology, Murchison issued a series of pronouncements collectively termed the 'General Gold Restriction Bill of Nature'.[42] He also plied Prime Minister Robert Peel with advice on the amount of gold potentially available in unexplored regions abroad.[43] 'Well may political economists and politicians now beg for knowledge at the hands of the physical geographer and geologist', he boasted, 'and learn from them the secret on which the public faith of empires may depend.'[44]

Murchison was convinced that only alluvial gold, which he believed to have been accumulated by cataclysmic floods redistributing Silurian detritus broken down by gradual erosion, could be profitably mined. He argued that gold was the 'last formed of the metals', and had been concentrated near the surface by emission in quartz veins. As mining experience seemed to testify, Murchison therefore maintained that auriferous veins in solid rock decreased in richness in proportion to their depth.[45] He even attempted to persuade the British Association to sponsor research to prove this contention.[46] Similarly, Murchison believed, history demonstrated that 'just in proportion to the time a country has been civilized, the extraction of the precious metal had diminished'.[47] An oversupply could never occur because the extent of Silurian formations capable of producing gold was finite and production from new mines would be absorbed by expanding commerce. He therefore hoped that the exploitation of alluvial gold in the colonies would serve 'the purposes of Providence in providing for a great augmenting population, and in converting wild tracts into flourishing hives of human industry'.[48] Murchison's confident prognosis that the gold-rushes contained the solutions to the very problems they raised proved true before the end of the decade; while anxieties ran high, however, he occupied a powerful position as the chief scientific bulwark against financial panic.

After the discovery of gold in New South Wales and richer strikes which followed in Victoria, articles, speeches, and successive editions of *Siluria* cemented Murchison's fame as 'The Goldfinder', the acknowledged expert on the world's gold supply.[49] Even the great Humboldt deferred to him on this point.[50] Popularisers further disseminated his views: the geographer James Wyld decided a pamphlet on the occurrence of gold to Murchison and the staff of the Bank of England was schooled in his theories.[51] At the same time, De la Beche offered a special lecture series at the School of Mines for gold-seeking Australian emigrants. These discourses, also printed for general sale, were illustrated by Murchison's maps and diagrams, and the speakers relied heavily upon his theories.[52] Among the experts called upon to offer advice was J. B. Jukes, who had returned from the Antipodes in 1846, secured a post on the home Survey, and published the first outline of Australian geology.[53]

As Murchison boasted in 1852: 'The *public men* think much more highly of me for having been the first who *worked out* mentally the Australian gold . . . and dwelling on it in successive years until the diggers discovered it'.[54] At the same time, he ensured that his friend Sir John Pakington, Secretary of State for the Colonies, understood full well that his predecessor Grey had complimented Murchison for 'having proved so true a prophet'.[55] To safeguard his reputation and the status thus won for his science, Murchison attempted to thwart charlatans posing as geologists from exploiting the public's mania about gold and its growing faith in science as a font of truth.[56] He also refused the presidency of an Australian gold-mining company, fearing his name would be used to mislead investors, though he admitted that 'if I did start such a thing I might *waddle out* of the Stock Exchange a much richer man than I am . . . If I had been 20 years younger I might in 1846 have backed up my prediction . . . & have secured *lands* which would now have been worth a colossal fortune.'[57]

Still, Murchison felt no compunction in recommending the geologist and mining engineer Friederich Odernheimer to the Australian Agricultural Company to evaluate the coal and gold potential of its holdings. Murchison had met Odernheimer while the German was engaged in a mineral survey of the Marquis of Breadalbane's Highland estates, and the appointment was probably arranged through his own connection with P. P. King.[58] Murchison later presented Odernheimer's results to the Geological Society and wove them into his gold theories.[59] He also influenced appointments to the new mints and assay offices set up to process the diggers' yield,[60] and mining promoters sent him ore specimens to publicise Australia's wealth and validate their speculations.[61]

Other British geologists felt the impact of the Australian gold-rushes as well. Adam Sedgwick was requested in 1853 by his former pupil George Stephen, younger brother of Sir James Stephen, who was Regius Professor of Modern History at Cambridge and former Permanent Under-Secretary at the Colonial Office, to recommend a geologist for his British Australian Mining Company.[62] It was unquestionably Murchison, however, who emerged as the great metropolitan expert on overseas gold resources, and the Australian prediction went far towards securing his appointment as Director-General of the Geological Survey in 1855. The colonial geologists and mining engineers thereafter trained at the School of Mines imbibed Murchison's ideas about gold, and proponents of his theories whom he had helped win university appointments taught courses in geology and engineering where students flocked 'to prepare themselves for the colonies and gold-hunting'.[63]

Murchison's stubborn adherence to his gold theories prevented him foreseeing – or, for a long time, even accepting evidence of – the impact on durations of yield produced by technological innovations opening up the exploitation of deep hardrock deposits. As the guiding hand of British overseas mineral reconnaissance, he consequently emphasised a search for fresh alluvial deposits rather than intensive development of existing sources. Murchison's geological explorers invariably sought high ground, for the 'Goldfinder's' constants dictated that regional 'dorsal spines' would exhibit the oldest strata and thus evidence of any coal or metallic ores. Paradoxically, however, most of Murchison's misconceptions actually facilitated mineral exploration: he proved right for the wrong reasons. Gold *is* usually found in mountain chains, whatever their direction, and though the age of deposits varies widely, quartz veins are more common in older rocks such as Silurian formations. The idea of yield decreasing with depth was also largely fallacious, yet recent evidence suggests that hydrothermal gold deposits tend to be concentrated near the top of intrusive igneous masses. Until the late 1850s, however, when revelation of the amounts of gold which might be won from venous deposits stimulated rapid technological advances, the prohibitive costs of quartz extraction justified Murchison's argument that alluvial gold was more profitable to mine.

Placers were also the easiest deposits to locate in unexplored countries where rivers often provided the only access, and they pointed the way to parent lodes which might be exploited once alluvial concentrations were played out. In their early stages, moreover, alluvial gold-rushes favoured individuals or small groups of miners who required no special skills or equipment. In the new era of low-cost ocean transport, rapid worldwide publicity, and unhindered emigration, the gold-rushes became gigantic lotteries in which men of all classes and nations might fairly compete. In these terms, Murchison's emphasis on alluvial gold made good sense for the initial exploitation of new regions, for placer gold-rushes rapidly peopled a country and generated economic activity.[64] Similarly, Murchison's belief that catastrophic inundations or 'waves of translations' were responsible for the detritus characteristic of gold-producing regions had little impact on the success of geologists and prospectors in discovering gold actually deposited by fluvial and glacial action.[65] Murchison's 'golden constants' – alluvia derived from Silurian rocks occurring in meridional mountain chains and exhibiting evidence of metamorphism and igneous intrusion – therefore provided a rough guide to the conditions in which gold could be expected to be found. His prestige as a geological authority and his influence on mining curricula may have slightly retarded the development of quartz mining, but innovators rapidly proved him wrong and younger geologists then jettisoned a set of assumptions based on the outmoded theory of catastrophism.

Before his appointment as Director-General gave him a direct role in Australian mineral reconnaissance, Murchison concentrated on promoting geographical exploration of the southern continent. As early as 1837 his advocacy had been essential in securing official support for an expedition in search of a new site for colonisation in Western Australia. George Grey, its leader, named the Murchison River and discovered Palaeozoic rocks in the hinterland north of Perth.[66] Addressing the RGS in 1844, Murchison encouraged further penetration of the interior to determine whether major river systems existed which might render the apparently forbidding central regions habitable. He also broached a topic to which he would repeatedly return – the need to establish permanent settlements on the north coast. Linked to the other Australian colonies by overland communication routes and to Britain's Far Eastern possessions by steamship lines, such ports could provide facilities for extending commercial domination in the East Indian Archipelago as well as for defending the Antipodean colonies and regional trade routes from French or Dutch aggression. Murchison was in this case advocating the promotion to Crown Colony status of the precariously maintained military outpost of Port Essington on the Cobourg Peninsula.[67]

This settlement had been founded in 1838 through the influence of George Windsor Earl, a visionary who saw himself as another Stamford Raffles, the founder of Singapore. Earl won the post of trade representative in Port Essington for alerting officialdom to the potential of an entrepôt in Torres Strait, and he was to emerge later as another rival for priority regarding the Australian gold prediction.[68] Sir John Barrow of the Admiralty also hoped to create a 'second Singapore' on the exposed north coast for strategic reasons. Having sponsored two similar but unsuccessful enterprises during the 1820s, he played a key role in bringing Earl's scheme before the Colonial Office. The hydrographers Francis Beaufort and John Washington had aided Barrow in the venture, co-ordinating their appeals to government through the RGS, whose Council was then dominated by naval officers. Despite such support, as well as official consideration of a scheme to use this outpost as the nucleus of a new penal colony, Port Essington was also abandoned in 1849 after Barrow's retirement eliminated its most influential advocate.[69] Murchison was thus expressing an institutional commitment to the survival of a colony which the Society had helped found. On a personal level, he was demonstrating himself as Barrow's lineal successor in the fields of Australasian imperial strategy and promotion of exploration.

In 1845 Murchison used his Anniversary Address to the RGS to praise Charles Sturt's recent exploration in South Australia. Citing naval hydrographer Captain Owen Stanley, FRGS, he also called for an exhaustive survey of the seas and islands north of the continent in order to stimulate commerce and the establishment of steam navigation links with Britain's Oriental possessions.[70] Stanley had participated in the establishment of Port Essington and in 1840 had prevented French settlement of New Zealand by proclaiming British sovereignty. Murchison's exposition of Stanley's exploring scheme represented an early example of his willingness to be guided by the opinions of naval surveyors eager to extend British dominance in the region.

The Australian sections of Murchison's RGS addresses of 1852 and 1853 were largely concerned with defending the priority of his gold prediction, but he also promoted exploration, geological research, emigration, extension of the settlement frontier, and colonisation of the Gulf of Carpentaria. Displaying cotton and silk samples donated by Stuart Donaldson, FRGS, a wealthy landowner and member of the Legislative Council of New South Wales, he argued that gold digging must give way to the more permanent occupation of agriculture. In 1853 he also announced that the proposal of an Austrian, Ernst Haug, to retrace the lost explorer Ludwig Leichhardt route across northern Australia was under consideration by the Expedition Committee of the RGS.[71] Haug stressed that his contemplated exploration south-east from the Victoria River offered a chance to take formal possession for the Crown, discover fertile new lands and navigable bays, accomplish a full range of scientific observations, and – no doubt to catch Murchison's eye – survey the 'northern mineral empire, where plenty of the precious metal will be found'.[72] The RGS recommended a modification of this plan concentrating on the region between the Victoria and the Gulf of Carpentaria to the Duke of Newcastle, Secretary of State for the Colonies. Murchison followed this up with a cost estimate of £2,500, arguing that land sales would repay the imperial government and that the scientific results of the expedition would justify its cost in any case. Newcastle provisionally agreed to the expenditure pending further information.[73]

Meanwhile a swindler who hoped to profit from a new gold-rush advertised the mineral potential of the far north, and the colonial promoter Trelawny Saunders, FRGS, published a pamphlet urging the government to support Haug in order to stimulate settlement in the Gulf of Carpentaria.[74] After perusing this document, Murchison promised its author that Haug would explore 'the very tracts you wish to colonize'.[75] At this point Haug infuriated

Murchison by letting it be known that the RGS 'had countenanced & recommended the project' before its final official acceptance. To limit possible damage from this indiscretion and dissipate the implication that, as was certainly the case, the RGS was working behind the scenes to secure government agreement before publicly requesting it, Murchison acknowledged support of Haug to several newspapers.[76] Knowing well that projects not pressed to conclusion could backfire embarrassingly upon their promoters, he invariably sought private assurance from key ministers that prospective expeditions would be supported before appeals were launched. Murchison also asked Herman Merivale, FRGS, Permanent Under-Secretary at the Colonial Office, to explain the issue to Newcastle and, seizing on the pretext of Haug's foreignness, began recasting the expedition as an entirely English affair.[77]

By 1854 the Admiralty and James Wilson, the expedition's geologist-designate, had also become suspicious of Haug, but, as Murchison ruefully admitted, the Society was now irrevocably committed.[78] Saunders, having decided to emigrate to the Gulf of Carpentaria in anticipation of an appointment as land commissioner in the settlement he hoped the expedition would found, now urged as its leader the naval surveyor John Lort Stokes, FRGS, who had charted much of the north coast and testified on the usefulness of maintaining Port Essington.[79] Stokes urged Newcastle at this time to found a convict colony in the Gulf of Carpentaria.[80] The Council of the RGS, hoping that the exploring project at least might be carried out as an extension of the navy's hydrographic work should Newcastle veto Haug's scheme, then forwarded Saunders' plan to the Colonial Office.[81]

After a proposal for Stokes and Sturt jointly to lead the expedition fell through and Haug finally withdrew,[82] Augustus Gregory, the Assistant Surveyor-General of Western Australia, accepted the leadership. Gregory was known to Murchison for his discoveries of coal, copper, and lead near the Murchison River and his confirmation of George Grey's suggestion of Palaeozoic strata in the region.[83] Meanwhile a deputation from the City of London waited upon Newcastle to express commercial interest in the mission, and when Sir George Grey (not to be confused with his namesake) replaced him at the Colonial Office Murchison renewed his pressure to 'get the N. Australian Ex. afloat'.[84] Through all of these vicissitudes Murchison kept Wilson, a civil engineer with Australian and Californian mining experience and Silurian views regarding gold, on hand as the prospective geologist.[85] Because of his desires to extend the scope of his stratigraphic system and prediction, as well as to ensure practical results from the venture, Murchison clung to the hope that the Victoria River region would prove auriferous.

With the advice of Merivale and Edward Eyre, Murchison, Sturt, and Captain John Washington, Assistant Hydrographer of the Admiralty and former Secretary of the RGS, worked out the organisation of the expedition.[86] Because of his earlier role in the founding of Port Essington, Washington offered advice on circumventing the Admiralty's reluctance to commit money and matériel, believing the project offered an opportunity 'to press the immediate occupation of Cape York – the French, as you know, are at Tahiti & New Caledonia and what are we doing?'[87] The Admiralty itself was simultaneously considering, and declining, the proposal of a Bradford merchant that New Guinea be annexed and Fiji and other islands in the region be taken under British protection.[88]

By early 1855 an appropriation of £5,000 had been secured for the expedition and its officers appointed. The scientific staff included Gregory; the geologist Wilson; Ferdinand von Mueller, Government Botanist of the colony of Victoria; the surgeon Joseph Elsey, with responsibilities for zoology and meteorology; the artist Thomas Baines, whose earlier explorations in South Africa had brought him to the attention of the RGS; and a specimen collector. While these officers sailed to meet Gregory at Sydney, Murchison used the British Association to maintain pressure on Palmerston's new cabinet to carry through with the obligation incurred by Aberdeen's administration.[89] The parliamentary subsidy was now raised to £12,000, and Sir William Denison, FRS, FGS, FRGS, Governor of New South Wales, began corresponding with Murchison on the expedition's progress. When Denison predicted that squatters would disperse the colony's population if the Victoria River region proved fertile, Murchison, drawing on the proposal of Stokes, suggested that a penal colony be founded.[90] Denison doubted that such a settlement could succeed, but prepared to found a free township in the Gulf of Carpentaria if Gregory's results were positive and plans to establish a steam mail service through Torres Strait materialised.[91]

The expedition began its work in the autumn of 1855.[92] The reports of Gregory and Wilson were read at RGS meetings, permitting Murchison to highlight the imperial significance of the expedition as tending 'to the establishment of a colony in North Australia which would materially strengthen the powerful position we already held in the East' as well as providing a terminus for a new land route to the southern colonies. Matthew Uzielli, a Fellow from the City of London, convinced by this harangue that occupation of the north coast was necessary to pre-empt Britain's rivals and safeguard her Oriental possessions, offered £10,000 for the project should the government refuse to act.[93] The presentation of a memoir by Stokes on the strategic value of a trading entrepôt organised as a convict colony again elicited Murchison's support for this solution.[94] The insubordinate Wilson was soon complaining that Gregory had frustrated his research and suspended him from service, but the leader pressed on to confirm the pastoral promise of the Victoria River region and northern Queensland, earning an RGS gold medal for his accomplishments.[95] Murchison's hope of gold discoveries in the far north was disappointed, however, and the expedition's cost overrun prejudiced the Colonial Office against future exploratory ventures.[96] Wilson's final reports were forwarded to Murchison via the Colonial Office and, being satisfied of their scientific value, he backed Wilson's conduct in the quarrel with Gregory.[97]

Though Murchison, now Director-General, declined Wilson's overtures for another government appointment, the geologist soon found other employment through his connection with the RGS.[98] Wilson's consolidated geological and geographical account, containing descriptions of Silurian strata, the newly-named Murchison Mountains, and exploitable soils and pastures, was refereed by Murchison for the RGS in 1858.[99] The President also requested further commentary from William Fitton, a former mentor conversant with the geology of northern Australia. Fitton remarked upon the parallelism of the region's meridional mountain ranges with the principal chains of the northern hemisphere and, despite the absence of fossil evidence, followed Wilson in assigning the ancient sandstones found there to Murchison's Devonian period.[100]

The reading of Wilson's account at the RGS provoked Sir George Everest, the Society's Vice-President and the former Surveyor-General of India, to suggest the establishment of a penal colony on the Victoria River for rebellious sepoys captured during the suppression of the India Mutiny. Trelawny Saunders, still intent on a civil appointment, countered with a plea for a trading entrepôt. William Lockhart, a missionary on leave from China, suggested alternatively that Chinese labour would be useful for developing a port and 'extending our dominion into the interior'. The orientalist John Crawfurd, however, who had testified to the Colonial Office in 1843 against the maintenance of Port Essington on climatological grounds, objected that neither Chinese nor sepoys would form permanent settlements and that the region was unsuitable in all respects for European settlement.[101] Murchison had already made Everest's point in 1857, arguing that since France had acquired a 'point d'appui' on Australia's flank by annexing New Caledonia, 'which our Cook discovered and named', Britain must establish naval stations on the north coast to defend her colonies and trade.[102] In 1858 he pointed out once more that this 'high political object' might best be secured by the transportation of sepoy mutineers.[103] Charles Sturt, eager to win appointment as governor of the prospective colony, also wrote to the Colonial Office in advocacy of this solution.[104]

New penal colonies, however, had become politically unacceptable by this era, and a settlement was not established on Australia's north coast for some years because of the continued routing of steamships via Singapore. Yet the North Australian Expedition reveals Murchison's tenacity in pursuing projects he believed to be of imperial importance. The opportunity to obtain new scientific data was subordinated to the larger goals of occupying the north coast, opening a new field for commerce and emigration, and establishing strategic bases for naval deployment and extended steam navigation. Murchison provided the continuity that saw the mission to conclusion through personnel shifts in the cabinet, the RGS, and the expedition itself – a feat all the more impressive for being accomplished in the midst of the Crimean War.

During the period 1859–61, Murchison also took part in a project

originating in Perth to settle Australia's north-western coast. The RGS had
been interested in such schemes since the 1830s, when Grey's explorations had
suggested the possibility of introducing herds and crops in the region, but
nothing had been attempted because subsequent exploration along the lower
Murchison River following a severe drought seemed to indicate that the entire
north-west was an arid waste.[105] In 1858, however, Francis Gregory, a brother
of Augustus, revealed vast tracts of excellent grazing north of the Murchison.
He also named two of the Murchison's upper tributaries the Roderick and the
Impey, fixed the location of Mount Murchison, and discovered Palaeozoic
formations as well as traces of useful ores. These new findings encouraged
Murchison to hope that the north-west might yet prove to be fit for
colonisation and to contain major sequences of his own systems' strata.[106]

Francis Gregory wished to explore the coast of this promising region, but
Sir Arthur Kennedy, Governor of the colony of Western Australia, refused to
sanction another expedition. To circumvent Kennedy, representatives of the
expansive settler interest approached the RGS for support with the Colonial
Office, for the discovery of exploitable resources would attract immigrants and
capital to the colony. To arouse Murchison's interest, it was pointed out that
for centuries the region had been termed 'Provincia Aurifera'.[107] This plan and
a similar scheme submitted by a Melbourne politician was considered by the
Society's Council during 1859. As Strzelecki pointed out, the time was
unpropitious for the founding of new colonies, but extensions of existing
settlements might be looked upon with official favour.[108] Realising that
Palmerston's Whig cabinet would be more open to such proposals than the
outgoing Tory government, Murchison recommended Gregory's proposal to
the Colonial Office.

In 1860 Gregory, several settlers, and the Surveyor-General of Western
Australia arrived in England to promote the expedition through the RGS.[109]
Murchison, now Vice-President, agreed to submit to the RGS Council
Gregory's application for repayment of personal funds advanced and authori-
sation to proceed. He also wrote to Newcastle at the Colonial Office and
appealed to the new RGS President.[110] Lord Ashburton, believing that the
RGS should 'bring to bear on Australian Exploration the united interests of the
Cotton Lords & of the settlers', and that 'the Colonial Off. will not much heed
the first suggestion of Sir Roderick's', urged Gregory to approach Sherard
Osborn, an influential naval explorer, to sound his Manchester friends about a
co-ordinated approach to the Colonial office.[111] In light of the cotton famine
resulting from the American Civil War and Grey's earlier reports that
north-western Australia might become a great cotton producer, this was a wise
tactic. Osborn's feeler apparently failed, however, for the Society itself was left
to stress the suitability of the country for intensive cotton production using
Indian labour. The RGS also maintained – and here the hand of Murchison was
evident – that 'men of science have come to the conclusion that this district
may be productive in gold and copper, as well as other minerals.'[112]

Newcastle agreed to this proposal, promising £2,000 if Western Australia put up a like amount. Gregory returned to Perth in 1860 to organise the expedition. Despite local fears that cotton cultivation would undermine the interests of stockmen, he raised sufficient funds within six months to launch his exploration.[113] Murchison had meanwhile kept the Colonial Office up to the mark by direct exhortation as well as public pronouncements.[114] At the Manchester meeting of the British Association he read a letter from Newcastle which stressed that the government was supporting Gregory because of its anxiety to develop new sources of cotton.[115] Murchison also passed on to the Geological Society the results of Gregory's 1858 exploration. This report stressed the mining potential of the far north-west and the probability that Silurian strata might soon be confirmed as the source of the region's traces of alluvial gold.[116]

In 1862 Murchison refereed Gregory's official report of the 1861 expedition for the RGS, and its reading prompted a promotional encomium from the Perth faction on the capabilities of Western Australia.[117] The Society's role in the expedition was recorded by Gregory's naming of rivers after Presidents Ashburton and Earl Grey, Secretary Norton Shaw, and Chichester Fortescue, FRGS, the Under-Secretary at the Colonial Office who handled the negotiations. Fortescue also received the society's gold medal from Murchison on Gregory's behalf.[118] The expedition revealed extensive agricultural and pastoral lands, but no valuable minerals. From the beginning, the RGS had proceeded on erroneous assumptions regarding the region: now the Society and the imperial government dropped the project of establishing a new cotton colony, probably because better opportunities beckoned in Africa and Fiji.[119] Pastoralists began arriving on the north-west coast in 1863, but nothing was done to establish cash crops. Mineral exploration also languished until the mid-1880s, when prospectors acting on the assessment of a government geologist made the discoveries in the Kimberley region which ignited Western Australia's first gold-rush and opened the exploration of one of the world's most richly endowed mineral provinces.[120] In promotional terms, however, such failures to produce promised resources were almost irrelevant. Since the examples of California and Victoria suggested that every wilderness region must harbour useful products, the Society's inveterate optimism constituted the best strategy for constructing the coalition of interest necessary to force the government to fund explorations.

Murchison also maintained close ties with scientific Australian governors and officials. In exchange for the provision of news and the dissemination of his own opinions, he publicised and aided their efforts to stimulate exploration and development. In 1859, for example, when Sir William Denison suggested that an encyclopaedia of the natural history of the empire would benefit colonial economies and reflect credit on the mother country, the Colonial Office requested advice from the Royal Society as well as from Murchison in his capacity as a geographer. Murchison extolled Denison's project as tending to increase imperial unity and both societies reported favourably upon the scheme. It fell to the ground, however, when Gladstone's Treasury attempted to limit the encyclopaedia to a series of unillustrated volumes, an expedient which defeated the purpose of furnishing field guides for the identification of useful colonial products.[121] Still, this nearly successful attempt to convince the imperial government to finance a vast, multidisciplinary research project offers evidence of the sophistication of scientists in presenting proposals and the influence which key savants could exert upon government decisions regarding such matters.

In 1860 Murchison formed a friendship with Sir Richard MacDonnell, FRGS, Governor of South Australia. MacDonnell provided Murchison with exploration details and backed his claim to have predicted the first Australian gold-rush. Murchison, in turn, supported MacDonnell's subsequent efforts to recoup financial losses by attempting to establish a new colony as chairman of the North Australian Settlement Company.[122] The same year Sir George Bowen, FRGS, the first Governor of Queensland, informed Murchison that while developing the resources of that colony he hoped to 'add new conquests to Geography and to Science'. He begged the RGS President 'to assist in making Queensland known in England', and the service was assiduously performed.[123] At the time of Bowen's appointment Charles Sturt requested Murchison's help in obtaining the governorship of Queensland himself, as he did later that of South Australia. Throughout the 1860s Sturt acted as an RGS adviser on Australian exploration, and as an agent for introducing Antipodean natural scientists into metropolitan circles.[124]

Murchison also maintained a link with Sir Thomas Mitchell, FGS, FRGS, Surveyor-General of New South Wales from 1828 to 1855. Mitchell, a fellow Scot and Peninsular veteran, provided Murchison with early indications of Palaeozoic strata in eastern Australia and named a mountain in his honour.[125] When Mitchell visited London in 1853 Murchison hosted him at the scientific societies and supported him in a dispute with his superiors. Murchison also asked Mitchell to provide introductions requested by Sir James Clarke, the Queen's physician and another Scot, for an emigrating protégé.[126] The King of Siluria's Australian network was thus consistently deployed to facilitate transplantation to the southern colonies of the connections of his wealthy friends. Murchison's Antipodean contacts also included Stuart Donaldson, a New South Wales politician who supported imperial loyalty. Donaldson, too, served as one of the RGS President's informers on Australian affairs until his retirement to Britain in 1859, when he became active at the Society.[127]

In 1854 Murchison helped secure the appointment of the palaeontologist Frederick McCoy to the chair of natural science at the new University of Melbourne. McCoy had aided Sedgwick in his attack on the Silurian System, and the professorship in distant Victoria offered a means of transporting him to a colony where he might, as one Silurian supporter put it, 'do some good work and not stir up mischief between old friends'.[128] Murchison's recommendation also helped McCoy win a second appointment in 1856 as Palaeontologist to the Geological Survey of Victoria. Despite this patronage, McCoy maintained his stratigraphic partisanship by attempting to prove that Australia's chief gold deposits occurred not in Silurian rocks proper, but in the formations Sedgwick claimed as Cambrian.[129] He likewise embroiled himself in a debate with W. B. Clarke over the age of Australian coal deposits. Clarke believed the coals to be Carboniferous, while McCoy maintained that they were younger Oolitic (Jurassic) formations. Since Carboniferous coals were considered of higher value as steam fuel, this stratigraphic battle, like the Murchison–Clark dispute over the occurrence of gold, had direct implications for colonial development. The quality of New South Wales coal was well established, but public perceptions of the value of this resource could be influenced by scientific pronouncements, especially in London, where Eurocentric notions about the inferiority of Mesozoic coals prevailed. The Palaeozoic assessment of most of the New South Wales coals gradually won acceptance, however, and Murchison advertised this fuel source as 'worthy of imperial notice'.[130]

The Geological Survey of Victoria had been founded at the height of the gold-rush in 1852 by A. R. C. Selwyn, an appointee from the British Survey with extensive experience among the Palaeozoic and slightly auriferous rocks of north Wales. Selwyn had explored the colony's gold resources with a single assistant until given an expanded staff and a direct link with the museum of natural history which McCoy established in 1856. A chemical laboratory was added to the Survey in 1864. The Victorian Survey rapidly achieved worldwide recognition for its precise mapping, and many of the geologists trained by Selwyn went on to senior survey and academic posts throughout the Antipodes. Because of bureaucratic conflicts, Selwyn resigned his post in 1869 to take over the Canadian Geological Survey, and the Victorian Survey was disbanded.[131] In the initial phase of Selwyn's work, the British Survey co-operated in identifying Victorian minerals. In 1853, for example, analysis in London of a 'black sand' from the gold diggings alerted Lieutenant-Governor La Trobe to the wastage of a rich source of tin ore which became the basis for a subsidiary mining industry.[132] This relationship was particularly close during the governorship of Murchison's friend Sir Henry Barkly, FRGS, a keen amateur geologist and botanist and the first President of the Royal Society of Victoria. Barkly also supported Murchison's drive for a northern colony, and Murchison reciprocated by securing his election to London's Royal Society.[133]

Selwyn's research into Victoria's gold deposits soon revealed grave discrepancies in Murchison's theories. In 1857 Barkly sent him two reports issued by a commission which had been established to look into the future prospects of the colony's mining industry. McCoy wrote these reports, but they were based largely on Selwyn's findings. McCoy stated that Murchison's theory of poverty with depth was born out by Victoria's falling gold yields. Though the Director-General assumed Selwyn's agreement with McCoy in drafting his official reply, Barkly and the Director of the Victorian Survey soon disabused him of this conceit. They argued instead that the exploitation of veins in Cambrian as well as Silurian rocks, along with buried placer deposits, could prolong gold mining in the colony almost indefinitely.[134]

The future of Victoria was actually at stake in this esoteric debate mediated by the Colonial Office. The report of the mining commission caused a faltering of public confidence in quartz-crushing companies until Selwyn repudiated the position of McCoy and Murchison. As Barkly observed in rebuking the 'received geological doctrines' of the King of Siluria, 'the question is indeed not one of authorities but of facts'.[135] This declaration of scientific independence from metropolitan control could probably only have occurred at that time in Victoria, where the prestige of Selwyn and Barkly and the Director-General's desire to promote the booming colony combined to persuade him to modify his views. The 'Goldfinder's' subsequent publications were accordingly revised to reflect Selwyn's conclusions.[136] And although Murchison sustained his old opinions on gold's formation, occurrence, and rapid exhaustion as an exploitable metal, his enforced shift of position on deep-level mining reopened the debate which his pronouncements had stifled during the very decade when the great rushes were transforming the situation.

Murchison nevertheless supported the Victorian Survey because it increased knowledge of Australian geology, benefited Victoria's economy, reflected prestige on the parent British Survey, and encouraged the foundation of similar institutions in other colonies. In 1861, for example, he campaigned at the British Association for the maintenance of Selwyn's budget by the Legislature of Victoria.[137] The following year, after awarding Selwyn a medal for his mineral display at London's International Exhibition, the Director-General arranged to supply the Victorian Survey with promising graduates of the Royal School of Mines.[138] In 1870 Murchison was approached for advice on founding a school of mines at Ballarat, Victoria. By the late 1850s the profits of alluvial mining were waning, and deep-level extraction demanded technological reorientation. In 1866 Murchison had similarly been requested to recommend a professor of geology for the University of Sydney capable of introducing 'the newest and best methods'.[139] He now advised a curriculum for Ballarat based on that of the Royal School of Mines, but modified to emphasise the practical techniques deemed appropriate to a colonial academy which did not aspire to challenge the scientific instruction offered by the University of Melbourne.[140]

In 1858 Murchison installed a geologist in Tasmania. As Lieutenant-Governor from 1847 to 1855, William Denison, aided by analyses carried out at De la Beche's Museum, had continued Franklin's efforts to stimulate scientific research regarding local resources and to promote coal mining.[141] When the mainland gold-strikes raised expectations of further discoveries, W. B. Clarke informed Denison that the metal was likely to occur in Tasmania. Denison attempted to suppress this news in the interests of maintaining public order, but Clarke published his prediction and informed Murchison in 1852 that discoveries were already reported.[142] The same year Murchison received the first specimens of Tasmanian gold sent to England.[143] In 1855 Lieutenant-Governor Sir Henry Young commissioned Selwyn to examine the island's coalfields and submitted the results to Murchison. The new Director-General recommended that the Colonial Office appoint a geologist, and on this advice the Tasmanian Legislature agreed to fund a limited survey. After a prolonged search for a geologist, Murchison chose Charles Gould, a graduate of the School of Mines employed by the British Survey and the son of John Gould, Britain's foremost authority on Australian birds.[144] Murchison used this appointment to advertise the imperial utility of his own Survey. Within the government, it bolstered an unsuccessful bid for independence from the Department of Science and Art; externally, it permitted him to boast that most colonial geological surveyors 'had either been brought up in the establishment which I direct, or recommended by my predecessor or self'.[145]

The Tasmanian government had by now offered a generous reward for the discovery of a paying goldfield, and Gould began his survey according to Murchison's instructions where the first specimens had been found. His observations, however, contradicted the Director-General's expectations and reinforced Selwyn's conclusions. In contrast to Victoria, Silurian alluvia were too scanty in Tasmania to promise much gold, but as in Victoria quartz veinstones had proved so auriferous that several mining companies were investing in crushing mills. The Van Diemen's Land Company also anticipated increased revenues, sending Murchison's Museum samples of auriferous sand from its vast holdings.[146] The expected boom never materialised, however, and Gould failed to find gold in the interior mountains. After producing a coal report, he began to complain that economic activities were hindering his scientific researches.[147] Murchison publicised Gould's coal findings at the British Association and eventually recommended their publication as likely to benefit Australian economic development.[148] Gould named a prominent mountain and a river in the west of Tasmania in his patron's honour and mapped much of this wild region as Silurian, but he made no major mineral discoveries. In 1869 his contract was terminated by a disgruntled colonial legislature and the Tasmanian Survey lapsed until 1882.[149]

Throughout the 1860s Murchison continued to promote Australian development. Like many of his contemporaries, he viewed the displacement or extinction of the Australian Aboriginals by the march of civilisation as an inevitable process, justified by the European technological and cultural superiority which it apparently demonstrated.[150] In the field of exploration, Murchison awarded the gold medal of the RGS to John MacDouall Stuart, the first person to cross the continent from south to north, and arranged special meetings of the Society to maximise the impact of his discoveries.[151] He defended Stuart's employers, the Adelaide pastoral and mining firm of

Chambers and Finke, from charges of mercenary behaviour by pointing out that 'it is by such bold and riskful methods of spending their capitals – a boldness which is peculiarly characteristic of the Anglo-Saxon race – that Geography owes many a bright discovery and Commerce many a useful end'.[152] The disastrous Burke and Wills expedition, mounted as a Victorian rival to Stuart's thrust in 1860–61, also won Murchison's support. Two of its members received RGS awards, one bestowed via Governor Barkly, whose colonial Royal Society had sponsored the expedition. Murchison also used the emotive value of Burke and Wills' fate, as he had that of Leichhardt, to promote further exploration.[153]

In the field of mineral reconnaissance, Murchison encouraged explorers to collect specimens and processed the data received through the Survey and the London scientific societies. Contemporaries considered that his judgement influenced the attraction of colonies to immigrants and investors. In 1856, for example, an official report from Western Australia suggested in light of evidence of the King of Siluria's 'constants' that a goldfield might exist on the Murchison River. Refereeing the paper for the RGS, Murchison allowed this statement to stand, hoping that it might stimulate an extension of his Silurian prediction.[154] In 1864 Edward Hargraves, now employed by the Western Australian government, contradicted this view. He too, however, shielded Murchison from any appearance of error, since by this time he owed him a debt of gratitude. Hargraves' report was also refereed for the RGS by Murchison, and at its presentation he changed tack to defend the famous prospector's negative assessment of gold potential. In opposition, one of the Western Australian activists who had recently promoted Gregory's north-west coast expedition protested that this publicity constituted an unfair attack on the colony's capabilities. The visiting Selwyn pointed out similarly that Hargraves' limited excursions hardly constituted grounds for concluding that the entire interior of Western Australia contained no gold.[155]

Other geological reports received included those of Richard Daintree regarding Queensland. Accompanied by the comments of W. B. Clarke, these reports were forwarded by Governor Bowen, who hoped they would attract English investment capital to develop the new colony. Daintree, a veteran of the Victorian Survey who had studied briefly at the Royal School of Mines, had pursued amateur research in Queensland from 1864 to 1867, when his discovery of gold ended a local depression and prompted his appointment as one of two government geologists. Both men lost their posts by 1870, but their discoveries accelerated the development of Queensland and eventually led to the permanent refounding of the colony's geological survey.[156] Daintree's evidence of Queensland's gold deriving predominantly from Silurian and Devonian strata gratified Murchison, who recommended the publication of these findings by the RGS. His extended observations and pioneering geological map of the colony were published by the Geological Society in 1872 – the same year in which the scientist was appointed Queensland's Agent-General in London.[157]

Murchison's steady promotion of Australian emigration even included advising a temperance society on founding a settlement.[158] His pleas for colonisation of the north coast, however, continued to meet opposition from John Crawfurd, who remained sceptical of the region's suitability for European settlement.[159] Murchison gave ground on this issue, but in his final address to the RGS he boasted that the establishment of Port Darwin and the construction of a telegraph line along Stuart's route from Adelaide fulfilled his predictions of a new northern colony.[160] Murchison continually advocated improvements in steamship and telegraph communications between Australia and Britain. Every new geographical discovery, outpost of settlement, and advance in railway or telegraph mileage was also recorded on a huge Australian wall map displayed at RGS headquarters as a record of the success of British enterprise in conquering the southern continent.

Australian imperial loyalty also received considerable comment while the American Civil War made secession a fashionable topic. In 1862 Herman Merivale of the Colonial Office explained to the British Association that Britain must remain on good terms with her existing colonies, as well as continue to found new ones, in order to maintain outlets for her redundant population. Murchison apprised Barkly:

Merivale has done good service in smothering that grovelling and unworthy sentiment of a few *doctrinaires* as to the inutility of our colonies. I am furious when I read their cold and heartless reasoning, and I shall take good care to show at our evening affairs how warmly the Australian colonists support the mother country, and sympathize with it.[161]

In 1856 Murchison also formed an association with the Australian politician Sir Charles Nicholson, FGS, a vociferous advocate of colonial loyalty. Murchison had Nicholson elected to the RGS and to the Athenaeum, despite opposition to his tendency to force too many colonials into the club. Nicolson reciprocated by hosting the President at London's first Australian reunion and defending the priority of his gold prediction.[162] Nicholson's *idée fixe* was the expansion of Australian influence into neighbouring islands. This concern, which meshed with Murchison's desire for north coast settlements, had been stimulated by the French occupation of New Caledonia in 1853. P. P. King and John Washington had railed to Murchison of Britain's failure to pre-empt this annexation,[163] and John Lort Stokes, who likewise had a personal interest in the matter, also urged the RGS President to press his promotion of north Australian settlement. Stokes hoped to circumvent the second Derby administration's distaste for colonial expansion and thereby 'prevent the French taking the wind out of our sails there, as they most certainly have done in New Caledonia'.[164]

Murchison publicly deplored the French move in 1857. While whipping up anti-French sentiment at the RGS the following year, he privately sounded Bulwer-Lytton of the Colonial Office on Nicholson's behalf about establishing a post in north Australia. He cited Denision and Stokes to emphasise that 'now that the French have occupied in force the New Caledonia of our Cook, the case seems to me to be one of absolute necessity'.[165] In 1859 Murchison's RGS address again raised the issue in phrases redolent of Nicholson, and two years later Queensland's Governor Bowen followed the President's advice in founding a settlement and coaling station near the tip of the Cape York Peninsula.[166] When Nicholson returned to England in 1862 Murchison installed him on the RGS Council, engineered his presidency of the Geographical Section of the British Association in 1866, and continued to help him promote north Australian settlement, penetration of New Guinea, and the maintenance of an adequate naval force in Australian waters. Nicholson further focused his imperial interests when he helped found the Royal Colonial Institute in 1868.[167]

Murchison's concern with imperial rivalry in the South Pacific also extended to Fiji. By the mid-1850s its annexation was being advocated in certain British commercial circles in order to forestall perceived French ambitions.[168] In 1857 the RGS published a memoir on Fiji by naval surgeon–naturalist John MacDonald which included geological observations and estimates of timber values.[169] When Consul William Pritchard visited London in 1859 to announce the cession of the islands to Queen Victoria by their king, the government despatched a fact-finding mission before taking a decision. This expedition included the Kew-trained botanist Berthold Seemann, who was charged with evaluating Fiji's cotton-growing capability. Murchison was approached through Prince Albert by a German geologist anxious to accompany the expedition but, caught out for once in complete ignorance of the undertaking, he was forced to report after making inquiries that the mission had already sailed.[170] Murchison recovered his equilibrium when the expedition returned. He elected Seemann to the RGS and organised a special meeting to consider his report, which dwelt upon Australian desires that Britain accept the cession, American and French threats to Fiji's sovereignty, and the suitability of the islands for European colonisation.

In the ensuing discussion the Reverend George Pritchard, a former Consul at Tahiti expelled by the French when they established a protectorate there in 1842, spoke in favour of Britain acquiring Fiji to curtail French expansion. He was supported by Rear-Admiral Sir Edward Belcher, another imperially-minded hydrographic officer who had been instrumental in effecting the cession of Labuan to Britain as a naval base.[171] In 1863 Murchison's warm praise for Seemann's further Fijian publications left no doubt of his opinion as to the desirability of colonising the islands.[172] Though Fiji was not occupied by Britain until 1874, the reasons cited in justification were precisely those aired at the RGS over a decade before. Seemann's appointment demonstrates the growing willingness of the government to rely on scientific evaluations of potential colonies. The treatment of his memoir illustrates Murchison's willingness to permit the RGS to be used as a venue for annexationist agitation by scientists, missionaries, Antipodean expansionists, and naval officers – all of whom shared his own suspicion of French motives in the Pacific.

New Zealand

Because New Zealand was colonised by a joint-stock company during British geology's 'golden age' and many of its early immigrants boasted scientific training, the colony demonstrated precocious awareness of the ability of geology to stimulate economic progress. Much of the South Island, especially, was first explored by government geological surveyors. Like Australia, New Zealand fascinated European scientists as an evolutionary laboratory: its archaic fauna survived in a geologically active landscape dominated by vegetation reminiscent of the Carboniferous swamps being brought to light by palaeontologists and illustrated in popular texts such as *Siluria*.[173]

Coal, iron, and other economic minerals were reported in New Zealand during the 1830s. But it was the arrival in London early in the next decade of skeletons of the *Dinornis* or moa, an extinct giant flightless bird, which first stimulated metropolitan interest in New Zealand's geology.[174] The simultaneous publication of an accurate description of the thermal region of the North Island by Ernst Dieffenbach, a German naturalist employed by the New Zealand Company, represented a milestone of scientific observation in the colony. This account, to which Murchison was alerted through the British Association, reported Silurian strata as well as exploitable copper and coal deposits.[175] Dieffenbach also gathered natural history specimens for the British Museum and another collection which was maintained in London by the New Zealand Company as an inducement to emigration.[176] Dieffenbach had been recommended to his post by John Washington, Secretary of the RGS; as a political refugee in London before sailing to New Zealand he had forged connections with several prominent metropolitan scientists.[177] His appointment typifies the flow of scientific talent into the British empire which resulted from political repression and lack of opportunities for distant travel on official service in the German states. With Prince Albert as a focus of aspirations for German scientists seeking such patronage, the British empire became the natural outlet for their ambitions.

As Dieffenbach's troubled relations with the New Zealand Company demonstrate, however, liberal-minded scientists working in the colonies could find themselves at odds with the exploitative goals of expansive capitalism. In 1841 Dieffenbach's employers had provided the RGS with a report that he had written on the suitability of the remote Chatham Islands for colonisation. The New Zealand Company purchased these islands to evade a prohibition of further land acquisitions in New Zealand itself, but Britain, which claimed the Chathams, awarded them to the government of New Zealand. The RGS published Dieffenbach's account while legal possession of the islands was still in doubt. Dieffenbach had advised his employers against colonisation of the Chathams because of the scarcity of resources, yet the expurgated version of his report issued by the Company gave a contrasting impression of a country ripe for European exploitation.[178]

In his presidential address to the Geological Society in 1843, Murchison solicited the transmission of more fossils from New Zealand to settle the palaeontological issues raised by the moa. The discovery of the first fragmentary moa fossils had shaken belief in progressive biological development and the stratigraphic system based on the fossil record, for they were believed to pre-date all known birds. As Murchison pointed out, the discovery of moa foot bones of verifiable age might settle a debate as to whether certain ancient fossil footprints in the United States might be taken as evidence that birds had appeared contemporary with the first terrestrial reptiles. Citing his own endeavours to extend 'the order of succession established in our own isles . . . eastwards to the confines of Asia, and westwards to the back-woods of America', Murchison also called for colonial contributions to complete this global project of taxonomic annexation.[179]

In 1845 the Scottish geologist Hugh Miller introduced Murchison to their countryman James Coutts Crawford, an employee of the New Zealand Company who submitted a collection of fossils for his examination.[180] Another source of information was the English geologist Gideon Mantell, whose son Walter, a government employee in New Zealand, sent home many moa fossils and published the first geological account of the South Island's east coast.[181] During the 1850s geological specimens from New Zealand began flooding into Britain. The New Zealand Company, for example, presented a series of economic minerals to the Museum of Practical Geology to advertise the colony's advantages.[182] While the naval surveyor J. L. Stokes charted the colony's coasts, his surgeon–naturalist Charles Forbes accomplished research whose details were conveyed to Murchison by W. B. Clarke and eventually published by the Geological Society.[183] Forbes' observations included the first accounts of Otago's fiords and coal deposits, and Stokes sent a more general description of these seams to the RGS.[184] In 1850 the Society also published a memoir received through the Colonial Office describing Thomas Brunner's explorations on the north-west coast of the South Island.[185]

The principal news of the decade, however, was the discovery of gold. The metal had been reported in the 1840s, but its exploration was ignored until 1851 when the government offered a reward for the discovery of a goldfield to stem the settler exodus to California and Victoria. A rush at Coromandel in the North Island resulted in 1852, but the alluvial deposits were rapidly exhausted.[186] W. B. Clarke wrote to Murchison claiming credit for predicting this strike as well as those in New South Wales and Tasmania.[187] Sir George Grey, the science-minded Governor of New Zealand who had discovered Western Australia's Murchison River years before, submitted specimens and reports by the new goldfield's Commissioner through Murchison to the Geological Society.[188] Grey delivered these data while on leave in England, dining at Murchison's Belgrave Square mansion and appearing at the inaugural dinner of his Geographical Club.[189] New editions of James Wyld's prospectors' field guide, probably on Murchison's advice, downplayed the extent of the colony's gold resources to avoid exciting expectations unwarranted by the slim returns from Coromandel.[190]

While Murchison took up his duties as Director-General of the Geological Survey, a second minor rush in the South Island's Nelson Province revived interest in New Zealand gold. At this time Murchison involved himself in launching the Austrian *Novara* Expedition, which inspired the founding of several geological surveys in the colony. This voyage of circumnavigation during 1857–59 was organised by the Imperial Geographical Society of Vienna, an institution that had been founded in 1856 on the model of the RGS. Its leader, Wilhelm von Haidinger, was, like Murchison, also the head of his country's geological survey. The founding of Vienna's Geographical Society, like that of similar institutions in other countries, represented a deliberate attempt by Austria's scientists, military men, and expansionist commercial interests to enter the international competition in overseas exploration, and the *Novara* Expedition was an ambitious opening bid in this high-stakes game.

Haidinger requested Murchison to provide the *Novara*'s scientists with introductions to British authorities abroad, a service for which he was eventually rewarded by the dedication of the English edition of the expedition's narrative.[191] Ferdinand von Hochstetter, the *Novara*'s geologist, spent nine months in New Zealand during 1858. He first examined a coalfield near Auckland for the colonial government at the request of Governor Denison of New South Wales, who sent Murchison the results of his own preliminary analysis of the fuel. After the RGS received Hochstetter's official report, Murchison advertised the coal as abundant and of sufficient quality for both steam navigation and manufacturing.[192] Hochstetter also examined various ore deposits in Nelson Province and constructed New Zealand's first geological map in south Auckland.[193] Accompanying him was Julius von Haast, a newly-arrived German geologist. Commissioned to explore Nelson for nine months in 1859 after Hochstetter departed, Haast discovered several new coal and gold deposits.[194]

At the turn of the new decade Murchison's involvement with New Zealand intensified dramatically. When the Admiralty became interested in sources of cheap, light armour plate for the naval race begun by France, he was requested to have specimens of New Zealand iron sand, which were believed to be of steel-making quality, analysed in his Survey laboratory.[195] Hochstetter meanwhile brought two Maori chiefs to England. Murchison proudly displayed them at the RGS and the British Association, as well as publicising Hochstetter's research and encouraging the sale of the *Novara*'s narrative.[196] The work of Hochstetter and Haast also stimulated the province of Otago in the South Island to consider systematic mineral exploration.

In 1858 the RGS published a memoir by Otago's Chief Surveyor which reported abundant copper and coal on the province's west coast and suggested, according to Murchison's theory, the presence of extensive goldfields.[197] In

1860 the provincial government applied to Murchison to recommend a geological surveyor. Dr James Hector, a Scot who had recently distinguished himself on the Palliser Expedition in British North America, was selected. Murchison also offered Hector a more lucrative post in Kashmir, but on the Director-General's advice that New Zealand was more likely to satisfy his scientific ambitions, Hector chose Otago. During 1861 Murchison, Hector, and the Edinburgh agents of the Otago government worked out the details of a three-year contract which included provision for a chemist as assistant. Murchison's instructions undoubtedly emphasised alluvial gold since official reports of a rush near Dunedin had been received just before Hector's departure. [198]

Hector arrived in Otago in 1862 and immediately began surveying the goldfields with his assistant Charles Wood, a former student of the Royal School of Mines. The visiting Scottish naturalist William Lauder Lindsay had undertaken a personal survey of the goldfields just before Hector's arrival, noting evidence of glacial action and the derivation of gold from what he correctly surmised were Silurian rocks. By predicting the discovery of extensive goldfields throughout the province in a lecture entitled 'The Place and Power of Natural History in Colonization', Lindsay had ensured Hector's good reception by the colonists. [199] Lindsay also visited Auckland before returning to Britain and delivering two papers to the British Association in 1862. In his Otago memoir, which contained the startling phrase 'the supply of gold is at present practically *unlimited*', he noted that the development of these vast resources would require systematic exploration, an extension of administrative and transport systems, and massive capital investment once alluvial washing gave way to shaft mining. In his Auckland memoir Lindsay mistakenly stated that the North Island's gold deposits, which are actually quite recent, also derived from Silurian strata, and he pleaded that geological surveys should be founded in every province of the colony. [200]

The government of Wellington, on the advice of Walter Mantell, had already requested Murchison to recommend a geologist for a two-year survey in 1861. To influence Murchison's choice, two reports from J. C. Crawford, who had also been supplying information to the Geological Society, were enclosed with this request. [201] Extrapolating from field observations, a thorough reading of *Siluria*, and the stratigraphic conclusions of a geologist in the neighbouring province of Hawke's Bay who had unsuccessfully applied to Murchison for a surveying appointment before emigrating, [202] Crawford also identified Silurian rocks as the main component of the North Island's mountains. On the basis of this erroneous assessment, Crawford predicted the discovery of extensive coal, copper, and gold deposits in these ranges. He also hinted that the axial chain of the South Island might contain gold. Highly satisfied at the views expressed in these reports, Murchison recommended Crawford's appointment as geologist, after ascertaining that none of his own Survey staff would accept the post. Maps and reports of Wellington's geology were soon being received by the Director-General as a result. [203]

Having made two appointments to the South Pacific colony within a year, Murchison boasted to Sir Henry Barkly that 'Geological surveys are all the fashion in New Zealand'.[204] In his *Annual Report* as Director-General he noted that Hector and Wood were extending the Antipodean influence of the British Survey initiated by Selwyn and Gould, and that the Wellington request likewise demonstrated the desire of British colonists for geological information. Justifying the fact that his imperial enthusiasm sometimes worked against more parochial national concerns, Murchison explained that 'in thus serving the colonies we occasionally lose, it is true, some of our best officers, and thus partially retard the progress of the home Survey, but the imperial interests of Britain are thereby essentially promoted, and the colonies are all surveyed on the same principles as those which regulate our own labours.'[205]

During the same year Murchison began a correspondence with Julius von Haast, now Surveyor-General and Geologist of Canterbury, who informed him of the naming of a mountain, glacier, and river in his honour.[206] Until his post lapsed in 1868, Haast was occupied with the exploration of this province. He discovered several goldfields, opened the Southern Alps, researched the effects of glaciation, and initiated other scientific investigations carried on in future years as Director of Canterbury's Museum.[207] Murchison urged Haast to verify rumoured sightings of live moas and commiserated with him about the Maori wars.[208] His own nephew had served in the army in New Zealand, and it is probable that he viewed the decline of the Maoris with the fatalism many scientists displayed in regard to native peoples drawn into competition with Europeans by the forces of imperial and commercial expansion.[209] The Murchison–Haast exchange benefited both participants. Murchison secured valuable geological and geographical data regarding a little-known colony brought into prominence by the gold-rushes, while Haast received stimulating encouragement from one of the giants of metropolitan science.

In 1864 Haast sent Murchison a map, memoir, and series of watercolours illustrating New Zealand's Southern Alps, including the Murchison Glacier.[210] When these were presented at a meeting of the RGS, a discussion ensued about the commercial value of the South Island's gold, coal, and timber resources.[211] This dilution of science in the interests of popularity and imperial development exemplified precisely what Joseph Hooker, Director of the Royal Botanic Gardens at Kew and the most persistent critic of Murchison's methods of management, detested about the Society. While admitting to Haast that his paintings were best exhibited at the RGS, Hooker sniffed: 'I hate the claptrap and flattery and flummery of the Royal Geographical, with its utter want of Science and craving for popularity and excitement, and making London Lions of the season of bold Elephant hunters and Lion slayers, whilst the steady, slow, and scientific surveyors and travellers have no honour at all.'[212]

All this notwithstanding, the results of Haast's research were absorbed by the President, for he acknowledged that the New Zealand evidence had compelled some modification of his inveterate opposition to glaciers as engines of erosion. In contrast to his rather shabby treatment of W. B. Clarke, Murchison also brought Haast's views prominently before the metropolitan scientific societies.[213] By challenging Murchison's gold theories, Clarke had threatened the sanctity of the Silurian System, while Haast merely opposed Murchison's anti-glacial position. Yet this too represented a subtle threat to the inner redoubt of Siluria, for the issues of glaciers and gold intermingled. As the New Zealand evidence was beginning to make clear, the distribution of alluvial gold in formerly glaciated countries was largely determined by the action of ice. Thus, for Murchison, acceptance of the glacial theory ultimately meant not only jettisoning the catastrophic assumptions underpinning his geological creed, but an enforced further alteration of his gold theory, which by the 1860s he was no longer prepared to contemplate. Still, Murchison was highly impressed with Haast's pioneering research, commenting to him disparagingly upon the insularity of many British geologists; '*here* they care little about practical geology in other Countries, whilst they are keen as hawks about beds of a few feet thickness *chez nous*.'[214] In 1867 Murchison and Hooker co-operated in electing Haast to the Royal Society, for Kew Gardens had also benefited from his work.[215] In the next four years Haast provided Murchison with further information on the Southern Alps, and though the President was considering him for an RGS gold medal at the time of his own death, the explorer was not awarded this honour until 1884.[216]

Hector, too, had been working at full stretch in Otago during the gold fever of the early 1860s, and he made discoveries which triggered several minor rushes. In 1863 he also led an exploring voyage to the western fiords. During this expedition he examined coal deposits, inquired into further reports of living moas, and named an entire mountain range in honour of his patron Murchison. Hector discovered no further goldfields on this journey, but he had ascertained that Otago's alluvial gold derived from veins in Palaeozoic – probably Silurian – schists. Because of the unsuitability of the glacier-scoured landscape to the large-scale formation of concentrated placer deposits, he decided that Otago's geological characteristics would preclude the maintenance of an extractive industry unless it was based on large-scale sluicing operations and deep shaft mines.[217] Hector prepared a report of his expedition for Murchison, but it was never sent.[218] The President fell back instead on a report from the Commissioner of the Otago goldfields to compose the remarks in his Anniversary Address of 1863 on the probable extent of Otago's auriferous deposits. He here concluded, much as had Hector, that the bulk of the gold was found in the alluvial detritus of quartz veins in Silurian rocks. But since in contrast to Victoria these parent formations rarely cropped out, the colonists of Otago, he argued, could not look forward to a permanent mining industry after their alluvial deposits were played out. This statement paid mute tribute to the modification of Murchison's views recently effected by Selwyn and Barkly. At the same time, it countered Lauder Lindsay's alarmist prediction of an 'unlimited' yield with the now-standard placation that the gold supply would last just long enough to set the colony on its feet without threatening monetary chaos.[219]

In 1864 the publication in the RGS *Journal* of an abridged version of Hector's official report on his west coast explorations, together with a description of Otago's lake districts and Haast's paper on the Canterbury Alps, constituted the first comprehensive map and description of the South Island's entire alpine zone.[220] The same year Murchison presented the British Association with a memoir from Hector on the geology of Otago. In 1865 he submitted this description to the Geological Society as well, for Hector's views corroborated his own surmises about the nature and extent of Otago's gold.[221] Though gold production represented 70 per cent of the colony's export earnings in 1863, the output, which was never large by world standards, peaked in that year. The companies organised later in the century to tap the rich veins of Auckland with shaft mines and exploit Otago's poorer placers by dredging faced tapering yields.[222] In 1865 Murchison also published another letter from Hector which supported his position in the disagreement with Haast about the power of glaciers to excavate lake basins in solid rock.[223]

At the same time Hector was appointed to direct a geological survey of the whole of New Zealand. Following the organisation of similar institutions in the United Kingdom and India, this was the third geological survey within the empire to be organised on a national scale.[224] Hector established his headquarters in Wellington where he soon took charge of the colonial museum, laboratory, meteorological service, and botanical and zoological gardens. When the New Zealand Institute, forerunner of the colony's Royal Society, was founded in 1867, Hector was also installed as its manager, and he served for many years as the first chancellor of the University of New Zealand. From this central position of authority, unparalleled in Britain or any of her other colonies, Hector controlled every aspect of New Zealand science until his retirement in 1903. In 1865 he published the first geological map of the entire colony; in 1879 the first edition of his influential immigrants' guide to the colony's resources; and in 1886 the first authoritative overview of the colony's geology.[225] Murchison continued to receive official reports from the New Zealand Geological Survey throughout the 1860s. He and Hooker were again instrumental in electing Hector to the Royal Society in 1866. Haast's election was in fact postponed in Hector's favour, but this was due to Hector's greater eminence as the colony's senior government scientist rather than to his support for Murchison's views on glaciers and gold.[226] Hector was not to receive the gold medal of the RGS until 1891, seven years after Haast.

Hector's original assistant Wood died after a transfer to Victoria in 1863, but once installed at Wellington Hector assembled an impressive research team. It included Frederick W. Hutton, FGS, a former army captain who had acquired his geological training during a posting at Sandhurst. There he had studied under Professor Thomas Rupert Jones, expert palaeontologist, editor of *Siluria* as well as the Geological Society's *Quarterly Journal*, and one of the dispensers of the system of military scientific instruction which Murchison helped foster. Hutton had previously held an appointment as a geological explorer in Auckland before being hired by Hector in 1871. A memoir which he sent to the Geological Society at this time prompted Warington Smyth, a Royal School of Mines professor who was also involved in the Sandhurst scientific lecture programme, to express his gratification 'that the new system of education of military officers was productive of such good results in a geological point of view'.[227] Hutton remained with the New Zealand Survey for only three years, but he went on to forge a distinguished scientific career in the colony. Following Murchison's death Hector also hired two other assistants recommended by Ramsay, the succeeding Director-General, one of whom was a School of Mines graduate. There was thus a good deal of direct input from the British Survey to its New Zealand counterpart. The structure of Hector's Survey, like those of other colonies, largely reflected the model provided by the London establishment, but when New Zealand's first school of mines was founded in 1876, it was attached to the University of Otago.

Practical and theoretical concerns were mutually served by the Survey, but serious errors could thus be perpetuated which hindered economic development and the pursuit of knowledge. Because of the assumption that New Zealand's stratigraphy would mirror Europe's, and the desire on the part of the authorities that seams of 'old' or 'true' coal be discovered in the colony, Hector's geologists wasted considerable effort seeking non-existent Carboniferous-age coal deposits. Lacking a palaeontologist, Hector consistently overestimated the age of younger fossils which he thought gave evidence of such formations.[228] In 1867, Murchison commented favourably to a Royal Commission studying the empire's fuel reserves on Hector's first report on New Zealand's Cretaceous and Tertiary lignites.[229] After Murchison's death, however, a dispute developed between Hector and Hutton over the dating of these deposits which, like the controversy over Australian coals, had implications for the industrial and maritime future of the colony.[230] Once the gold-rushes which had prompted so much exploration subsided, the flow of data from New Zealand also abated and public interest moved elsewhere. The scientific institutions which Murchison had helped establish, however, and the tradition of independent local research which they in turn fostered, provided the colonists with the expertise and organisational structure to help solve the ongoing problems of adapting to a new environment.

Murchison's promotion of geographical and geological exploration as concomitant instruments of development was most consistently demonstrated in Australia and New Zealand. By roughing out the basic features of unknown territories – terrain, climate, resources, inhabitants – exploring expeditions provided data for capability assessments preliminary to occupation. Refined scientific inventories would subsequently aid systematic development. Murchison regarded the Antipodes as an exclusively British preserve wherein a long-term experiment in the scientific colonisation of an alien environment might be played out in splendid isolation, untainted by the meddlings of rival powers or competing cultural influences. For this reason he jealously guarded the inviolability of Australia's coastline, remained suspicious of French, Dutch, and American intentions in neighbouring islands, and advocated improvements in communications that would knit the colonies into a cohesive whole and bind them permanently to the mother country. Because of his own role in initiating development of their vast potential, Murchison remained fascinated with the South Pacific colonies. In this respect, as in his role as an unofficial representative of Australian interests in London, he followed in the tradition of Banks and Barrow.[231] The coincidence of his rise to authority, the saturation of the official mind with confidence in science as an engine of progress, the stimulus given mineral exploration by the gold-rushes, and the beginning of the vogue for exploration as a form of national drama gave Murchison unparalleled opportunities for action in the Antipodes. His impact on the exploration, settlement, mineral development, and establishment of scientific institutions in Australia and New Zealand was consequently greater than in any other part of the formal empire, as the number of natural features bearing his name still attests.

3

The Americas

Murchison displayed an opportunistic attitude towards the Americas throughout his career. In British territories, he urged the installation of scientific institutions to foster economic development and generate data. He encouraged similar undertakings on the part of sovereign nations, using the resulting information to refine and expand the range of his geological systems and theories, to maintain public interest in the RGS as a font of sensational discoveries, and to extend Britain's commercial and financial reach. Murchison considered North America as Western Europe's natural hinterland, but because the continent's most desirable real estate comprised the territory of the United States, his capacity to operate freely in this theatre was restricted. Murchison demonstrated pride in the accomplishments of the United States as a progressive nation sprung from British stock. Mixed with this fellow-feeling was suspicion of American assertiveness and military strength, for the revolutionary history and threatening expansionism of the former colony little endeared it to such an imperial patriot.

A spirit of heartfelt competitiveness on behalf of his country thus dominated Murchison's activities with respect to the United States – competition for political control of western North America, for influence in Latin America and the Pacific, for domination of westward trade routes to the Orient, and for international prestige in the patronage of science. In Latin America, by contrast, where the cultural tradition was distinct, no serious commercial rivalry loomed, and Britain's territorial interests (excepting Guiana) were not in question, Murchison's policies exhibited less anxiety and aggression. Here his concerns were scientific exploration, reconnaissance of resources for commercial development, the discovery of coal deposits to facilitate naval deployment, the extension of riverine and railway communications to open inaccessible areas, and the construction of an isthmian route to the Pacific colonies and the Orient.

Canada

Murchison's involvement in the exploration of British North America dated from the inception of official geological research in the eastern colonies. His recommendation, along with those of Sedgwick and Darwin, helped secure the appointment of Joseph Beete Jukes to survey Newfoundland in 1839. The colonial legislature, however, deriving no immediate benefit from research which included evaluation of coal and copper indications, terminated the contract in 1841.[1] Jukes' published journal provided the first accurate description of the island. Murchison praised the book for helping to raise Newfoundland's status from that of a mere fishing station to a resource-rich colony. He argued that Jukes' reconnaissance, which suggested a major exposure of Palaeozoic rocks, should be followed up by a detailed survey. Jukes' Newfoundland work thus established a precedent for research in the other North American colonies as well as helping to secure his appointment on the H.M.S. *Fly* expedition to Torres Strait.[2] Before sailing for Australia, Jukes spoke to Murchison at the Geological Society and at a soirée in Belgrave Square, and he would eventually be elected to the Royal Society with the help of the Silurian chief.[3]

By establishing an abundance of Carboniferous coal deposits, a simultaneous reconnaissance of New Brunswick conducted by William Jory Henwood, FRS, FGS, a Cornish mining expert, gave Murchison hopes of more Palaeozoic discoveries.[4] George Featherstonhaugh, a British geologist whom Murchison helped to secure an appointment with the American government, also provided early evidence of Silurian rocks in the United States.[5] When Feathersonhaugh later served on the Maine-New Brunswick Boundary Commission, whose official report was a highly partisan document replete with 'natural frontier' arguments based on bogus topographical features,[6] he also reported to the British Association on this colony's geological formations. In an incident illustrating the imperial sentiment of many British scientists, Featherstonhaugh had to be called to order for alluding to the political issue of the disputed territory.[7] Similarly, the coal researches in Nova Scotia of Charles Lyell and William Logan stimulated stratigraphic discussion and interest in economic development.[8] Logan was a Canadian geologist who had worked in the south Wales mining industry and, while aiding De la Beche's Survey, distinguished himself mapping coal measures and researching the fossil fuel's formation. Murchison was particularly concerned with Logan's Canadian work because it established a transatlantic extension of his new Permian System.[9]

In 1842 Logan was appointed to conduct a geological survey of Upper and Lower Canada – an institution which became the first permanent colonial survey in the empire.[10] Earlier attempts had been made to found a Canadian survey, yet no action was taken until Sir Charles Bagot, a progressive new Governor-General, reconsidered the question in light of the research in the maritime colonies and similar undertakings in the United States. Bagot persuaded the Provincial Legislature to finance a geological survey that 'might not only make known the hidden mineral riches of this country, but might also serve to elucidate the scientific theories in regard to the formation of this continent.[11] Logan was recommended to the post by Lyell, Sedgwick, De la Beche, and Murchison. The latter, having used Logan's Welsh maps in *The Silurian System,* backed him from the chair of the Geological Society as certain to 'render essential service to Canada, and materially favour the advancement of geological inquiry'.[12] Logan arrived in Canada in 1842 with Alexander

Murray, an assistant from the British Survey, and introductions from Murchison.[13]

Since Logan's initial appointment would terminate in two years, he set out to demonstrate the utility of geology by surveying coal and copper reserves. He also founded a museum modelled on De la Beche's in which he displayed impressive specimens of economic minerals. Arguing that his Survey's provision of accurate knowledge of mineral resources was responsible for attracting development capital and pointing out where it might be invested, Logan secured funding to hire a chemist and found a laboratory. His Survey gradually increased its staff and budget until as a permanent institution it was producing a full spectrum of maps and memoirs. These publications, along with Logan's ore displays at international exhibitions, advertised Canada's mineral potential and encouraged the surveying of other colonies. Logan's geologists only infrequently discovered economic minerals because they lacked the time for detailed prospecting, but their mapping of formations favourable to the occurrence of ores contributed to the opening up of several major mining regions. Negative results were also valuable. By proving that Canada lacked profitable deposits of 'true' coal, Logan prevented the speculative losses which Murchison estimated as having been sufficient in Britain alone 'to effect a general geological examination of the whole crust of the Globe'.[14]

Logan's Survey also functioned as a general-purpose science and exploration agency. As members of the 'first scientific arm of the Canadian government',[15] Logan's surveyors were the first officials to appear in many districts. The ethnological, natural history, and meteorological data they gathered eventually formed the nucleus of Canada's National Museum, and the original mapping required to record geological data proved invaluable for developing water transport, routing railways, and evaluating Crown lands. The training and research opportunities provided by the Survey, together with Logan's personal support of academic and sociate science, fostered the growth of an independent tradition of inquiry in Canada.

It was initially assumed that the Director-General in London would supervise Logan's geologists. As in India and Australia, however, the Canadian Survey soon achieved independence because local authorities controlled its funding and the problems encountered required a local data base. Still, the two surveys continued to exchange specimens and publications, and the Museum of Practical Geology helped Logan distinguish and date unfamiliar strata.[16] Logan also transmitted data to the Geological Society and, before temporarily running afoul of the Canadian Parliament for his involuntary involvement with mining entrepreneurs during the Lake Huron copper boom of 1847, even advised De la Beche on taking advantage of Canadian mining stock flotations.[17] Until his retirement in 1870 Logan ran his survey as a private empire in much the same manner as Murchison, Selwyn, and Hector controlled their respective establishments. His scope grew with Canada's federal authority. In 1864 he seconded Murray to survey Newfoundland; in

1868 he acquired New Brunswick, Nova Scotia, and Prince Edward Island following the passage of the British North America Act; in 1869 his writ expanded again when the Hudson's Bay Company surrendered its rights to the new confederation; in the 1870s, under pressure to explore the west, the Survey was caught up in the nation-building debate under Logan's successor, Selwyn, former head of the Victoria Survey.[18]

Logan's early research on Canadian Palaeozoics particularly endeared him to Murchison. Murchison displayed his evidence in *Siluria* and dedicated its fourth edition to him for applying Silurian nomenclature to 'vast regions of British North America' and proving that Laurentian Gneiss, the oldest stratified rock series then known, underlay even the earliest strata of Sedgwick's Cambrian system. In his native Scottish Highlands, Murchison won the reciprocal honour of first discovering this same 'foundation-stone' in Britain. He initially designated these formations as Fundamental Gneiss, but soon recognised their equivalence with Logan's system and used them as an anvil for his Silurian hammer in the attempt to obliterate Cambria. The discovery of primordial fossils in the Canadian Laurentians during the 1860s also pushed the origin of life back before the Cambrian period, just as an earlier find in Ireland had robbed Murchison of his claim to a Silurian beginning for organic creation.[19]

Murchison praised Logan's work as a model to other colonies and as an example of the interdependence of geology and geography. For this latter reason Logan's annual reports were forwarded to the RGS as well as the Geological Survey.[20] When Logan visited London to organise Canada's mineral display at the Great Exhibition of 1851, Murchison secured his election to the Royal Society and used specimens he provided to illustrate a lecture on gold.[21] In 1856, the year Logan was knighted, Murchison accepted the Geological Society's Wollaston Medal on his behalf and arranged his election to the RGS.[22] Logan later reciprocated by contributing his advice on campaigning in the Canadian forests to the RGS's *Hints to Travellers*.[23] In 1867 the *Siluria* dedication and the bestowal of a Royal Society medal helped Logan win an extension for his Survey throughout the new dominion.[24] During Logan's visits to London in the 1860s to oversee the publication of his *Geology of Canada* and act as Canadian Commissioner at several European exhibitions, he consolidated his relationship with Murchison,[25] who provided introductions to many Europeans visiting Logan's museum in Montreal.[26] Murchison also maintained relations with other key Canadian scientists, notably J.W. Dawson, FGS, Principal of McGill University, whose son was a student at the Royal School of Mines and later became Director of the Canadian Geological Survey.[27]

As Director-General of the Geological Survey, Murchison remained keenly interested in the discovery of Canadian gold and coal because of their Palaeozoic associations. In 1861, when a gold-rush occurred in Nova Scotia, Murchison was able to cite the metal's occurrence in curious formations of

Silurian age as evidence for his auriferous theory, while his friends mused whether he had predicted the discovery.[28] In 1866, reporting on colonial coal reserves for a Royal Commission, Murchison predicted Newfoundland's potential to be as great as Nova Scotia's and recommended its publicity to encourage exploitation. Nova Scotian financial interests, however, complained that the development of Newfoundland's coal would be delayed for years because of that colony's under-financed survey, and that meanwhile investment in Nova Scotia would be unfairly hindered. Murchison accordingly disavowed the statement, for by promoting both colonies' development he neither contradicted his original assessment nor unduly favoured specific parties.[29] Other economic minerals also received his attention. Murchison's advice to Logan concerning the competitiveness of Canadian copper ores in the British market *vis-à-vis* newly developed Namaqualand sources demonstrates that the metropolitan and colonial surveys co-operated in using the empire's centralised data network to advise colonial investors about the profitability of opening new mines.[30]

The Arctic

Beside the geographical lure of the North-west Passage, Murchison's interest in the Arctic sprang from his efforts to promote the scientific investigation of British North America. Knowledge of the structure of 'a region which we have almost made our own', he believed, would provide a bridge between European research and that of Canadian and American geologists.[31] Arctic exploration offered opportunities to extend British stratigraphic nomenclature and gather data for theoretical debates about the history of the earth. It also provided publicity for the RGS, potential economic and strategic benefits for Britain, and peace-time exercises for the Royal Navy. Murchison had received his first instructions in geology in the company of the Arctic explorers Franklin, Richardson, and Back when they were preparing for an expedition in 1825.[32] The scientific results of this mission included the discovery along the Mackenzie River of extensive tracts of ancient sedimentary strata later classified as Silurian. The explorers also reported coal, copper, and lead deposits on the shores of the Beaufort Sea. Accumulating evidence from such Admiralty expeditions which suggested a warmer Arctic during the Palaeozoic era sparked controversy regarding climatic change, and succeeding editions of *Siluria* charted the growth of geological knowledge of the region.[33]

Murchison remained closely associated with the group of officers and scientists constituting the 'Arctic Lobby', and after his assumption of command at the RGS he was centrally involved in nearly every expeditionary scheme regarding the region. He even participated vicariously, lending a geological hammer to George Back for use during two expeditions in the 1830s.[34] The dramatic search for Franklin was ready-made to generate public sympathy, and Murchison exploited it brilliantly as the champion of Lady

Franklin's forlorn cause. A clique of scientists employed this issue to win financing for expeditions undertaken long after Franklin had been given up as lost. As RGS Vice-President, Lord Ellesmere remarked to Murchison that such projects 'should be started with the fiat of the experienced and sustained with the money of the ignorant.'[35] Murchison promoted the despatch of the various Franklin search expeditions, helped draft their instructions, rewarded their leaders, received privileged access to their geological collections, and even had a sound in Greenland named for him – an honour he defended with characteristic vanity.[36]

Among the results of the Franklin searches were valuable mineral discoveries. In 1853 Murchison's announcement to the RGS that specimens of the rich copper ores from western Greenland submitted by Captain Edward Inglefield indicated 'that a very large portion of this region may prove to be metalliferous' prompted an expedition by a British speculator.[37] Murchison also used his influence to advance the careers of the officers and naturalists of these voyages. He helped secure an appointment as Geological Surveyor of Natal, for example, for the Scottish surgeon-naturalist Peter Sutherland, who had discovered Silurian fossils in Baffin Bay.[38] In 1860 Murchison also recommended a naturalist and provided geological instructions for an Admiralty expedition sounding the North Atlantic preparatory to the laying of a submarine cable by a private company.[39] Yet the government was never zealous enough in the cause of Arctic discovery for Murchison's taste. When a bid for renewed exploration failed in 1865, he disingenuously blamed the authorities for failing to support a project which he had characterised in private as an 'Arctic adventure in the cause of science & the Fine Arts' because it promised no 'political or monetary profit.'[40] No further opportunities for publicly funded polar research presented themselves during Murchison's lifetime, but he continued to encourage work by individual scientists, such as a visit to Greenland by the glaciologist Edward Whymper.[41]

The far west

While promoting exploration in the Atlantic colonies and the Arctic, Murchison maintained a vigil over American expansionism. In 1842, no doubt influenced by Featherstonhaugh, he denounced the Webster–Ashburton Treaty for conceding northern Maine – 'a ravelin indented . . . upon the Southern bank of the River St. Lawrence!' He also criticised Peel's government for 'knuckling down to our Trans-Atlantic cousins' in regard to the treaty.[42] During the Oregon boundary dispute Murchison similarly informed the RGS that Britain's claim was valid and that a proposed canal across the Isthmus of Tehuantepec would offer 'a short cut to the Pacific and all colonies, *in esse* and *in futuro,* whether on the Oregon or elsewhere, to which British enterprise may extend.'[43] The Oregon question also affected Murchison financially, for he was concerned to sell his holdings of the notorious Pennsylvania state bonds before they became utterly worthless. When the war for Oregon failed to materialise, he retained the bonds but became worried again about how to 'act accordingly with his *Pens*' in regard to possible British involvement in the Mexican War.[44]

When California achieved statehood in 1850, Murchison noted in the *Quarterly Review* that the Hudson's Bay Company had known of the gold thirty years before and might have secured a mineral lease of the entire territory from the bankrupt Mexican government. This arrangement would have given Britain the wealth of the Sacramento, the port of San Francisco, and an inviolable claim to the entire region south of the Columbia River. Still, the scientific efficiency of Frémont's occupying expedition won Murchison's praise. Content to see the United States expanding west and south rather than north, he also expressed his opinion on mineral discovery as a spur to annexation by advocating the American purchase of Baja California to legalise the forays of Californian miners into the Mexican Sierras.[45] A minor gold-rush in the Queen Charlotte Islands illustrates the sort of buccaneering Murchison feared Americans might attempt on the north-west coast. Gold was discovered in these little-known islands in 1850. The authorities in London and Vancouver Island were forced to make rapid preparations to defend British sovereignty against the threat posed by the arrival of several hundred prospectors from California in 1852. An extension of the authority of Vancouver Island's Governor, the deployment of a naval vessel, and a diplomatic request that the American government restrain its citizenry averted any attempts at a filibuster, but the rapidly dwindling gold output did at least as much to defuse the situation.[46]

Murchison was himself tangentially involved in these events. In 1852 a British firm trading out of San Francisco sent an expedition to the islands which found a gold vein in Mitchell Harbour. One of the partners subsequently travelled to London to interest investors and apply to the Colonial Office for a mineral lease. When the firm's representative solicited the opinion of the eminent mining consultant Richard Taylor, FGS, FRGS, Taylor advised Secretary of State for the Colonies Newcastle that the ore could be profitably worked only by a large-scale, mechanised operation. In consequence, he recommended the granting of a sixty-year lease in exchange for a 5 per cent royalty to attract the capital required. The Colonial Office then consulted De la Beche. He agreed that mineral exploration of the islands should be stimulated but cautioned that a long lease might hinder development of the harbour, which could prove strategic to the developing Pacific trade.[47] To promote the speculation, the applicant next offered specimens and a memoir describing the discovery to Murchison, who as RGS President publicised them through that society and the British Association.[48] This enterprise failed, as did other attempts to mine gold, coal, and copper. In the late 1860s the RGS was still calling for development of the islands' minerals, samples of which were displayed in the Museum of Practical Geology.[49]

In the late 1850s, when another border dispute erupted over the San Juan

Islands, Murchison provided the Foreign Office with the latest American maps from the RGS library[50] and developed a distrust of the American senator William Seward for his apparent desire to annex all of British North America. When, as Secretary of State, Seward purchased Alaska in 1867, Murchison worried that the gambit presaged a new pincer attack on British Columbia: the territory appeared to contain little of value except traces of the gold whose abundance Murchison had predicted in the hope of profiting his friend the Tsar.[51] The American Civil War caused Murchison another financial crisis linked to the threat of United States' expansion: he lost upwards of £2,500 liquidating his Pennsylvanian bonds and nearly took a 75 per cent loss on his Massachusetts investments. A long war, he reasoned, would bankrupt the North and render his holdings mere paper. If the war proved short and the South achieved independence, the two American republics would fall on Canada and Mexico to recoup their losses, precipitating war with Britain and Federal repudiation of British-held securities. Murchison hoped instead that peace could be rapidly effected, and that the South, the North, and the West might separate to Britain's benefit.[52]

Murchison also did what he could to combat the growing American presence in Hawaii, which by the early 1850s was considered a potentially important market and port. Visiting the islands in 1861, Lady Franklin employed Murchison to persuade Palmerston that Queen Victoria should stand as godmother to King Kamehameha IV's infant son. She thus hoped to thwart the influence of American and French missionaries who she and the King believed were bent on annexation. Demonstrating the global reach of Murchison's name, she also informed him that the King planned to write to him about cotton cultivation in Hawaii.[53] As Palmerston agreed with Murchison that a Hawaiian royal visit to England 'would be a good thing', Lady Franklin arranged Queen Emma's visit to coincide with the International Exhibition of 1862 in order to augment further Britain's prestige among the powers.[54]

Against this background of suspicion of American expansionism, Murchison took part during the later 1850s in the organisation of John Palliser's British North American Exploring Expedition and Lieutenant-Colonel J. S. Hawkins' North-West American Boundary Commission. These enterprises sought to assert a British presence in western North America, link this region with Canada, and assess the Hudson's Bay company lands turned over to the Crown in 1858.[55] Palliser's plan to explore the Canadian prairies and Continental Divide had originally been submitted in 1856 to the RGS, whose Expedition Committee then employed Murchison to recommend it to the Colonial Office. The Expedition Committee refined Palliser's aims to comprise tracing the course of the Southern Saskatchewan, evaluating the region for settlement, and exploring the Rockies for a southerly pass to British Columbia. The RGS estimated the cost at £5,000 and suggested that the expedition could accomplish its goals in two years. The Society also recommended that trained subordinates be appointed to make accurate scientific appraisals, and Murchison emphasised the probability of discovering coal and other economic minerals. Injecting a note of international rivalry, Murchison likewise cited the parallel American railway surveys in the far west and Russia's explorations in the Amur region to emphasise that Britain had a duty to fund such researches.

Herman Merivale, Permanent Under-Secretary at the Colonial Office, approved the project on political grounds but distrusted the cost estimate, 'recollecting what passed when the N Austr. expn. was undertaken'. Assistant Under-Secretary John Ball – a friend of Palliser's, keen botanist and alpinist and, like Merivale, Fellow of the RGS – adroitly circumvented such objections to secure official sponsorship for the expedition. Ball argued that the margin of error regarding the cost of Palliser's venture was much less than that of Gregory's Australian expedition. On his advice Secretary of State for the Colonies Henry Labouchere approved the proposal in order to provide objective information upon which to base the decision as to whether to renew the chartered privileges of the Hudson's Bay Company.[56]

While the RGS had transformed Palliser's plan for an exploring adventure into a scientific undertaking with political and economic overtones, Ball further amplified the project in line with his own interests. He then consulted Murchison and the other directors of government scientific establishments, as well as a committee of the Royal Society consisting of these same men, for recommendations as to research desiderata and personnel. Murchison was ill pleased to learn that an expedition proposed by the RGS had thus been taken over by the Colonial Office and its scientific planning entrusted to the Royal Society. His request that the RGS be permitted to communicate directly with Palliser in the field was vetoed, though Ball promised that the Society's wishes would continue to be considered and Labouchere ordered that the results of the expedition not considered politically sensitive should be transmitted to the geographers as received. Despite this loss of authority, Murchison made the best of what remained a promising opportunity. He offered his advice on behalf of the RGS, approved the Royal Society's suggestion that a botanist, terrestrial magnetician, and geologist be appointed, and noted the geological recommendations which had been solicited from other scientists. As Director-General of the Survey, Murchison chose a young Scot, James Hector, to accomplish the geology and zoology.[57]

Announcing Palliser's departure to the RGS, Murchison emphasised the importance of the expedition to British national interests, noting that it sought to establish a direct line of intercourse between Canada and the valuable ports and resources of Vancouver Island.[58] American settler infiltration into the tempting British prairies was proceeding unchecked, and the only sizeable British settlement in central North America, the fractious Red River colony, looked south rather than east for lack of a transport route. Palliser's survey represented the preliminary move in a new policy of colonisation and route construction to check this trend. Murchison understood the urgency of the work, for in 1852 he had announced to the RGS that an American proposal for a transcontinental railway as 'the shortest road to the richest countries of the East' had been countered by Captain Millington Synge, FRGS, of the Royal Engineers, with a rival British project involving canals, river improvements, railways, settlements, and mining development. This scheme emphasised that the Canadian route would turn the 'tide of emigration, labour, and commerce' northward from the United States.[59] Murchison cautioned, however, that a complete survey was required since Lord Selkirk, proprietor of the Red River colony and a stockholder of the Hudson's Bay Company, had assured him that little of the west was fit for cultivation and its topography presented 'almost insurmountable' barriers to a commercial route.[60] Synge fumed that this comment, by contradicting earlier remarks by Murchison on the feasibility of the route and the mineral wealth of the Rockies, made 'all the difference between *possible speculation* or assured investment in the eyes of men of capital & business . . . who cannot judge in the matter for themselves'.[61] The following year Synge reported to the RGS that he had secured the support of Newcastle and Gladstone in Aberdeen's new cabinet, and in 1855 he thanked the Society for helping him keep the scheme before the public.[62]

The issues raised by Lord Selkirk surfaced again in regard to the Palliser Expedition, for opponents of monopoly contended that the Hudson's Bay Company had exaggerated the inhospitableness of the prairies, sequestered knowledge of easy southern passes across the Rockies, and played down the resources of its Pacific coast territories in order to maintain them as an exclusive preserve. In 1857 the RGS cross-examined Sir George Simpson, retired administrator of the company's territories, on these points in order to suggest instructions for Palliser, whose aloofness from the controversy had weighed heavily in the Colonial Office decision to sponsor his expedition. Yet the questions which the Society put to Simpson were hardly disinterested, for they had been formulated by the Reverend Charles Nicolay, an academic geographer and member of the Aborigines Protection Society who had emerged as a critic of the Hudson's Bay Company. Nicolay had in fact sponsored Palliser's election to the RGS shortly before his plan was proposed.[63] Parties concerned with depriving the Company of its monopoly in order to achieve commercial gain also attempted to influence the aims of the expedition through Murchison. William Kernaghan, a partner in a Great Lakes steamship company desiring free access to western British North America, testified against the Company before a parliamentary committee in the same year in which he provided introductions for Palliser and advised Murchison on the route the expedition should follow.[64] Political and commercial interests in the Pacific settlements also wrote to Murchison pressing the necessity for Palliser to discover a pass which might break the seal of isolation imposed by the Company to guard its trading sphere.[65]

The RGS thus became an alternative court of appeal in which the stockholders, politicians, commercial promoters, and philanthropists appearing before the Select Committee of the House of Commons on the Hudson's Bay Company attempted to shape the pending decisions regarding the Company's territories. To give point to these issues, a memoir on the endowments of Vancouver Island and the efforts of the Company to hinder its development by private interests was read at the RGS soon after Palliser's departure. Its author was an army officer who hoped to be appointed the first Crown colonial governor of the island.[66] A similar memoir offered to the British Association stressed the government's duty to develop the island in order to ameliorate the condition of its natives.[67]

Palliser had several interviews with Murchison before he departed from England.[68] The geologist James Hector, however, a medical graduate of Edinburgh University, received not only precise instructions as to economic and theoretical research, but a complete outfit of books, maps, and instruments to aid his inquiries.[69] West of the Great Lakes, Hector was to look out for Silurian and other Palaeozoic strata, especially Carboniferous coal seams. Murchison hoped such a fuel source for the railway and steamship lines being planned to open the prairies might be found because of accumulating fossil evidence brought together by a critic of the Hudson's Bay Company whose results he had helped publish.[70] In the Rockies Hector was to search for the Director-General's 'golden constants', and as a result of the controversy concerning glaciation, note evidence of ice action. On the Pacific coast he was to examine the coal deposits of Vancouver Island and search for similar deposits on the mainland. Vancouver Island's coal seams had been known to the British Geological survey since samples were forwarded by the Admiralty in 1847 during De la Beche's Coal Enquiry. They had been desultorily mined by the Hudson's Bay Company for local use and export to the Californian goldfields. The Royal Navy, however, was anxious that these deposits be exploited under official controls similar to those governing the coal mines of Labuan in order to facilitate warship deployment and prevent the fuel being leased by American steamship interests intent on dominating trans-Pacific commerce. Vancouver Island's coal thus became an issue to opponents of the Company's monopoly as well as a primary influence on the Admiralty's decision to create a major naval base at Esquimalt Harbour.[71]

While the Palliser Expedition began its field work, the Canadian government sponsored a second exploring party to survey routes between the Great Lakes and the Red River settlement. This expedition included the geologist Henry Youle Hind, whose tracing of Silurian and Devonian strata across a broad region west of Lake Winnipeg was avidly seized upon by Murchison.[72] At the same time, the North-west American Boundary Commission was organised by the Foreign Office to survey the 49th parallel from the Pacific to the Rockies in co-operation with an American team. The geographers were uncomfortable with the 49th parallel as an international boundary because, in ignoring natural features, it violated the logic of their science and 'cut off from either side portions of territory which will therefore be valuable to neither'. It was bad enough, they felt, to endure this 'absurdity' east of the Rockies, but among the tortuous watersheds and massed peaks of the Pacific side they hoped to prevent its implementation altogether. Such was the nature of the extraordinary protest, so reminiscent of Featherstonhaugh's attempt to influence the Maine–New Brunswick boundary with the false positioning of watersheds, which Nicolay requested the RGS Council to address to the Foreign Office, and which constituted nothing less than a bid for authority to attempt a border rectification in Britain's favour by way of finesse in the field.[73]

In January of 1858 Clarendon, the Foreign Secretary, requested Murchison as RGS President to contribute scientific suggestions for the Commission. The staff would comprise Lieutenant-Colonel Hawkins, RE, FRGS, a geologist, a surgeon–naturalist, and two officers trained as astronomical and magnetic observers.[74] Similar letters were sent to Kew Gardens and the Royal Society. Before answering, Murchison obtained Hawkins' instructions from the War Office. Drawn up by Henry James, Superintendent of the Ordnance Survey and a member of the RGS Council, these directed Hawkins to evaluate British Columbia as a colony and explore for passes across the Rocky Mountains in British territory. The orders also reiterated the RGS position in stating that the choice of strictly following the parallel or searching for a natural frontier was the Foreign Secretary's alone.[75] Murchison now made the point even more forcefully on behalf of the RGS Council, emphasising that Hawkins should fix the positions of permanent features along the border so the two governments might obtain if possible the advantage of a natural boundary between their respective territories 'instead of the mere straight line defined by the 49th N. Latitude'.[76]

As they had for Palliser's Expedition, Sir William Hooker of Kew Gardens and Sir Edward Sabine, RE, of the Kew Observatory chose and instructed, respectively, the botanist and terrestrial magnetism observers. Murchison provided geological orders for his appointee Hilary Bauermann, a graduate of the School of Mines employed on the British Survey.[77] Bauermann's instructions recapitulated Hector's. He was to examine Vancouver Island's coal tracts, explore the mainland between the Pacific and the Rockies, and search the mountains for Silurian, auriferous, and glacial evidence.[78] The Director-General was clearly desirous of another gold-rush to set what he hoped would soon be a new colony on its feet and render a transcontinental link necessary. Within two months his expectations were justified by arrival of the news of the Fraser River gold discoveries, which he claimed to have anticipated though they were made by prospectors rather than his geologists.[79]

During 1857 and early 1858, Hector had meanwhile discovered Silurian strata near the Red River settlement and coal on both branches of the Saskatchewan. Murchison used this information and Palliser's reports to advertise the potential of the Canadian prairies, the value of geological exploration, and the imperial role of the RGS.[80] Six months later Hector sent

Murchison a route map, provided details about the coal formations now dated as Cretaceous, and announced that he would explore the Athabasca River during the winter while Palliser crossed the Rockies to meet Hawkins.[81] At the same time, the Foreign Office alerted the RGS about the latest U.S. Army surveys of Pacific railway routes.[82] Palliser was confidentially instructed by the Colonial Office, now under the direction of Bulwer-Lytton, to establish the feasibility of a Pacific railway leading west from Red River, whose development as a separate colony was under consideration.[83] Palliser had already advised Labouchere that the Hudson's Bay Company monopoly should be abolished and that Canada could not control the far west because the cost of constructing transport routes would remain prohibitive until mineral discoveries increased the incentives. He suggested the creation of new colonies centred on Red River and Vancouver Island, but these views were politically unacceptable to Canadians, who now saw their national destiny in westward expansion.[84]

   The Fraser River gold strikes dramatically altered the situation and gave Palliser's explorations fresh immediacy. By coinciding with the debate over the Hudson's Bay Company and introducing a flood of American prospectors into the region, the discovery of gold precipitated the assertion of Crown authority over Vancouver Island and the adjacent mainland in August 1858. Even a new name had to be found for the colony which became British Columbia, and Murchison was consulted by the Colonial Office on this matter. Harking to Highland associations, he proposed the title of 'Fraserland' for the territory hitherto designated as New Caledonia, but nothing came of the suggestion.[85]

   By January 1859, Murchison had received reports regarding the crossing of the Rockies, Hector's discovery of Kicking Horse Pass, the route eventually chosen by the Canadian Pacific Railway, and his naming of peaks after his mentors. Murchison boasted of the honour to his friends and directed that the account of Mount Murchison be read at the RGS.[86] He now requested the Colonial Office to permit Palliser and Hector to examine the interior of British Columbia. With the gold discoveries attracting more American than British immigrants, it was important to ascertain if easy communication might be established between the new gold regions and the east, and Hawkins' Boundary Commission was too far south to determine this issue.[87] The RGS meeting to discuss the discoveries was held in February 1859. Murchison borrowed Hector's and Palliser's maps and despatches from the Colonial Office and presided over a spirited debate between John Ball, who had now left the Colonial Office, Sir George Back, the Arctic explorer, George Hills, the Bishop-designate of British Columbia, Edward Ellice, the Deputy Governor of the Hudson's Bay Company, and Lord Bury, MP, former Superintendent of Indian Affairs in Canada, future founder of the Royal Colonial Institute, and an advocate of Canadian federation and independence. Bury dominated the meeting, attacking the Company's administrative record and urging a new programme of road building and colonisation to link Britain's North American possessions.[88]

Murchison wrote to Hector immediately after the meeting, relating how the expedition's reports had given great satisfaction and filled the Society's meeting room. 'The people (few in number) who are interested in the Hudson Bay Company's occupation are of course untoward', he continued, but his own summation of the proceedings had countered their protests by emphasising the possibility of the new discoveries 'laying the ground work for a profitable eventual intercourse between the region of the Saskatchewan & British Columbia'. Supplementary geological instructions were conveyed at the same time: 'The recent accounts of the increasing abundance of gold dust in the Fraser riv. & *its affluents* & the statement that there are slate rocks with quartz veins in the adjacent ridges make me very desirous that you should prove the original matrix to be a true Silurian rock as the Gold rocks of Victoria in Australia & elsewhere.' The Director-General's eagerness to add a province to his stratigraphic empire and claim another colonial gold-rush to the credit of his gold theory was clearly evident in his plea that 'a single trilobite or graptolite will suffice to explain' the age of the rocks forming the axis of the Rockies.[89]

By the Anniversary meeting of May 1859, Murchison had been warned by Hector that the imminent arrival of thousands of American prospectors lent urgency to the need for an all-British route across the Rockies to the diggings. Despatches were also received from the Governor of British Columbia extolling the colony's climate and forwarding gold nuggets for Murchison's Museum.[90] In presenting the RGS gold medal to Palliser through the Earl of Carnarvon, Under-Secretary of State for the Colonies, Murchison praised Canadian efforts to compete with the United States in its westward surge. He expressed his hope that Palliser's explorations and the gold-rush would stimulate the construction of a transcontinental railway, 'creating fresh centres of civilization, and consolidating British interests and feelings'.[91] In his Anniversary Address, Murchison predicted the discovery of valuable coalfields in the prairies on the basis of Hector's finds and urged the adoption of Synge's railway project. He continued to warn against a too hasty beginning by 'men of ardent minds', but believed that telegraphic communications should be established with the Pacific colonies. He went on to review their attributes, including Vancouver Island's lignites, which Bauermann reported were fuelling steamers connecting the island's capital of Victoria with the Fraser goldfields, and to infer that the gold formations of British Columbia probably extended northward into Russian America and eastward into the Columbia River drainage.[92] The gold strikes in the Cariboo country on the upper Fraser in 1860 and on the Stikine River during 1861 and 1862 would soon offer what seemed to be yet more evidence of Murchison's prophetic powers.[93]

In December of 1859, at the reading of a memoir jointly authored by the Chief Justice of British Columbia, a Lieutenant of Engineers stationed at the capital of New Westminster, and a naval officer engaged in charting the surrounding waters, the RGS hosted another discussion on the colony. This meeting devolved into a political debate about the dispute which had erupted between Britain and the United States regarding the sovereignty of San Juan Island, the largest in the archipelago lying between Vancouver Island and the American mainland.[94] By this time Hector had reported his failure to find another pass providing a more direct route to the goldfields. Yet he soon established that the mineral province known to yield gold in the Washington Territory continued north of the border, and discovered what were probably auriferous Silurian slates in the Rockies, through no confirmatory fossils presented themselves. Hector also noted the 'stiff-necked' attitude of American military forces towards the San Juan crisis.[95] Palliser meanwhile reiterated to Bulwer-Lytton his conviction that Canada could not be given responsibility for governing western North America, and that a series of independent British colonies presented the logical alternative.[96] In June an RGS meeting devoted to reports from the expedition was transformed into another attack on the Hudson's Bay Company, which still retained control of vast stretches of British North America.[97] At the meeting of the British Association in 1860 Murchison again held forth on the discoveries of the Palliser Expedition and the utility of Synge's railway proposal, noting that, should Hector's passes prove impracticable, territory should be exchanged with the United States to secure the best route linking British Columbia with the Red River settlement and Canada. This transcontinental artery was to be effected by stages – a telegraph would pioneer the route in the wake of the explorers, roads would follow to open commerce, and the culminating railway would weld the colonies permanently together.[98]

The same year Palliser's and Hector's reports were published in the *Parliamentary Papers*, where a contribution by the expedition's terrestrial magnetician warned that, without the immediate construction of east-west routes, Britain would lose its prairies to the United States.[99] In 1860 Hector also returned to Britain, where he completed his scientific work and presented his collections to the public museums.[100] Murchison elected Hector a Fellow of the RGS and the Geological Society and soon secured him a new appointment in New Zealand. He also invited Hector to compose the remarks regarding British North America which he himself issued on behalf of the RGS in 1861. In encouraging colonisation of the prairie region, Hector here contrasted the fertile British plains with the corresponding 'desert' in the United States. This factor, he argued, coupled with the disturbances of the Civil War, would 'indefinitely' delay construction of an American transcontinental railway. A Canadian line could succeed, but only 'as part of a great national enterprise' to extend the settlement frontier and link Canada with the Pacific colonies. The mineral wealth of the far west offered the best inducement to build a railway across the Rockies, and the harbours and coal of Vancouver Island designated it as the natural western terminus. Pressing the concept of a new route to the Orient, Hector also reminded Murchison's audience of the coal deposits of Japan and Formosa.[101]

As Merivale had feared, the £13,000 cost of the Palliser Expedition was nearly three times the original estimate. Yet in terms of the results achieved, and in comparison with the far greater sums expended by the United States on transcontinental railway surveys, the enterprise had been a bargain. The general topography of a huge swath of territory had been confirmed and corrected, the first accurate map of the British Rockies laid down, several passes discovered, and substantial new knowledge produced concerning the geology and natural history of the region. Reports meanwhile continued to arrive from Bauermann, but he had been unable to explore the gold districts visited by Hector, and Murchison consequently pressed Palmerston's government to provide British Columbia with a regular geological survey. This suggestion was declined, and Bauermann's appointment terminated in 1863 after he arranged his collections and reports under Murchison's supervision.[102]

British Columbia and Vancouver Island continued to receive publicity from the RGS, including memoirs by colonists eager for development.[103] Reports and specimens of British Columbian ores also flowed into the Museum of Practical Geology.[104] The Civil War in the United States temporarily dulled British anxieties about American expansionism in the far west, but the isolation of British Columbia from London for six weeks during the *Trent* affair of 1861 (see below, p. 87) provided Newcastle, Bulwer-Lytton's successor as Secretary of State for the Colonies, with renewed incentive to promote a Canadian rail link. A memoir read before the RGS which demanded this service prompted Murchison to predict on the basis of the gold strikes that British Columbia 'was about to open out to us a complete new California'.[105] Thomas Baring, MP, FRGS, also organised a petition to the Colonial Office from fellow bankers that called for the abolition of the remaining Hudson's Bay Company monopoly of trade and transit in central British North America. This petition also demanded the establishment of communications with the Pacific colonies in order to thwart American settlement of the prairies and ensure British retention of the lucrative China trade, in which Baring Brothers was heavily involved.[106] Synge offered yet another paper to the RGS on the same theme, and an overland transit company established in London subscribed £500,000 as a result of the euphoria created by the new gold discoveries in British Columbia.[107]

Newcastle was prevented from complying with the growing pressure to act in the matter by the retrenchments of Gladstone's Exchequer. The lack of either economic incentives or political will sufficient to mobilise capital investment continued to prevent the transformation of Palliser's exploration into a transport route uniting Britain's North American possessions. The initiative for new colonies on the prairie likewise foundered on the question of finance, and no steps were taken to buy out the residual rights of the Hudson's Bay Company. Financial trouble plagued the Pacific colonies as well: a depression caused by falling gold yields led to the unification of British Columbia and Vancouver Island in 1866 on the grounds of administrative economy.

In 1864, however, Murchison was still advertising Vancouver Island as a 'rising colony' and commercial entrepôt in his Anniversary Address.[108] He also refereed a memoir for the Society which described British Columbia's bright future as a locus for European settlement. The author, one of the officers of the Royal Engineers who were playing a prominent role in organising the new colony, closely followed Murchison's theories in predicting that, as the alluvial gold deposits which were peopling the country became exhausted, capital investment in quartz mining would stimulate further development.[109] Though the Cariboo rush had eliminated the demographic dominance of Americans in British Columbia, the discussion generated by this paper caused Murchison to lament that two-thirds of the population reaping riches from the colony's gold were United States citizens. The similar remarks of other Fellows regarding the recent San Juan dispute and the need for a railway helped to create an atmosphere of intense anti-Americanism and imperial loyalty.[110] In 1868, stimulated by a memoir discussing the need for a railway link to develop British Columbia and acquire the lion's share of North America's growing Oriental trade, which was being captured by the American rail terminus at San Francisco, Murchison demanded the exertion of 'a strong imperial will' to construct 'lines of communication between our widely-separated provinces, which otherwise will be absorbed one by one by our energetic neighbours of the United States'.[111]

Such fears were justified, for in early 1870 the British Minister in Washington alerted Clarendon that Hamilton Fish, the American Secretary of State, had proposed the cession of all British possessions in North America to settle the United States' claims to compensation for the depredations committed by the British-built Confederate raider *Alabama* during the Civil War. The minister warned further that the American public was 'hankering after' such annexations. President Grant received a petition at the same time from certain inhabitants of Vancouver Island requesting his intervention with Her Majesty's government to permit their territory's joining the Union, and it was rumoured that an American military expedition was being sent to British Columbia to report on the colony's resources and the political proclivities of its inhabitants. The cross-border Fenian raids, Riel's Red River rebellion, American delays in signing the San Juan Convention, rumours of an impending swoop on Bermuda by the Federal Navy, and American connivance in filibustering expeditions against Cuba and Santo Domingo only complicated matters.[112] British Columbians were indeed anxious for a political solution to their problems, but the majority remained unenthusiastic about formal connection with the United States. In 1871, the year of Murchison's death, and shortly after the Hudson's Bay Company was deprived of its remaining territories, British Columbia joined the Canadian federation.

While the future of the prairies and the Pacific colonies had hung in the balance, both British imperial purpose and Canadian national aspirations had found timely vent at the RGS through the active agency of Murchison.

The West Indies

Murchison began promoting exploration of the West Indian region in 1844, when in his first address to the RGS he reported the progress of Robert Schomburgk in surveying the boundary of British Guiana. Schomburgk's career exemplifies the close relationship between metropolitan science and the government departments concerned with imperial affairs. A Prussian-born disciple of Humboldt, he had already received the gold medal of the RGS for explorations conducted in Guiana during the 1830s on behalf of the Society. On the subsequent boundary delimitation expedition, the RGS had secured permission from the Colonial Office for Schomburgk to deviate from his official work for scientific purposes.[113] Among Schomburgk's latest contributions, Murchison announced, were the discovery of valuable stands of wild cocoa and a new yam which might prove a cheap food source for plantation labour.[114] Schomburgk's mineral specimens were donated to the Geological Society, while his memoir on British Guiana's structure suggested the probable presence of gold in the colony's rivers.[115] The next year Murchison entertained the returned explorer and again praised his accomplishments, which were recorded in geographical memoirs published by the RGS and the British Association.[116] Schomburgk continued to transmit information to the RGS from subsequent consular posts in Santo Domingo and Siam, and his explorations in Guiana remained an example of the Society's usefulness to the imperial government. When the RGS launched a campaign for financial support in the 1850s, Schomburgk's discovery of a valuable hinterland in the colony was cited by one Fellow as having 'led to the Govt. laying claim to a considerable portion of Territory well worth alone anything they can give us in repayment'.[117]

In the mid-nineteenth century the geology of the Caribbean islands remained almost unknown. In 1855 Trinidad's Governor submitted samples of iron supposedly containing gold to the Colonial Office for analysis, along with an application from a local place-seeker to survey the island's minerals. Murchison, as the new Director-General, joined Charles Lyell in encouraging instead the founding of a geological survey conducted by 'a really intelligent agent' to expose any fraudulent claims by mining speculators and benefit the colony 'for various weighty reasons beyond the search for gold'.[118] When the specimens in question proved worthless, Murchison's move to maintain the home Survey's monopoly over colonial appointments and analyses won the support of the Colonial Office. In consequence, all the West Indian colonies were canvassed about co-operating with the imperial government in financing a survey.[119]

Labouchere, the Secretary of State for the Colonies, agreed to the Director-General's demands for a salary of at least £600 and an assistant for the surveyor. He expressed concern, however, about the selection of someone 'immune to pecuniary interests of a commercial kind which may be involved in procuring favourable official reports of the advantages offered by particular Colonies or Districts'.[120] Jamaica, Grenada, Barbados, St Kitts, and the Virgin Islands agreed to host the surveyor after Trinidad, while Antigua and British Guiana refused. Murchison recommended George Wall, a graduate of his School, as the geologist, while cautioning that Trinidad alone would take years to survey. James Sawkins, a mining engineer at work in Jamaica whose previous research on Hawaiian volcanoes had impressed Murchison, was appointed assistant surveyor.[121] While Wall and Sawkins began their fieldwork in Trinidad during 1856, Murchison used their appointments to advertise the imperial utility of his establishment in his first *Annual Report* as Director-General.[122] The surveyors' thoroughness was not appreciated by the colonists, however, and the legislature restricted them to a rapid reconnaissance of economic minerals. Murchison pleaded that they be allowed to examine Tobago and the Venezuelan coast in order to link Trinidad's formations to the mainland. As events were to prove in Jamaica, he was determined to maintain high research standards even at the risk of alienating short-sighted local economic interests. In this case the Colonial Office compromised, sanctioning the Venezuelan visit but vetoing the Tobago survey.[123]

In 1857 Wall and Sawkins surveyed a deposit of lignite coal in Trinidad whose steam potential De la Beche had already analysed, but which had now gained in value because of advances in boiler technology permitting the efficient burning of lower-grade fuels.[124] Murchison publicly praised this discovery, convinced the Colonial Office to publish the results of the Trinidad Survey in the British Survey's *Memoirs,* shared them with the Geological Society, and used them to encourage the examination of other colonies.[125] At this juncture, Murchison's establishment performed another imperial service in the Caribbean by ascertaining the commercial worthlessness of specimens of marble from the Bay Islands north of Honduras.[126] Since the signing of the Clayton–Bulwer Treaty in 1850, the United States had demanded that Britain vacate these islands, which were strategically placed in relation to one of the proposed interoceanic canal routes. Palmerston, however, refused to be intimidated, and before relinquishing the islands to Honduras in 1859 ordered their examination for any resources whose exploitation might justify retention or postponed cession.

Wall resigned his post following the completion of the Trinidad Survey in 1858. By this date, precedent had established the principle that geologists appointed to lead colonial surveys should be graduates of the School of Mines – a factor which increased the patronage powers of the Director-General. But because Murchison refused to recommend another appointee from his Survey or School at a salary which had proved grossly inadequate,[127] the Colonial

Office consulted John Phillips, Professor of Geology at Oxford. Lucas Barrett, a protégé of Sedgwick's as curator of Cambridge's Woodwardian Museum, was thus selected for the post.[128] The preference shown this young academic over the veteran Sawkins reflects the determination of metropolitan geologists to place the most highly qualified scientists available at the head of colonial surveys. Yet Murchison's insistence on a higher salary for Sawkins, whose practical experience as a miner he believed indispensable to the success of the palaeontologist Barrett, also demonstrates their concern with economic aspects.[129]

The West Indian Survey now commenced work in Jamaica. Palmerston had consulted Murchison in 1841 about the value of newly discovered Jamaican copper deposits relative to the rich Cuban ores Britain was already importing, but he had turned this inquiry over to De la Beche, who had once owned a sugar estate in the island and whose knowledge of its geology was unrivalled in Britain.[130] A minor copper boom during the 1840s raised Jamaican hopes in the desperate economic climate created by the sugar slump. Following the great mid-century rushes, even gold was reported. Murchison himself received copper and what were thought to be gold and coal specimens from a proprietor anxious to save his failing estate through the intervention of scientific expertise.[131] Barrett and Sawkins had been at work only six months, however, when the copper interests lost patience with their systematic approach, demanding in a parliamentary petition supported by Newcastle at the Colonial Office that the geologists report on the mining district in order to encourage sceptical British investors.[132] Murchison temporised that the surveyors had been delayed by the necessity of making a topographical map until he received assurances from Barrett that copper deposits were being examined, though without the preparation of reports on individual properties. He then decisively throttled this challenge to the scientific basis of the West Indian Survey by instructing Newcastle that 'mining details form no part of a Geological Survey, properly so-called', and that geologists should not be called upon to give opinions on 'the probably relative value of this or that lode, so as to reassure the public, and impart confidence to those who speculate in mines'.[133]

While Barrett established the age of Jamaica's copper ores, Sawkins examined phosphate of lime deposits in the Anguilla Islands. Murchison had long advocated Britain's development of an independent source of fertiliser to break the Peruvian monopoly, and once guano deposits off the south-west African coast were exhausted, the American discovery of a rich phosphate source in a Caribbean islet suggested that the Anguillas might be similarly endowed. The Governor of St Kitts, Sir Hercules Robinson, provided Murchison with specimens while visiting London, and was in turn elected to the RGS. The samples proved rich, and Murchison urged their survey through appeals to the RGS, the Colonial Office, and the Royal Agricultural Society, arguing that the issue was interesting both 'to the geographer and geologist, whilst it is very likely to become very valuable to the merchant and the shipowner'.[134] Sawkins was thus diverted for the task – also briefly examining St Kitts – but the phosphate deposits were found to be unfavourable for commercial exploitation.[135] The Jamaican Survey continued to receive publicity from Murchison, but it suffered setbacks following Barrett's death in 1862. The Director-General recommended Arthur Lennox, a former student of the School of Mines and the son of an old friend, as Sawkins' assistant. Lennox resigned after six months because of poor health, and was replaced by another student, Charles Brown. Next, the Jamaican Legislature twice suspended the Survey's annual appropriation. Despite Murchison's protest, the geologists were only allowed to complete their work by Governor Eyre's manoeuvre for a snap vote in the House of Assembly.[136]

Being informed that the smaller islands no longer felt able to afford their services, Sawkins and Brown, after home leave to prepare their Jamaican report,[137] proceeded in 1867 to British Guiana, which had reversed its original decision. The surveyors were now provided with a copy of Schomburgk's outline map of Guiana, but Murchison warned once more that the necessity of constructing a working topographical map would delay progress.[138] Again, the colony's sudden desire for the survey stemmed from hopes of gold discoveries. In 1850 alluvial gold had been found in Venezuela on a tributary of one of British Guiana's larger rivers. Further discoveries took place near the area disputed with Britain since the unsuccessful border negotiations of 1840–44. Worried that a gold-rush in the colony might precipitate French or American intervention on the pretext of defending private concessions sold by the Venezuelan government, the Foreign Office renewed pressure for a border settlement. Britain wished to exchange its claim at the mouth of the Orinoco for Venezuela's claim in the interior in order both to protect the indigenes and to effect a settlement which would 'bring the English into, or close on, the gold district, and probably lead to further valuable discoveries'.[139] The Colonial Office therefore needed accurate evidence of the disputed territory's resources, and when the results of several small-scale expeditions sponsored by Guiana's government proved inconclusive, Labouchere turned to Murchison for an expert opinion.[140] The 'gold-finder' aired these expeditionary reports at the RGS, but without scientific assessments he could only offer pessimistic theoretical predictions until the West Indian geologists arrived in Guiana.[141]

Sawkins and Brown began an immediate survey of the gold district, but mining activity had given out after a brief rush in 1857 – ostensibly because the uncertain boundary stifled investment, but actually because unprofitable amounts of gold were being recovered.[142] The gold tracts were simultaneously being examined from the Venezuelan side by another Royal School of Mines graduate employed by an American company, whose report was published by the Geological Society.[143] The surveyors' grudging opinion that gold might eventually be found in some of Guiana's streams was proved misleadingly conservative by subsequent events, but their identification of commercial deposits of kaolin, the clay used in porcelain manufacture, and soils suitable for the extension of plantation agriculture offered more valuable contributions.[144]

Central America

Murchison's interests in Central America revolved around the discovery of economic minerals to increase British trading opportunities and the establishment of interoceanic communication for strategic and commercial purposes. Science was served in the process of researching these practical issues. Random mineral specimens had been reaching the Geological Society for some time from British merchants and scientists visiting the region, and in the late 1840s De la Beche's Museum had analysed Panamanian coals for the Admiralty as well as gold specimens from the Mosquito coast for Palmerston.[145] In 1844 Murchison hosted two memoirs on trans-Isthmian canal projects at the RGS. One advocated a Nicaraguan route while the other, by William Wheelwright, founder of the Pacific Steam Navigation Company, backed a Panamanian route. Wheelwright's scheme was based on a survey completed in 1825 by Captain J. A. Lloyd, FRS, Bolivar's chief engineer, who had also collected mineral specimens for the Geological Society. Murchison remained neutral towards these plans, but demanded more exhaustive surveys and emphasised the need for a canal.[146] The following year he lauded the formation of an international consortium to construct a canal by a third French-surveyed route across the Isthmus of Tehuantepec. This scheme, he believed, seemed likely to provide a short cut to Britain's Pacific colonies and open out 'a grand field for commerce'.[147]

When the United States was granted a right of transit across the Isthmus of Panama in 1846 by New Granada, and began building a railway, British interest in Panama sharply intensified. The Clayton-Bulwer Treaty of 1850 neutralised control of any future canal, but Britain hedged her bets on American sincerity by pushing forward surveys for a canal to countervail the railway, which was completed in 1855. Agitation in Australia and New Zealand for an extension of steam mail service to the South Pacific focused discussion on the relative merits of the various routes.[148] In 1850 The RGS almost became directly involved in these undertakings when Lloyd and Admiral Robert Fitzroy – one-time commander of H.M.S. *Beagle*, RGS gold medallist, and former Governor of New Zealand – proposed that the RGS launch a public subscription to survey a Panama canal route. President Smyth demurred, arguing that such activity was irregular for a public society, but he submitted the proposal to Palmerston and Fitzroy's memoir was published by the Society.[149]

The Panama route had been partially explored by Dr Edward Cullen, an Irish adventurer who hoped to promote emigration to New Granada through the offices of the RGS. Cullen sent the Society a memoir in 1850 which dwelt on the ease of constructing a canal, the fortunes to be made from reopening abandoned Spanish gold mines, and the brilliant prospects offered by New Granada.[150] At the same time, Murchison's pronouncements on the global distribution of gold, which characterised the entire region from Mexico to New Granada as a vast goldfield connecting the cordilleran troves of the Andes and Rockies, lent some credence to Cullen's exaggerations.[151] In 1852 the engineering firm of Fox and Henderson sent two employees to survey the Panama canal route. They went to work armed with maps provided by Murchison, and their first optimistic results were transmitted to the RGS by the Foreign Office and read before the British Association.[152] Another agent hired to arrange Indian treaties also consulted Murchison before sailing for Panama.[153]

Cullen had meanwhile negotiated an Indian treaty of his own, secured a land concession from the government of New Granada, and laid plans to take out 800 Irish emigrants.[154] He also induced a British planter in New Granada to send the RGS evidence of the country's colonisation prospects. In simultaneously citing the opinion of the British Minister in Bogotá on the eligibility of his estate to receive colonists, and quoting the remark made to him by Sir William Hooker, Director of Kew Gardens, that they were both 'working under the same flag', this writer clearly revealed the spirit of patriotic endeavour in which even enclave colonisation in sovereign nations was undertaken by capitalists, scientists, propagandists, and officials alike.[155] With the publication of the full report of the British engineers in 1853, the Atlantic and Pacific Junction Company was formed to build the canal. Murchison promoted the concern in his Anniversary Address and praised a new book by General Thomas Mosquera, ex-President of New Granada, about his country's resources.[156] The same year Fitzroy evaluated the various canal and railway proposals,[157] while Murchison threw open the RGS map room to the officers of foreign navies suddenly interested in the approaches to Panama.

A more detailed survey carried out with joint financing by the governments of Britain, France, and the United States now reported the construction of a canal to be impossible.[158] The Junction Company had mismanaged its funds, and by 1855 Cullen among others was financially ruined by participating in its affairs. The RGS organised a special meeting in 1856 to discuss the results of this new survey and its bearing on Australian communications.[159] Murchison argued that the Society should press the British government to survey all of Central America in order to establish the feasibility of every possible canal route.[160] In his Anniversary Address of 1857 he expressed his hope that Britain and France might join the United States in financing a survey of a newly proposed American route. His desire was not only to internationalise the endeavour in the interests of science and progress, but to forestall a *de facto* monopoly of the route arising from more energetic American activity.[161]

During the same year, two Englishmen anxious to people a two-million-acre land concession acquired from the Nicaraguan government offered the RGS a square mile and a hundred guineas for every three emigrants it could induce to make the passage to their estates.[162] The Society declined to accept, but the incident illustrates the public's perception of the unique status of the RGS as an arbiter of overseas development schemes.

In 1858 Murchison announced that a trans-Honduran railway projected by a Birmingham Member of Parliament offered a new opportunity for extending knowledge of the region. Besides a staff of surveying engineers, the company sent out a geologist–naturalist to explore mineral and agricultural potential, while a Colonel of the Royal Engineers signified official interest. Murchison's comment that 'there will, therefore, not long remain any doubt respecting the capabilities of Honduras' reflected the attitude of national purpose in the Society's promotion of such strategically, commercially, and scientifically useful private enterprises.[163] The following year the President announced that another concessionaire had laid before him plans for a Nicaraguan canal in the hopes of generating interest among investors.[164]

The *Trent* affair of 1861, during which the United States prevented Britain from communicating with her Pacific colonies by blocking her use of the Panama railway, lent new urgency to the search for routes across Central America. Bedford Pim, for example, a naval officer who had been involved in the search for Franklin, proposed to the RGS a British-controlled railway across Nicaragua. Pim emphasised the potential of Nicaragua for cotton production, while at the same time stressing that his projected railway would tie Britain's Pacific colonies closer to the mother country and direct the growing commerce of that ocean into British channels. British financiers whose investments in western South America stood to be enhanced by development of the Panama route attended the discussions on Pim's proposal to urge instead the completion of the long-envisioned canal and the British government's defence of the bondholders of the Panama railway.[165] Murchison, who helped Pim organise his surveying expedition, permitted these political views to be aired. At one of the debates to which he had invited General Mosquera, now New Granada's minister in London, he again demanded a decisive government survey of the region.[166] In 1867 the RGS published the results obtained by Pim's surveyors, and Pim himself, who had donated gold specimens collected by his agents to Murchison's Museum, pressed the viability of his route at the meeting of the British Association by stressing its proximity to the new gold-mining district of Nicaragua.[167] In his address to the same audience as President of the Geographical Section, Sir Samuel Baker reiterated the vital importance of an interoceanic route to Britain in a grand panagyric linking geography, technology, commerce, colonies, and military strength as interdependent factors of Britain's greatness.[168] During the remainder of his life Murchison's steady promotion of projects to traverse the Central American isthmus echoed the same imperial theme.[169]

South America

Because British ambitions in South America were limited by political considerations to the fields of commerce and finance, Murchison's activities in the continent focused on exploration of resources, markets, and trade routes rather than on the acquisition of colonies. South America had claimed Murchison's attention even before he had taken up science. The financial crisis which changed his career in the 1820s was partially caused by losses from the default of a notorious loan raised by the Potosi, La Paz, and Peruvian Mining Association to rejuvenate production at the mines of Potosi. The failure of this company, of which he was offered a directorship, [170] permanently jaundiced his attitude towards South American mining ventures. Though he remained interested in establishing the extent of the continent's mineral wealth, he never again invested in the region, and even wove the popular wisdom regarding the poverty of Andean mines into his theory on the occurrence of precious metals. [171]

Murchison's earliest interest in South American geology stemmed from his efforts to establish the world-wide occurrence of the Silurian System. Joseph Barclay Pentland, a naturalist hired by Foreign Secretary George Canning to assess the resources, economic potential, and political stability of the new republic of Bolivia following the wars of independence, provided early evidence of what would be classified as Silurian fossils high in the Andes. [172] When Pentland was subsequently appointed Consul-General in Bolivia during 1836–39, he carried out extensive scientific observations on behalf of the British Association[173] and was nominated by Murchison for the Athenaeum Club. Pentland later served the RGS as a referee of memoirs and an adviser to Andean travellers.

Another of Canning's agents, Sir Woodbine Parish, British representative at Buenos Aires from 1823 to 1832, was an active Fellow of the Royal, Geological,and Geographical Societies. After joining the RGS in 1833, he served for nearly fifty years as the Society's chief South American adviser. As he later observed: 'I foresaw its importance to our Commercial Community, and felt sure of its eventual success, even when its existence was thought by some to be in peril.'[174] Parish frequently helped Murchison compose the South American sections of his RGS addresses. In so doing, he kept Britain's free-trade interests well to the fore by rigorously sifting out the covert political content of memoirs submitted to the Society by citizens of nations perceived as commercial rivals. [175] As early as 1839 Parish employed Murchison to evaluate an Argentine memoir for the RGS, and the latter's comments betray the interests of a geologist as well as of a once-incautious investor in publicising the details of gold and silver mines. [176] Charles Darwin also discovered fossils initially considered to be Silurian in the Falkland Islands, which whetted Murchison's appetite for further extensions of his stratigraphic realm in that quarter. [177] After his first tour in the Urals, Murchison's friends speculated that he might next 'Silurianize' the Andes in the wake of his hero Humboldt, but the trip never eventuated.[178]

Murchison used the Geological Society, the RGS, the *Mining Journal*, and his own correspondents to stay abreast of advances in South American geology. In 1850, for example, he chaired an RGS Council meeting which decided to publish an Admiralty notice on lignite coal seams discovered in the Straits of Magellan. De la Beche's favourable analysis of this fuel, which encouraged development of a port and road to facilitate exploitation of the deposit, was printed at the same time.[179] Murchison also collated South American evidence received from miners and scientists such as W. J. Henwood, manager of a Brazilian gold mine in the 1840s, to support his theories on the formation and occurrence of gold.[180]

At the RGS, Murchison urged the exploration of the continent's rivers as highways for British commercial penetration of hitherto inaccessible or closed markets. In 1852 W. B. Baikie, future explorer of the Niger, won Murchison's support for an abortive expedition up New Granada's Magdalena River, and the President called for the opening of Paraguay and Bolivia to foreign shipping in his Anniversary Address the following year.[181] His 1857 address announced that the opening of the Atrato River to steam navigation was facilitating British imports to the New Granadan interior, advocated similar development of the Orinoco system, and urged its linkage via canal to the Amazon's tributaries.[182] In 1865 an RGS gold medal was awarded for a survey of the Peruvian River Purûs to determine its navigability. Murchison lavished praise on its recipient, William Chandless, as well as the government of Peru, for continuing these researches.[183] Memoirs submitted to the Society which promoted European emigration to South America and development of the continent's resources were read at British Association meetings held in the great manufacturing and shipping centres.[184]

The RGS also processed mineral reports. Murchison received through Prince Albert and Lord Clarendon several memoirs from J. A. Lloyd, now British Chargé d'Affaires in Bolivia, describing mineral resources in the Andes and Bolivian water communications with the Amazon.[185] This information was presented to the British Association as well as the Society, with the geological data subsequently being incorporated in *Siluria* and the geographical memoir recommended for further use by the Foreign Office.[186] Once Murchison was installed at the Geological Survey, his efforts to evaluate and publicise South America's mineral potential achieved new coherence. He henceforth received all pertinent consular and naval reports, oversaw analysis of accompanying specimens, and laid data of commercial or scientific significance before the Geographical or Geological societies. Reports by foreign scientists of economic mineral deposits were also publicised at the RGS, and the Museum of Practical Geology displayed specimens of ores and coals donated by scientists, travellers, officers, and mine owners.

It was from this official position that Murchison attempted in 1860 to secure the appointment of the metallurgist David Forbes as a representative of the British government in Bolivia. The younger brother of palaeontologist Edward Forbes, David Forbes had spent ten years superintending a mining colony in Norway for the nickel refiner and merchant Brook Evans. During this period his research on the occurrence of metals in Silurian rocks brought him to Murchison's notice. Forbes passed the years from 1857 to 1860 exploring Chile, Peru, and Bolivia for nickel and cobalt ores for the Evans firm, returning to England with palaeontological evidence of Siluria's extension throughout much of the Andes.[187] Forbes' data also seemed to corroborate Murchison's theories that Silurian rocks 'constitute the chief matrix of the gold and other metals so extensively worked along this great chain', and that gold veins invariably decreased in value with depth.[188]

As a result of this felicitous research and the request of influential mine owners in Bolivia, Murchison supported Forbes' bid for appointment as British Chargé d'Affaires in Bolivia in 1860.[189] The republic had gone without British representation since Lloyd's recall in 1853, and the British mining and commercial community as well as Bolivia's President hoped to re-establish diplomatic relations to defend British interests against the aggressive American presence. In the event, however, the Foreign Office declined Forbes' proposal.[190] Murchison's acquiescence in the wishes of private capitalists during this affair exemplifies his willingness to attempt to influence official policy regarding countries exporting essential industrial minerals, especially when doing so could pay off a personal scientific debt. Nor was his attitude inconsistent: the Bolivian case was dissimilar to that of Jamaica because an official geologist was requested to aid the entire British mining effort against co-ordinated foreign competition, rather than to raise the credit of specific firms by certifying the value of their holdings.

Certain other activities at the RGS representing a congruence of scientific, financial, and official interests support the contention that an informal British imperialism operated in South America during the nineteenth century. In 1859 two officers of the Ecuador Land Company, a group of merchants in the City of London who had been granted 4.5 million acres in Ecuador in partial payment of that government's debt, joined the RGS to promote exploration and development projects that might raise the value of their holdings. Isadore Gerstenberg, the company's secretary, used the Society to garner parliamentary and Foreign Office influence. At the same time he provided news of a private expedition to the enclave led by James Wilson, the former Australian exploration geologist, and a related excursion by Dr William Jameson, assayer at the Quito Mint and professor of chemistry and botany at Quito University.[191] Six years later when gold was discovered near the company's property and announced by the Foreign Office through the Geological Society, Jameson was appointed by the Ecuadoran government to investigate the find and draft a report encouraging speculation by foreign capitalists.[192]

Even before joining the RGS, Gerstenberg had attended its South American discussions to attempt to mobilise influence in favour of his company's interests. In 1858 he demanded that the Society pressure the British government to support Ecuador in its war with Peru in order to prevent monopolisation of navigation and disruption of development in a region rich in gold and cotton producing potential. In chairing this discussion Murchison commented that it was too political, but he allowed it to proceed in the interests both of Britain and of the Society's own popularity as a forum for debate on peripheral regions.[193] In 1859 Gerstenberg notified the RGS that his company had convinced the India Office to attempt to transplant the quinine-producing cinchona tree from its wild home on the eastern slope of the Andes to plantations in India.[194] This celebrated and illegal operation, organised by the India Office in conjunction with Kew Gardens and carried out by Clements Markham, Fellow and future President of the RGS, ensured the dependable supplies of cheap quinine necessary to sustain the health of the expanded British establishment maintained in India after the Mutiny of 1857.[195] Murchison gave Markham's exploit wide publicity through the RGS, analysed his Andean rock specimens at the Geological Survey, introduced him to John Murray, who published his account of the cinchona transfer, and secured his election to the Athenaeum Club.[196] Markham's devotion to the Society also caused Murchison to elect him to the Council in 1862 and appoint him Honorary Secretary the following year.

In 1862 Gerstenberg joined the agitation at the RGS for a Panama canal which had been brought on by the *Trent* crisis. The opening of direct communications with Europe by the new route would greatly appreciate his company's holdings as suitable for agriculture devoted to export crops, and would preclude the necessity of developing a trans-Andine route to the navigable Amazonian tributaries.[197] Throughout the rest of the decade he continued to encourage the development of free steam navigation between Europe and Ecuador via both these routes. He also urged the RGS against the wishes of Murchison to sponsor expeditions to settle the disputed frontiers of Brazil, Peru, and Ecuador and provided geographical and geological information garnered by the surveyor Wilson.[198] Gerstenberg was clearly attempting to exploit the Geographical Society in order to influence British foreign policy for his own financial benefit. While it appears improbable that his efforts had any impact on government, they did stimulate interest in South American trade and development. Murchison generally condoned the financier's agitation as contributing to the fame of the RGS, the advancement of geographical knowledge, and the extension of Britain's commercial reach. As long as such propaganda equated financial self-interest with Britain's national interests, Murchison was prepared to interpret the Society's rule against political comment with considerable flexibility.

Other South American projects of benefit to Britain also won Murchison's support. The Welsh colonists who founded a settlement in Patagonia in 1865 consulted him about favourable locations.[199] William Wheelwright's British-financed railway schemes in Argentina received backing from the RGS, while providing new data for Murchison to present to the Geological Society.[200] The exploration of Brazil remained a priority. The Amazonian researches of the naturalists A. R. Wallace, Richard Spruce, and H. W. Bates were closely followed by the RGS, and the investigations of Richard Burton while serving as British Consul at Santos received Murchison's blessing. The President urged the Brazilian government to found a topographical survey, and the RGS advised the Brazilians on the preparation of the first general map of their country.[201]

Information on Brazilian minerals deemed important to the deployment of Britain's steam navy and merchant marine, or offering opportunities for commercial development, was likewise publicised. In 1866 Murchison advised Clarendon to instruct the British Ambassador to Brazil to examine newly opened coalfields known to be of Carboniferous age.[202] When, in consequence, reports, specimens, and fossils arrived with the Ambassador's recommendation that a coal viewer be sent out, Murchison proposed instead that a geologist be despatched to produce 'a scientific and trustworthy report' of the deposits 'for British national purposes'.[203] The Brazilian coal data almost immediately found its way into *Siluria* to support Murchison's theory of a globally uniform climate during the Palaeozoic era.[204] A Foreign Office announcement encouraging investment in the railway and port facilities required to develop the mines was also published by the Geological Society.[205] Brazilian scientists visiting Britain were shown the facilities of the Geological Survey and encouraged to investigate the natural resources of their country. When the Brazilian Emperor came to London in 1871, he too paid tribute to Murchison's prestige by not only touring the Royal School of Mines and attending an RGS meeting, but visiting his home in Belgrave Square.[206]

Throughout the Americas, the activities of Murchison and the institutions which he controlled closely reflected British policy. In the British possessions, expansion of settlement, economic development, and imperial security were systematically promoted. A watchful stance was maintained in relation to the United States, coupled with a willingness to profit from the economic opportunities which its own exploratory and developmental initiatives revealed. In the nations of Central and South America, because of the greater strength of British influence and lack of direct political competition, a more aggressive policy of promoting British commercial and financial interests was pursued, with an added dimension of concern for the development of interoceanic communication routes. Woven through all of this activity was Murchison's interest in the advancement of science, which both served and benefited from British expansion in the Americas.

In the various projects in which he was involved in the Americas, Murchison's organisations acted in the vanguard of public opinion. While supporting official policy, they urged the government towards a more active role in encouraging a widening and deepening of British involvement throughout the two continents. Since much of the Americas had long been occupied by Europeans, Murchison had fewer opportunities to promote primary exploration than in regions never before visited. Only in the interior of South America and the Arctic did such blank spots still beckon geographers, and here Murchison's interest in winning major laurels for the RGS was correspondingly keen. In the remainder of the Americas, he fell back on supporting detailed secondary exploration. In these areas his goals were to redraw the maps with new precision, examine natural resources of potential commercial or strategic value, collect fresh data contributing to the progress of science, and survey routes for steamship and railway lines. These technological arteries would penetrate the territories being opened up and form links in the transport and communication network being extended around the globe by the Western powers.

4

The Middle East

Because of their biblical and classical associations, importance as expanding markets, and strategic position straddling the shorter routes to India, the countries of the Middle East had long attracted the attentions of British travellers and scientists. Politically and commercially, the region lay within Britain's informal empire. While Britain's special relationship with the Sublime Porte gave her scientists privileged access to the Ottoman hinterlands and Persia likewise received periodic visits, officials at the Foreign and India Offices were increasingly willing to profit from the intelligence they generated.

In 1835 Murchison took steps to extend the range of British stratigraphic nomenclature across Anatolia. At the behest of William R. Hamilton, FRS, a diplomat, co-founder of the RGS, and collector of antiquities, Murchison gave Hamilton's son his first field lessons in geology. Hamilton senior wished young William J. Hamilton, who already combined employment at the Foreign Office with the duties of Secretary of the Geological Society, to be trained in natural history 'so that he might explore foreign countries to advantage.'[1] Murchison also introduced Hamilton to the naturalist Hugh Strickland, Junior, and at Murchison's suggestion the two began a geological tour of Asia Minor, the first detailed British scientific exploration in Ottoman lands.[2] The Geological Society published several papers resulting from these researches, and in 1842 Hamilton brought out a complete account which became a standard reference on the region.[3] The following year he was awarded a gold medal during Murchison's first presidency of the RGS, and he served as President himself in 1847–48. Hamilton was also a Fellow of the Royal Society, twice President of the Geological Society, a Conservative MP between 1841 and 1847, and for some eighteen years chairman of the board of directors of the Great India Peninsular Railway Company. In this latter capacity he took part at Murchison's invitation in the RGS debate on sites for the new Indian capital projected in 1867.[4]

While Hamilton and Strickland were at work in western Anatolia, the Euphrates Expedition commanded by Captain Francis Chesney, R.A., and Lieutenant Henry Lynch of the Indian Navy was surveying Mesopotamia for the Indian government during the years 1835–37. This lavishly funded expedition was primarily intended to examine the feasibility of opening the Mesopotamian rivers to steam navigation as a new route to India. Subsidiary goals were warning Russia away from Ottoman and Persian territory, opening the country to Western commerce, gathering scientific and political data, and locating the sites of ancient cities.[5] Chesney had organised support for the expedition from the London scientific societies, and his staff included the geologist William Ainsworth, an original Fellow of the RGS. Chesney accomplished all of his objectives except the principal one, for the treacherous channel of the upper Euphrates and the highlands separating it from the Mediterranean proved disappointing barriers. Technical progress also seemed to indicate that seagoing steamships offered the best means of accelerating communications with the East. Still, a vast accession of data resulted from the Euphrates Expedition, and the RGS had the honour of first publication. Ainsworth constructed geological sections across northern Syria and the Taurus Mountains, discovered numerous deposits of economic minerals in Mesopotamia and Anatolia, and reconnoitred a substantial portion of southeastern Persia.[6]

Other significant consequences followed from the Euphrates Expedition. It was the means by which McGregor Laird, whose family's shipbuilding firm provided Chesney's steamships, inaugurated a long association with the government and the RGS as a promoter of steam navigation and riverine exploration. Several individuals, in fact, saw the possibilities inherent in this demonstration of river steamers as a means of deploying imperial military power and penetrating unknown countries. John Cam Hobhouse, President of the Indian Board of Control and an RGS founder, used the example of Chesney's expedition to persuade the Indian government to introduce steam gunboats on the rivers of India, Burma, the Persian Gulf, and China.[7]

Lynch continued Chesney's work by surveying the Tigris system during the period 1837–42. In the course of this exploration he also opened a trading company with his brother, Thomas Lynch, who later served as Consul-General for Persia in London. Henry Lynch used the RGS to encourage the extension of British commerce and influence in the region. At the same time, he extolled the efficiency of river steamers to Captain John Washington, Secretary of the RGS, who as Admiralty Hydrographer later organised the purchase of Laird steamers for the Niger and Zambezi expeditions of the 1850s.[8] Writing to Washington from Baghdad soon after an Anglo-Persian crisis in 1839, Lynch observed that 'there could be no difficulty in marching armies along these rivers in the present day'. He begged his correspondent to rouse officials at home to the necessity of a forward policy in Mesopotamia, 'or they will be losing an opportunity that may never recur – a few thousand would throw the flank of Asia open to us.'[9] Lynch continued his riverine explorations in Persia during the 1840s as Assistant Superintendent of the Indian Navy. In 1862 the Lynch brothers founded the Euphrates and Tigris Steam Navigation Company. Because this firm constituted the only concrete representation of British interests in its operational zone, it received protection from the Foreign Office.

Such activities, which used the interdependent agencies of technology, science, commerce, and official influence to strengthen and extend British interests, were expressions of the policy elaborated by the Foreign Office to deal with the Eastern Question. As the chief architects of this policy, Palmerston, Clarendon, Stratford Canning, and A. H. Layard sought to promote Ottoman and Persian reform and development in order to maintain these countries' internal political stability, independence from hostile foreign powers, and accessibility as markets for British manufactured goods.[10] Once scientific exploration revealed the commercial potential of these lands, it was believed, the ensuing penetration of foreign entrepreneurs would provide outlets for native produce. This mutually beneficial exchange would in turn stimulate cultivation, raise living standards, increase tax revenues, and contribute to the maintenance of peace and order.

In this atmosphere of governmental support for research and commercial activity in the Middle East, the RGS, in conjunction with the Society for the Diffusion of Christian Knowledge, despatched a two-man expedition to Armenia and Kurdistan in 1838. The expedition was commanded by Ainsworth, and Murchison was a member of the committee which oversaw its activities. The ostensible purpose of the undertaking was to establish contact with remnant colonies of Nestorian Christians in order to purchase or transcribe ancient religious manuscripts in their possession. It is clear, however, from the correspondence of Chesney, who acted as intermediary between the RGS and the government in organising the expedition, that its real motives were far more than antiquarian curiosity.

Chesney advised the Society that Ainsworth should map the eastern tributaries of the Tigris and Euphrates, examine ancient ruins, and use his visits to isolated monasteries as pretexts to explore the topography and geology of the region's mountain valleys.[11] After consulting the President of the Royal Society and Lord Glenelg, Secretary of State for War and the Colonies, Chesney noted that Lord Ponsonby, Britain's Ambassador to the Porte, should use the opportunity to ascertain the political state of a country rich in mining and commercial potential where Britain might 'form links which would *exlude others*'.[12] On this advice Glenelg helped the RGS convince Palmerston, the Foreign Secretary, to support the mission. The Foreign Office granted £300, and Ponsonby furnished £200 on the secret understanding that Ainsworth should provide intelligence regarding any political issues which fell under his purview.[13]

Palmerston shared with the RGS the reports of Ainsworth's movements received by the Foreign Office.[14] The two-year expedition, however, proved an embarrassing failure. Ainsworth's downfall resulted from espionage conducted for Ponsonby, who opposed the Egyptian occupation of Syria as an opportunity for French meddling. In attempting to observe the battle of Nezib, by which Mehemet Ali seized control of northern Syria in 1839, Ainsworth was arrested by the Turks. Ponsonby secured his release, but recriminations followed between the RGS and the Foreign Office about the political versus geographical nature of the expedition and the consequent financial responsibilities of its sponsors.[15] Ainsworth's expenditures also contributed to the financial crisis which nearly brought an end to the Geographical Society in the 1840s, and to its cessation of the practice of funding expeditions from its own resources.

The Kurdistan Expedition demonstrates the willingness of the leaders of the RGS to allow the Society to become a vehicle for enthusiasts like Chesney anxious to extend pre-emptive British influence under the guise of scientific or philanthropic motives. Conversely, it illustrates the government's – especially Palmerston's – readiness to use the Society both as an espionage bureau and an instrument for projecting a British presence into sensitive foreign territories. Murchison is revealed as playing a prominent early role in the planning of such operations: he even felt compelled to defend the scientific nature of Ainsworth's mission from the taunts of French geologists rendered suspicious of British motives by the maps and plans captured from the explorer at Nezib. [16]

Ainsworth's appointment established a precedent for the employment of geologists to conduct official explorations in territories where the discovery of profitable mines was a priority. At Palmerston's request, De la Beche appointed the geologist William Kennet Loftus to the Turko-Persian Boundary Commission in 1849. Again, the subsidiary object of this Commission was the collection of intelligence to stimulate trade, improve Turkish administration, and stabilise an area strategically near the route to India. [17] Loftus conducted various scientific investigations during the course of the four-year survey and a subsequent appointment to explore ruins for the Assyrian Excavation Fund. He corresponded with De la Beche, gathered intelligence for the War Office, and collected antiquities for the British Museum as well as advising Lynch & Co. on cornering the trade of southern Persia with river steamers. [18] Clarendon sent Loftus' preliminary geological report from the Foreign Office to the Geological Society for publication; his geographical memoirs were published by the RGS, which had provided his scientific instruments. [19]

In his quest to extend the bounds of Siluria, Murchison received many fossils during the 1830s and 1840s from General William Monteith of the Indian Army, an original Fellow of the RGS who had seen extensive diplomatic service in Persia. [20] As President of the RGS, he simultaneously encouraged the charting of the Mesopotamian rivers as valuable to science and commerce and urged further exploration of Arabia. [21] In 1851 Sir Henry Ellis, former British Minister to Persia, requested Murchison's advice on behalf of a friend about the geological feasibility of a commercial project to supply water to Tehran and Tabriz. [22] This overseas extension of his services as a consultant for the mineral development schemes of landed friends prefigured Murchison's imperial work at the Survey, illustrating the convergence of official, financial, and scientific interests in exploiting opportunities created by British influence in the informal empire. Historical geography was also utilised in attempts to rediscover valuable mineral deposits. In 1852, for example, the British Association published a memoir analysing cuneiform references to ancient copper mines in the mountains of Asia Minor.[23] Likewise, Murchison's interest in this region partially stemmed from its fame as a source of gold to early Mediterranean civilisations.

In the late 1840s and early 1850s, Murchison actively encouraged the Turkish geological researches of Pëtre Chikhachëv, a Russian naturalist who had proved the existence of Palaeozoic rocks in the Altai Mountains of southern Siberia. Chikhachëv now extended the accomplishments of Hamilton, Ainsworth, and others in Asia Minor using Russian army maps of Turkey and British Admiralty charts of the Black Sea supplied by Murchison.[24] Murchison published the Russian's letters from the field at the Geological Society, fêted him in scientific circles during a visit to Britain, encouraged the sale of his book, and elected him an Honorary Corresponding Member of the RGS.[25] In 1853 Murchison was rewarded by Chikhachëv's announcement that he had made 'new conquests' in eastern Asia Minor in favour of the King of Siluria's 'extensive estates'. He warned also that a war to dismember Turkey was imminent and that Russia was well prepared to take advantage of this 'substantial repast'.[26] Murchison's friendship with Chikhachëv survived the Crimean War – he continued to praise the Russian's results in *Siluria*[27] – but the struggle between his homeland and the nation where he had accomplished his greatest foreign geological triumphs proved a traumatic event.

Murchison deplored the Crimean War as an unnatural conflict between old allies: he told the Tsar afterwards that it 'touched me to the heart'.[28] He followed the course of the war with anxiety, keeping informed through friends privy to the latest official news. As 1854 began, for example, the physician Sir Henry Holland posted him on the likelihood of war as revealed by Aberdeen's daily conversation. At the same time Holland relayed to the Prime Minister Murchison's advice on Russia's limited naval coal production and put pressure on him, as a fellow original member of the RGS, to award the Society an annual grant-in-aid for its services to the nation. Explaining to Murchison that it was a poor time to request aid, he neatly summed up the pretensions of the RGS as an instrument of foreign policy: 'If we had discovered 2 or 3 safe harbours in the Black Sea, the Society might have approached with a demand rather than a supplication.'[29]

Murchison had more at stake in his hunger for advance news than concern for Anglo-Russian amity, however, for he was attempting to rearrange his investments in anticipation of shifts in the values of British and Russian government bonds.[30] When Baron Brunnow, the Russian Ambassador, left England in February 1854 as the Anglo-French fleet concentrated at Sevastopol, he praised his old friend Murchison for remaining faithful to Russia.[31] Residing in Germany for the duration of hostilities, he used Murchison's friendship with Clarendon to maintain contact with the British government and negotiate the treatment and return of war prisoners. For his part, Murchison used this channel to assure the Tsar that he was doing all in his power to work for peace and combat the anti-Russian sentiment fanned by the British press.[32]

When British troops were sent to the Black Sea, the London scientific societies resolved to request that scientists be appointed to accompany them in order to collect natural history data and advise on such questions as water supply and field entrenchments. They chose Murchison to appeal to Lord Raglan, commander of the expeditionary force. His request was refused on the grounds that civilians would encumber the army, whose officers could perform such functions. Warington Smyth, a School of Mines professor acquainted with the geology of the Black Sea region, nevertheless lectured departing officers on choosing bivouac sites according to the drainage characteristics of underlying formations.[33] By the summer of 1855 the Royal Engineers were belatedly begging Murchison, now Director-General of the Geological Survey, to advise them on boring wells to supply the suffering troops besieging Sevastopol. In providing recommendations, Murchison noted that a couple of geologists 'would have been worth a cartload of useless boys, & etc. classed *extra ADC's*' to the expedition. Referring to the scientific success of the American Fremont expedition of 1848, he concluded that '*wherever* an Army goes for the first time, it ought ... to be accompanied by some useful "savans"'. For years afterwards, he continued to chastise the army for having ignored the scientists in the campaign.[34]

Murchison believed that the Russians would prove much tougher adversaries than did the majority of his countrymen. Following the line of Major-General James Shaw-Kennedy, whose strategic views he relayed to his friends Clarendon and Sydney Herbert in the cabinet, Murchison opposed the choice of Sevastopol as the best location for attacking Russia. He argued instead that a threatened Allied invasion from the Balkans could intimidate the Tsar into suing for peace.[35] In early 1854 Murchison had the RGS print a related memoir by Baron Jochmus, a German general in the Turkish war ministry, on the possibility of a Russian attack across the Danube, in order to bring the concept before the British authorities and gain timely publicity for the RGS.[36] Murchison also influenced military appointments, such as that of Lieutenant-Colonel J. A. Lloyd, FRS, FRGS, the former British Chargé d'Affaires in Bolivia, secretly sent to raise the Circassians in the British interest.[37] Throughout the war, Murchison put his first-hand knowledge of the Tsar's character and Russian military and industrial capabilities at the disposal of Clarendon. At the same time, however, he laboured to maintain trust and respect between the antagonists until peace might restore normal relations.[38]

In public, Murchison's well-known Russian leanings provoked comment. In the autumn of 1854, news of the battle of the Alma reached England while he was attending the meeting of the British Association, where a thorough discussion of the Crimea had been organised in the Geography Section. When the victory was announced at the country house where Murchison was staying, he refused to toast the British success and warned his fellow guests that British statesmen would soon regret the war as a political mistake.[39] After this incident Murchison refused to dine out at all until Sevastopol fell in September 1855. Murchison also ridiculed the panic-stricken inhabitants of New South

Wales for fortifying Sydney Harbour against a possible Russian attack.[40]
When the Australian geologist W. B. Clarke, who had a nephew at the Alma,
congratulated Murchison on the victory, the Peninsular veteran thundered in
reply:

Considering that you are a clergyman, you seem to me very bellicose . . . the sooner Mr.
Bull & Co. can be persuaded to be moderate & not be so arrogant as to think that they
can ruin, humiliate & dismember "Rooshia" we shall have no chance of a reasonable
termination of the present deplored state of affairs, which is fatal to science & letters, as
well as to everything, save Contracts, Jobbing & Protection.[41]

Murchison blamed the war on 'the artifice of the French Emperor, and the
too skillful diplomacy of Canning'.[42] He confidentially begged John Murray,
editor of the *Quarterly Review*, to extricate himself from the Russophobic
influence of A. H. Layard and abstain from 'the hue & cry for dismembering
Russia' or occupying the Crimea – a goal which he considered the secret object
of Louis Napoleon and 'absolute insanity'.[43] Murchison urged the *Quarterly* to
encourage instead the preservation of Turkish independence and the exaction
of free navigation on the Black Sea and the Danube. Despite these strictures,
Murchison invited Layard to a dinner of the new Geographical Club, for as a
celebrity intimately acquainted with British policy in Turkey, he was sure to
bring the RGS publicity. Because of his divided sympathies, Murchison also
faced a dilemma about the propriety of continuing to display his Russian
orders in public. After consultation with Clarendon and Aberdeen, however,
he grudgingly abstained from wearing them for the duration of the war.[44]

Appointed Director-General of the Survey in May 1855, Murchison had
thenceforth to be more circumspect in his behaviour. His recent usefulness to
government in estimating Russia's fuel reserves may have been a factor in his
selection, for immediately upon assuming office Murchison implemented a
scheme to reconnoitre the coal resources of Turkey in the interests of the Allied
war effort. This project illustrates his success in promoting the use of geology
as an instrument of policy in the penumbra of informally controlled or
influenced states surrounding the official empire. At the same time, it
represents one of the first experiments on the part of the British government in
employing science to solve problems directly arising from wartime exigencies.

Desultory British interest in exploiting Ottoman coal had existed for some
time, with private and official motives sometimes overlapping. In 1846 the
Consul-General at Tripoli had consulted Adam Sedgwick on a speculative
scheme, under authority of a concession to be secured from the Porte, to mine
coal believed to exist beneath the Sahara for sale to English steamers at
Malta.[45] In 1850–51 Palmerston had sent the Geological Society two reports
from the Consul at Erzurum describing a newly discovered coal seam in Turkey
itself.[46] As Turkey declared war on Russia in 1853, a friend who had visited
the Allied fleet near the Dardanelles sent Murchison a sample of clinker from a
steamer supplied with coal from Eregli on Turkey's Black Sea coast. This mine
had recently been discovered by Captain Thomas Spratt, RN, a hydrographer
with a penchant for geological research.

Explaining that Britain might soon require abundant naval fuel supplies in the area, Murchison's correspondent requested him to analyse the specimen and report to Stratford Canning, Britain's Minister at Constantinopole, as to whether the coal could be turned to account or a superior fuel be found nearby.[47] Even before his appointment to the Survey, Murchison's geological skills were thus enlisted to further the deployment of Britain's steam navy in support of her foreign policy. The palaeontological staff of the Museum of Practical Geology were at the same time dating the Eregli coal as Carboniferous by means of fossils transmitted by Spratt. But the transaction involving Murchison, in bypassing De la Beche and the bureaucracy, illustrates the efficiency of informal Victorian acquaintance networks in mobilising expertise to accomplish national goals.

By the spring of 1854 the fuel situation had become desperate. The Earl of Ellesmere, President of the RGS, wrote to Clarendon: 'I have letters from the Black Sea howling for *coals*.'[48] Canning, now Viscount Stratford de Redcliffe, soon secured permission from the Porte to mine Turkish coals despite French suspicions of Britain's motives. By January 1855 he reported to the Foreign Office that an English coal-viewer despatched to examine the Eregli mines had promised that they could be 'worked without limitation'.[49] Though production at Eregli was expanded, this coal lay 150 miles east of Constantinople and could only be loaded from an open roadstead. Consequently, when the British Consul at Bursa reported the discovery of seams in the Sea of Marmara's sheltered Gulf of Nicomedia, a scant fifty miles west of the capital, the Allied fuel problem looked like being solved. Clarendon passed this report to the Geological Society via Murchison, whose appointment as Director-General was simultaneously being confirmed.[50]

After consulting W. J. Hamilton, who agreed that the Nicomedia coal probably represented an extension of the larger deposit at Eregli, Murchison advised the Foreign Office that a mining engineer should be despatched to reconnoitre the seams, which he believed might prove 'of vast importance both to Turkey & to the cause of the Allies'.[51] When Murchison began his official duties in May 1855 he recommended the appointment of Henry Poole, a mining engineer with experience in the coalfields of Nova Scotia, to survey the Nicomedian deposits and commence mining should their quality and accessibility justify it. The proposition was accepted. Stratford was ordered to obtain aid for the expedition from the Porte while Murchison arranged with Clarendon and Musurus, the Turkish Minister in London, that a Turkish student at the School of Mines accompany Poole as interpreter and assistant.[52]

Murchison provided Poole with maps, Admiralty charts, tools and instruments, and a briefing from Hamilton on his destination.[53] The Director-General instructed the surveyor to search for outcrops of coal, open out promising seams to judge their quality and extent, and ascertain the cost of mining and transporting the fuel to the coast. He was to collect both fossils and inorganic rock specimens to determine the stratigraphic succession, and construct geological maps and sections. If as Murchison anticipated the coal

seams proved of Carboniferous age, Poole was to establish their strike to test
the hypothesis that they were a prolongation of the Eregli deposits. Lastly, if
the coal proved of more recent age, Poole was cautioned not to 'cease your
labours until you have ascertained whether the combustible be of any
commercial value', because steamers with specially designed boilers were by
now successfully using better grades of lignite.[54]

Poole sailed on 20 June, and within a month had accomplished a good deal
of the survey. He reported to Murchison that he had examined several coal
outcrops, but that fossil molluscs indicated them to be lignites. Combustion
tests likewise revealed poor burning characteristics and unacceptable percent-
ages of ash. The country was so broken, he announced, that even if good coal
could be found, it could not be worked economically. Regretting that he had
had no success, Poole forwarded specimens, sections, and sketches and
promised to give Murchison detailed observations when he completed his
explorations. The Director-General instructed Poole to undertake further field
researches before abandoning his survey, but his subsequent reports were
equally pessimistic.[55] As Murchison pointed out to Clarendon, though such
negative results were of less value to military strategists than to scientists, the
survey had been worthwhile because 'the chances . . . of a better result were
very great & the expenditure is not large'.[56]

Thus encouraged, Clarendon found other employment for Poole. A gun-
powder shortage had developed and, as with the search for coal, the discovery
of deposits of its constituents might help the war effort as well as
contribute to Ottoman economic development. Palmerston privately ascertained from an English traveller that there were good indications of saltpetre
on the shores of the Dead Sea, and he advised Clarendon that it might be
worthwhile to send out a geologist. 'A mine of sulphur or saltpetre', he
commented, 'would be a source of wealth to the Porte, & might prove useful to
us.'[57] In consequence, Poole was ordered to Palestine in the winter of 1855 on
another survey, but again his results were negative. By the time Poole's reports
and specimens reached the Foreign Office, peace had been concluded.
Clarendon therefore directed Murchison to disseminate these data through the
scientific societies.[58]

Clarendon retained Poole for further service during 1856, but when his
exploratory ambitions began to outweigh the immediate usefulness of his
researches, the Foreign Secretary reluctantly dismissed him on the grounds
that the cost could not be justified.[59] Murchison had done very well at
exploiting the scientific opportunities generated by the resource demands of
the war, but even under the aegis of a sympathetic cabinet minister, science
had to pay its way abroad with practical results. Still, Poole's coal survey
represented an auspicious beginning in this line for the new Director-General,
who used it to advertise the usefulness of his geological institution in his first
Annual Report.[60] The Nicomedian coal survey also had repercussions on British
policy regarding the Ottoman empire. It re-emphasised the potential value of

new mines, which might be discovered by scientific survey and financed and worked along European lines, to the progressive administration which British statesmen were attempting to create. In 1857 James Brant, the Consul at Erzerum who had announced the coal discoveries before the war, drafted a memorandum calling for the encouragement of European capitalists to replace the corrupt and inefficient government monopoly in developing Turkey's 'inexhaustible' mineral riches. Since opening new mines would require the building of roads and ports to facilitate export, the attempt to revolutionise the mining sector fitted in with the entire spectrum of initiatives for reform and development advocated by Britain.[61]

When the Crimean War ended, Murchison rejoiced that scientific communication between Britain and Russia was restored and that Britain could once more turn her attention to overseas exploration. He also believed that the renewal of British coal exports to Russia would provide 'a *fundamental* reason for the continuation of peace'.[62] Friends who had taunted him for his Russian leanings during the war now praised his prescience regarding the peace. He congratulated Clarendon on its terms, and advised him on the ability of the new Tsar Alexander II and his cabinet. When the Russian Ambassador Brunnow returned to London and invited Murchison to St Petersburg for the coronation of Alexander, Murchison begged Clarendon to send him in an official capacity in order to help 'dissipate the residual irritations' of the war.[63] Instead, he found himself representing Russia at several London civic feasts, but he found other occasions to mix science and politics. Chairing a discussion at the RGS about a dispute between Russia and Turkey over possession of an island at the mouth of the Danube, Murchison remarked that 'if the structure of rocks was to be the ground for the construction of empires', this diplomatic issue could be decided by the island's geological affinity with the Bulgarian shore.[64]

The British expedition that occupied the port of Bushire in order to force the Persians to evacuate the Afghan city of Herat commanded the attention of the RGS during 1856 and 1857. Murchison had fixed on the topic as one sure to bring the Society publicity as a venue for debate between government officials and civilian authorities. At a meeting which he chaired, Sir Henry Rawlinson – RGS gold medallist, ardent Russophobe, and recently retired British Consul-General in Turkish Arabia – threatened annexations of Persian territory if Britain's demands were not met.[65] This was an extreme statement for a public official to make, but Palmerston himself, in permitting the RGS to print edited despatches regarding Persia during the several crises with that state in the first half of the century, had used the RGS as a subtle means to assert British interests against those of Russia. A confidential Foreign Office minute accompanying a report in 1837 on the Bushire–Tehran road warned the Society to omit 'every passage which has a tendency to shew that the writer had any military or political objects in his observation of the face of the country'.[66] The Foreign Secretary, however, knew that the *Journal*'s international audience could read between the lines. The alacrity of the Foreign Office in passing information to the Geological Society regarding Persian mineral deposits similarly suggests that this alternative outlet was being used to alert British investors to potential speculations.[67]

Murchison, who was selective in enforcing the RGS rule against political discussions and supported the punitive expedition to Persia, praised Rawlinson's strategic comments on the movements of the British army. He likewise expressed approbation for the related remarks of A. H. Layard and General William Monteith, the retired Persian diplomat (now FRS) who contributed a memoir to the Geographical Society emphasising the routes by which invading armies might advance through the country.[68] Meanwhile, however, Murchison acted as an envoy for Tsar Alexander and his ministers, expressing to Palmerston their distaste for the wording of the blue books describing Russia's role in the Persian imbroglio.[69] When Rawlinson went to Persia as British Minister in 1859, Murchison encouraged him to emulate his predecessor in promoting scientific explorations and reporting their results to the London societies.[70] At the same time, he influenced appointments to a second Turko–Persian Boundary Commission.[71]

Throughout the 1860s, while keeping abreast of developments via private contact with British officials, Murchison used his annual addresses to the RGS to maintain public interest in Persian exploration and to put pressure on the government to publish the results of the boundary surveys.[72] His RGS Councils always included a specialist on Persia or the Ottoman empire, and Persian memoirs by British diplomats appeared regularly before the Society. The explorations of Major Frederick Goldsmid, FRGS, in surveying a telegraph route to India through southern Persia and Baluchistan during the early 1860s, also received considerable attention because of their bearing on imperial communications.[73]

In the Ottoman empire, meanwhile, Francis Chesney surveyed the route for a railway to link Antioch with the Euphrates during 1856. He had first proposed this line to the British Association in 1852.[74] While possessing the double merit of validating Chesney's earlier navigational researches and providing Palmerston with a Middle Eastern route to India which might avoid the strategic complications a Suez canal would entail, the scheme failed to take hold. Chesney obtained a concession from the Porte, and the British government at first expressed willingness to guarantee a loan, but diplomatic considerations forced Palmerston to give ground to a rival French plan for a direct railway from Constantinople to Basra. The promotional success of De Lesseps' Suez canal ultimately doomed the project.[75] Still, Murchison supported it while it was under consideration and urged the government to complete publication of the now timely reports on Chesney's early explorations, arguing that if the Ottoman government would guarantee the safety of the route, 'all difficulties may vanish before the will of Englishmen'.[76] Working through his connections in Aberdeen's cabinet, he also used the proposal of a project related to Chesney's – of connecting the Red Sea with the Mediterranean by flooding the Dead Sea depression – to urge the survey of Palestine by the Royal Engineers.[77] In 1858 Murchison likewise called attention to the explorations of Cyril Graham in Syria for their value in pointing out the barbarising effects of Turkish 'misrule'.[78] Graham later served on the Dufferin Commission of 1860–61 inquiring into religious massacres and insurrection in Syria, and he acted as Foreign Secretary of the RGS from 1866 to 1870, in which capacity he helped establish the Palestine Exploration Fund.

In 1861 Murchison himself was involved in repercussions from the intervention by the European powers to restore order in Syria. Lord John Russell, the Foreign Secretary, apprised him that Under-Secretary Layard was anxious that Robert Etheridge, senior, Assistant Naturalist of the Geological Survey, should be granted leave of absence to 'report on the resources of Syria'.[79] This plan, which meshed with Palmerston's policy of developing the Middle East for strategic reasons, had originally been concocted by Sir Culling Eardley, FRGS, an evangelical philanthropist who had obtained important improvements in religious toleration within the Ottoman empire. As a member of the British government's Syrian Committee, whose brief was the recommendation of reforms to maintain peace and Turkish authority in the province, Eardley had decided with Etheridge's encouragement that a geologist should be sent out to examine mining potential, the suitability of the soil for cotton cultivation, and the condition of the peasant dwellings. His scheme testifies to the Victorian faith in the power of statistics and applied science to solve social, political, and economic problems, as well as to Britain's need for alternative cotton supplies during the American Civil War.

Murchison, who might have been expected to support such a project as a further opportunity to deploy geology on government service, viewed it instead as an attempt to interfere with his establishment and conduct a superficial survey that would bring discredit upon science. He fumed to T. H. Huxley, Etheridge's superior at the Museum of Practical Geology:

When Sir Culling came to me about the Syrian Humbug, I shovelled him off by saying that no one under my direction could think of undertaking such a thing – that a mineral survey of a vast region like Syria is an affair of years & etc. Nothing daunted you see . . . that he has wormed his way like a true son of Exeter Hall into Lord John & of course carries Layard with him "in re Syria".

Agreeing with Murchison's view that the scheme was a wasteful 'piece of absurdity' and that Etheridge was in any case unqualified for the work and insubordinate in entering into such negotiations, Huxley helped Murchison convince the Foreign Office to abandon it.[80] Their success demonstrates that the ability of established special interests to sway official policy could be countered by the growing influence of the government's own cadre of scientific experts when they felt that projects they were called upon to implement compromised their professional ethics. In disciplining Etheridge for attempting to create an alternative niche for himself by invoking political influence

Plate 4　The Museum of Practical Geology in Jermyn Street, facing Piccadilly

beyond the hierarchy of their institution, Murchison and Huxley were at the same time defending the status which the reputation of the Survey had conferred upon geology and defining with new precision the contractual basis on which the science served government.

Serious proposals for Ottoman research, however, did receive the Director-General's support. In 1865 Arthur Lennox, a veteran of the Jamaica Survey, solicited his aid for a proposal to conduct a comprehensive geological survey of the Ottoman empire. His plan emphasised the economic benefits of such surveys and the cultural stigma attaching to Turkey as one of the few European powers lacking one. He proposed to begin in the most productive mining regions of the Balkans and then extend operations to the entire empire. Lennox secured a promise of aid from Sir Henry Bulwer, British Ambassador at Constantinople, before approaching the Porte for a commission, but his plan proved too ambitious 'to suit the views of the Turkish authorities, so apathetic generally in matters of science'.[81] A visit to Turkey and a discussion with Murchison convinced him that a more modest proposal limited to European Turkey stood a better chance of acceptance.

Murchison then advised Layard at the Foreign Office that Lennox be recommended to the Turkish government for this limited appointment, but he cautioned that establishing a geological survey for the entire Ottoman empire would require a director of greater experience and a large staff.[82] Layard apparently backed the plan, as did Musurus, the Turkish Minister in London who had helped Murchison organise the Nicomedia expedition,[83] but again the Porte declined to act and Lennox fell back to request Murchison's recommendation for a third, even more dilute, proposal. Though Lennox failed to win appointment, his demand for '*"carte-blanche"* to work where I like' and his remark that 'if the Porte leave me to myself it cannot be said that *it* is responsible for the success of the enterprise' reflect the sovereignty-infringing status which Europeans enjoyed in Turkey.[84] These comments also typify the vision which British geologists held of career opportunities in the informal empire. The assumption on the part of Murchison and Layard of a virtual right to dictate appointments and arrange resource explorations likewise demonstrates that science was an integral component of Britain's unequal relationship with Turkey.

In 1865 Murchison also supported the foundation of the Palestine Exploration Fund, whose aim was to map the Holy Land. The activities of the Fund provide another illustration of geography and geology in the service of informal empire. An earlier version of this organisation, the Palestine Association, had been founded in 1804, but after sponsoring a single unsuccesful expedition it had merged with the Geographical Society in 1834. At the British Association meeting of 1865, the Geographical Section appointed a committee consisting of Rawlinson, Murchison, and the biblical scholar the Reverend H. B. Tristram to promote the Fund. The committee persuaded the RGS to donate £100 which the War Office matched to enable the Fund to begin operations, and Lord Stanley, Foreign Secretary in Derby's cabinet, lent his official support. Also present at the same meeting was Charles Wilson of the Royal Engineers. After serving as Secretary on the North-west American Boundary Commission between 1858 and 1863, Wilson had surveyed Jerusalem for two years on behalf of the philanthropist Baroness Burdett-Coutts and Sir Henry James, Director-General of the Ordnance Survey and head of the War Office's secret Topographical and Statistical Department.[85]

Wilson spent the next five years exploring Palestine and Sinai for the Fund. In 1869 he joined the RGS and succeeded James as head of the renamed Intelligence Division of the War Office, which functioned during the rest of the century as a centre for imperial military strategy.[86] Significantly, many of the officers of this elite corps were Royal Engineers, scientific soldiers trained in surveying, mathematics, and, partially as a result of Murchison's efforts, the rudiments of geology. Several of them served with the Palestine Exploration Fund before taking up their staff duties, and nearly all were Fellows of the RGS. Wilson himself sat on the Society's Council in the late 1880s when he had been promoted Director-General of the Ordnance Survey. The Fund thus functioned as a training exercise for the 'military mind' of imperialism, for in research regarding the topography of the Holy Land it offered a plausible justification for sharpening observational and cartographical skills in a strategically vital region.

For patriotic as well as scientific reasons, then, Murchison promoted the Exploration Fund, advertising its progress at the RGS while its reports appeared at British Association meetings.[87] The Director-General's protégé Archibald Geikie even consulted him about leaving the British Survey to join the Palestine expedition as its geologist.[88] Murchison attended meetings of the Fund's General Committee and served as co-trustee with Sir John Herschel, Sir Henry James, and a clergyman, to organise the loan of instruments, brief the departing explorers, and lend his name to fund-raising campaigns. Before Murchison's death the RGS had also provided around £150 in direct financial aid to the Fund.

In Arabia, too, sporadic exploration continued. In 1851 Murchison had arranged that the Russian Imperial Geographical Society, the RGS, the East India Company, and the Foreign Office join in providing aid for a Finnish professor to travel in the Arabian interior. The explorer's salary demands ruined the project, though Murchison's continued supplications to the British authorities on the strategic usefulness of such research resulted in the financing of Richard Burton's celebrated journey to Medina and Mecca.[89] In 1864 the RGS gave wide publicity to the central Arabian explorations of William Gifford Palgrave.[90] This explorer's exciting narrative brought large crowds to the Society's meetings, won praise from scientific geographers alienated by Murchison's popularity hunting, and stimulated further investigations on the part of the Indian government's Political Resident in the Persian Gulf.[91] The promotion of Palgrave through the RGS ensured a wide sale for his narrative and contributed to the launching of his career as a diplomat. He continued to provide the Society with information in subsequent years from his various postings, some of it of a highly political nature.[92] Murchison also received geological reports and specimens from the Arabian Red Sea coast sent by Hilary Bauermann, former geologist to the North-west American Boundary Commission, who spent the years 1867 to 1869 privately surveying the region's mineral deposits because of interest excited by the opening of the Suez Canal. Bauermann discovered iron and manganese ores near Suez as well as archaeological remains of ancient turquoise workings and copper-mining sites in the Sinai Peninsula. Fossils which he sent to the Geological Society were correlated with those collected by the Palestine Exploration Fund to further develop the understanding of the region's structure.[93]

Persia and the Asiatic provinces of the Ottoman empire remained a major focus of Murchison's exploratory interest throughout his career. This concentration resulted as much from his robust sense of imperial patriotism as his desire to win the support necessary to exploit research opportunities created by Britain's vested interest in the region. His own identification of these opportunities, however, went as far towards creating them as did the imperatives of British foreign policy. He was in effect laying open fresh fields of endeavour in which, by means of judicious investment, new capital in the form of scientific, strategic, political and economic information could be accumulated. Murchison's management of these data enhanced his own reputation both as a scientist and as a man of affairs, and their presentation to the public and government in the form of practical results facilitated and in some sense justified further activities in the region. Even independently generated projects fell under his purview as they were processed through the RGS, the Survey, or the Geological Society. In this complex 'feedback loop', where Murchison simultaneously influenced policy and implemented it, helped to shape public attitudes and gave them expression, his emphasis on the Middle East represented more than a simple product of British involvement in the region. It was also a factor which promoted the deepening and rationalisation of that involvement.

5

The Indian empire and Central Asia

The Indian empire

The natural sciences had enjoyed the patronage of the East India Company for many decades before Murchison's era. Ruling over an alien subcontinent with absolute authority and a shrewd sense of commercial profit, the Company was more willing to fund applied research than were *laissez-faire* governments in Britain. As the amateur activities of Company servants gave way to directed research and development during the nineteenth century, India became a gigantic laboratory for governmental experiments in the use of science to achieve economic, political, and social progress by organising the rational exploitation of the country's resources. Based on a pragmatic programme of solving specific problems rather than implementing established theory or policy, this interventionist model of science and technology being used to maintain European rule would later be applied throughout the dependent empire. In India the universally applicable laws of Western natural science, like those of jurisprudence and political economy, were used to bolster British political authority. Science generated knowledge that reflected credit on the East India Company while increasing its dividends, and it also helped to impose unity on the diverse territories under the Company's control. Furthermore, the Indian experience produced a cadre of scientific soldiers and civil servants who, by applying their organisational skills and promoting the centralised direction of research during retirement in Britain, strengthened the links between science and the imperial government.[1]

The Great Trigonometrical Survey of India, founded in 1800, conferred great prestige upon the East India Company as an enlightened sovereign power, for in scale and accuracy it equalled or surpassed even the most exacting European mapping operations. The Company also published the results of its explorations beyond India, such as the surveys of the Mesopotamian and Persian rivers. This research directly benefited Britain, for as A. H. Layard noted, the Company's explorers in Persia obtained 'great influence over the wild inhabitants of the countries in which they were employed. Through it they could promote to no inconsiderable extent the interests of their country.'[2] Reports of these expeditions and the coastal surveys carried out by the Indian Navy also appeared in the journals of the RGS and the Bombay Geographical Society. Founded by officers of the Company's Navy in 1832, this latter society started as a branch of the RGS and then functioned independently until its absorption by the Asiatic Society of Bengal in 1873.[3]

If precise maps were of immediate necessity for military and administrative purposes, geological and botanical investigations as a basis for extractive industries were also required. In 1780, accepting the proposal of an assistant surgeon who had studied science at Edinburgh University to search for exploitable ores, the Company funded a one-year mineralogical survey of Bengal. Hotchkis, the surveyor, became one of the oriental correspondents of Charles Greville, a friend of Joseph Banks, who attempted for economic reasons to establish a national mineralogical collection at the Royal Institution and also recommended the first mineralogist appointed in New South Wales.[4] The fledgling Geological Society of London also encouraged Indian research: in 1816 the government of Madras reprinted the Society's *Inquiries* to stimulate observation and collection.[5] This pamphlet was largely written by George Greenough, the Society's founder, who in 1854 produced the first geological map of India from data provided by Englishmen serving in the East.

In 1818 the Asiatic Society of Bengal, a focus for science in British India modelled on the Royal Society, planned to co-operate with the Geological Society in order to promote economic development.[6] From this period geologists were attached to the Great Trigonometrical Survey, employed in surveying specific districts of India, and appointed to diplomatic missions throughout the Company's sphere of interest.[7] Geology likewise began to be taught at the Company's Indian colleges.[8] English geologists also used the medium of the British Association to encourage the Company to collect and publish geological data.[9] British geology in India lagged only slightly behind that practised in the home islands during the nineteenth century. Specimens shipped home by amateur and official collectors were shared between the British Museum, the Geological Society, and the museum which the Company itself established in 1801.[10]

Murchison thus encountered a well-established tradition of inquiry when he was first exposed to Indian geology in the form of spoils from a campaign of imperial annexation. John Crawfurd, FRS, FGS, orientalist and East India Company Resident at Singapore, collected an important series of plant and animal fossils during a mission to Ava in 1826 to negotiate the treaty ending Britain's First Burmese War. When these sensational finds reached Murchison as Foreign Secretary of the Geological Society in 1827, he helped arrange for their classification and public display and established a lifelong friendship with Crawfurd, a fellow Highlander.[11] The next year he organised the collection of more specimens by correspondents in the newly conquered Burmese territories.[12] At the beginning of his scientific career, therefore, Murchison discovered the scientific benefits to be gleaned from imperial military expeditions.

As President of the Geological Society in 1831–32, Murchison persuaded his brother Kenneth, East India Company Resident at Penang in the Straits Settlements, to commission a survey of that outpost. He published the results in the Society's *Proceedings*, arguing that similar surveys in every colony 'would prove of inappreciable value'.[13] Murchison also helped integrate researchers from India into metropolitan scientific culture: he proposed for membership of the Athenaeum Club, for example, Alexander Christie, a company surgeon who had published on the geology of central India.[14] When Alexander Burnes visited Afghanistan, Persia, and Central Asia during a diplomatic mission in 1832, Murchison introduced his results to the Geological Society. Burnes collected organic remains on a subsequent mission to Kabul, and his distribution of copies of the Society's geological instructions resulted in Murchison's receipt of fossil shells gathered in the Himalayan foothills by two army officers.[15] In 1840 Burnes also reconnoitred coal deposits to facilitate the deployment of steamboats on the Indus, and in 1843 Murchison used the example of his work to promote the geological exploration of Afghanistan by the British army of occupation.[16] During the same period Murchison helped arrange for the British Museum to acquire the magnificent series of mammalian fossils collected in the Siwalik Hills of north India by Proby Cautley of the Bengal Artillery and Hugh Falconer, head of the Saharanpur Botanical Garden.[17]

A technological innovation in India soon permitted him to turn from such opportunistic endeavours to more systematic promotion of research. In 1836 the government of Bengal formed a Coal Committee to ascertain the location and quality of the presidency's coal reserves. It was hoped that the provision of such information would stimulate private enterprise to produce the fuel required by the steamers being introduced on the Ganges at costs sufficiently low to obviate the need for further importation or reliance on the monopolised existing local supply. The Committee's Secretary was Dr John McClelland, a Company surgeon and enthusiastic naturalist who had accomplished geological research while involved in the introduction of tea cultivation in Assam.[18]

After a preliminary investigation undertaken by surgeons and engineers, the Committee approached Charles Lyell and Murchison in 1841 to persuade the Company's Court of Directors that a long-term survey conducted by specialists was required to develop India's coal resources. The survey was projected to range as far east as the Burmese provinces of Arakan and Tenasserim, where mines had been opened during the first Opium War to supply warships in China, and as far west as Natal and the Zambezi River, where the Portuguese had offered the Indian government exclusive coal mining rights.[19] Lyell stated that the examples of Newfoundland, Canada, and the United States proved the economic advantages of geological surveys; Murchison replied as President of the Geological Society that the Indian government's apathy was 'marvellous and lamentable', and that he would do all in his power to help prosecute a survey.[20] In 1843, Murchison having publicly reiterated the economic and strategic value of a survey to India, the government of Bengal requested that he be asked to select a geologist.[21]

After evaluating applicants and consulting William H. Sykes, FRS, a retired Indian Army officer of diverse scientific interests who was one of the Company's Directors, Murchison turned the search over to De la Beche as head of the British Survey.[22] In 1845 Murchison held forth once more in his Russian book on the regrettable fact that 'In Hindostan, so eminently British, . . . no well-defined and precise labours have yet been devoted to the older rocks': without such researches, India's rulers could not possibly evaluate or rationally develop the country's coal resources.[23] During the same year De la Beche recommended the appointment of David Williams, a subordinate with extensive experience in the South Wales coalfields.[24] Williams died in 1848, but his reports influenced the planning of Bengal's mineral and transportation development. The Indian Survey was permanently refounded in 1851 under Thomas Oldham, former Director of the Irish Survey, and its scope extended to Burma and the other two presidencies of Bombay and Madras.[25] Oldham's museum, laboratory, and library superseded a Museum of Economic Geology which had been founded in Calcutta in 1841 by the Asiatic Society on the model of De la Beche's establishment to stimulate mineral exploration, iron smelting, and soil analysis for the introduction of plantation crops.[26]

Williams' survey in Bengal demonstrates that the Indian authorities, like the imperial government, were susceptible to co-ordinated pressure from field scientists and metropolitan savants arguing that scientific research could solve problems of technological deployment and economic development, and that the Company's international prestige was at stake in funding such research. The goals of the two groups of scientists are also evident – the desires of the men on the spot for secondments offering promotion and intellectual recognition, and of the men in London to increase their powers of patronage and the scope of publicly funded inquiry. For both the scientists and the Company, these negotiations were conducted within the framework of advancing British imperial interests. That the British Survey was initially bypassed in the process of arranging the coal survey illustrates the informal relationship between science and government that obtained in the first half of the century. The predominance of the Geological Society as the repository of metropolitan expertise yielded only gradually to the official authority of the Survey.

The coal survey also highlights the Company's willingness to depart from reliance on free market forces to call forth required production. This Utilitarian solution, however, served to break a monopoly, encourage private development of resources, eliminate the financial waste of uninformed speculative sinkings, and substitute revenue-generating local supplies for costly imports that drained the industrial strength of the mother country. A scientific and technological multiplier effect was initiated as well. In the course of his survey Williams discovered deposits of copper and lead as well as coking coal, iron, and limestone – complementary minerals that encouraged the Company to develop a local iron industry. As usual, extraction and transportation costs were the dominant considerations in the decision to open mines, and again the expertise of the geologist was critical.

Since Bengal's most promising coalfields were not convenient to the river transport they were required to fuel, Williams recommended the building of one of India's first railways to deliver coal to Calcutta at a competitive price.[27] He was also called into service, before a permanent river surveyor was appointed, to recommend navigational improvements necessitated by the introduction of the steamboats.[28] Steam technology provided a vector for Western penetration, but its deployment depended on the installation of scientific institutions which accelerated progress in related sectors.[29] An entire new spectrum of developmental possibilities was thereby suggested to the Company's planners. Lastly, sub-imperialism is evident in the Indian government's expansionist attitude toward coal procurement. Not content with providing fuel at Calcutta, its desire to survey deposits in the Bombay presidency, Assam, Hyderabad, Burma, Malaya, Natal, Mozambique – and later even Borneo, Labuan, and New South Wales – showed a willingness to reach out along communication routes throughout its sphere of interest and dominated strategic fuel supplies.[30]

During his first presidency of the RGS Murchison also served as the conduit by which the first fossils from Sind, collected by an officer of Sir Charles Napier's conquering army, were introduced to Europe.[31] Again, his praise for the military campaigns which made them available emphasised the growing co-operation between scientists and imperial planners.[32] The resemblance of the jaw of an extinct gavial from this lot to that of a prehistoric British form prompted anatomist Richard Owen to twit Murchison for his imperial sentiments: 'Would this go at all toward establishing a prior claim to the territory and justifying the occupation?'[33] The RGS simultaneously published a political officer's account of the defeated Baluchi tribes which Murchison considered 'of great practical importance now that Scinde is annexed to our eastern possessions'.[34] When Captain Vicary, the geologist of Sind, provided more fossils collected during the Punjab annexation, Murchison commended him for obtaining his information 'in an arduous campaign, which led the British forces into regions ordinarily inaccessible to geologists'.[35]

In 1846, the year of his British Association presidency, Murchison persuaded the Earl of Ripon, President of the Indian Board of Control and former RGS President, to select graduates with comprehensive scientific training for the Indian medical service.[36] Given the state of university science curricula in Britain, this private arrangement constituted a preferential measure for recruiting yet more Scottish medical graduates into the East India Company's employ, where they were already extremely well represented. And though Murchison was primarily actuated by his desires to advance Britain's related scientific and imperial interests, he was no doubt also attempting to systematise the creation for fellow Scots of lucrative Eastern opportunities such as that which had saved the fortunes of the Murchisons in the previous generation.

One of the resulting appointees was Dr Andrew Fleming, the son of John Fleming, a geological ally and Professor of Natural History at Aberdeen

University. Helped by the influence of Lyell and Sykes, Murchison secured young Fleming's place as an assistant surgeon in Bengal.[37] From 1848 to 1852 Fleming conducted a geological survey of the Punjab and parts of Kashmir, giving particular attention at Murchison's request to the Palaeozoic strata of the Salt Range. He examined coal and fossil wood as fuels for Indus steamers, surveyed deposits of gold, copper, and other minerals, and advised the government on problems concerning transport, irrigation, and miners' strikes.[38] Fleming's fossils and maps were sent home to the Geological Society, the India Museum, and his father, who relayed some of his portion of the collection to Murchison for classification by other specialists.[39] Fleming senior also passed on a series of reports which his son had written especially for Murchison, and these were published by the Geological Society.[40] In 1853 the Punjab Survey was subsumed by Oldham's expanding Indian Survey, but Fleming continued to prosecute his geological researches while on medical duty and Murchison used his results to link India's Palaeozoic stratigraphy with the European succession.[41] At the same time, as one aspect of Utilitarian experimentation in the Punjab, a professorship of geology was founded at the Roorkee Civil Engineering College to create a corps of mineral surveyors and channel native energies into practical endeavours which would strengthen administrative control and promote economic development.[42]

Immediately upon succeeding De la Beche as Director-General, Murchison begged his friend Viscount Canning, the Governor-General of India, to carry out the research he deemed essential to the well-being of the Raj by deploying geologists in every presidency.[43] Canning's acquiescence allowed him to boast later that his 'warm intercession' had led to the expansion and permanent endowment of Oldham's Survey.[44] Oldham never forgot this obligation, and for the next fifteen years his personal relationship with Murchison, besides the official information exchanges between the two surveys, remained a strand of metropolitan domination of Indian science. Oldham relayed scientific news and promoted RGS recruitment in India, while Murchison publicised Oldham's work as worthy of increased funding and advised the India Office on improving his museum.[45] The English geologist W. D. Conybeare suggested aiding Oldham by using a review of *Siluria* to 'furnish proofs of the great utilitarian importance of geological enquiries' for developing the resources, employing the idle capital, and establishing the manufacturing potential of the Indian empire.[46]

When Oldham supervised India's mineral display at London's International Exhibition of 1862, Murchison awarded him a medal; on a private hint from his friend Sir William Denison, now Governor of Madras, he also put pressure on Oldham to ascertain the origin of curious laterite deposits on the Malabar coast.[47] Chafing under his own subordination as Director-General to the Committee of the Privy Council on Education, Murchison looked longingly to Oldham's relative independence and ability to accomplish large-scale projects with efficiency. He viewed the authoritarian power structure of British India as analogous to that of Tsarist Russia, Imperial France, or the Royal Botanic Gardens at Kew, which he believed owed its excellence as a national institution to the autocratic power of its Director, Sir William Hooker.[48]

Oldham earned Murchison's ire, however, for luring both veteran Survey geologists and promising graduates of the School of Mines to more lucrative jobs in the East without consulting the Director-General. He complained that the haemorrhage of talent caused by this 'temptation of rupees' had inflicted 'a *serious blow* upon the well being & progress of the Survey of the British Isles', even while admitting that this impasse had resulted from 'a demand for India which I myself have to a great extent caused.'[49] Yet he managed to turn the Indian staff drain to account, using it to press for higher salaries for his subordinates and to advertise the imperial utility of his metropolitan establishment.[50] As the resignations continued, however, Murchison's justifications began to wear thin with his colleagues. Jukes, Director of the Irish Survey, remarked to his English opposite number A. C. Ramsay in 1866 that the unsatisfactory progress of which the Treasury was complaining 'is not our fault. When we remonstrated ten or twelve years ago against all our best men going to Australia, India, etc., were we not told that it was good for the Colonies, and that the Mother Country could afford to wait?'[51]

Throughout the 1850s and 1860s Murchison was involved in various other schemes using scientific expertise to foster India's development. School of Mines graduates staffed the Indian mints and telegraphs,[52] and Murchison continued De la Beche's policy of offering instruction to East India Company officers at half price. By 1856 nearly 100 officers had undergone this scientific training.[53] Murchison also occasionally secured Indian Survey appointments to reward worthies such as William Kennet Loftus, former geologist of the Turko–Persian Boundary Commission.[54] When the introduction of the *ryotwar* rent system in Madras necessitated a precise assessment of cultivated lands, Murchison was requested to appoint a geologist to evaluate soils.[55] Canal building, too, with its salinisation problems, forced the Indian government to turn to the Museum of Practical Geology for analyses of soils and waters and recommendation of a chemist to carry on this work in India.[56] When Ceylon's pearl fishery failed in 1864 the Colonial Office also turned to the Director-General who, together with his subordinate Thomas Huxley, recommended the appointment of John MacDonald, FRS, a naval surgeon distinguished for microscopic studies of marine invertebrates, to solve the problem.[57]

In 1867, during a scare over the exhaustion of Britain's coal reserves, Murchison urged the publication of Oldham's report on Indian coal to allay public fears. At the request of the Secretary of State for India, he also recommended the appointment of a specialist to the Indian Survey in order to stimulate coal production by introducing new mining methods.[58] The next year Murchison took part in an unsuccessful attempt to install a geological survey in Singapore and the Straits Settlements. Since the administration of this possession had just been transferred from the India Office to the Colonial Office, the governor was anxious to discover new sources of revenue to render the colony fiscally self-sufficient. The tin resources of the Malay Peninsula had long been known and, as Murchison himself believed, the region's limited gold production suggested the possibility of larger deposits awaiting discovery. Requesting a survey to stimulate mining, the governor proposed that funding be shared by the colonial and imperial governments on the model of the West Indian Survey. He also reported that the Maharajah of Jahore had agreed to underwrite a survey of his dependent territory. When Murchison demanded a salary of £1,000 to induce a qualified geologist to undertake this hazardous short-term assignment, however, the colonial Legislative Council balked and the negotiations collapsed.[59]

Despite his constant reiteration of the permanence of empire, Murchison's reaction to the India Mutiny demonstrated that he felt financial prudence to be the better part of patriotic valour. As early as 1843 he had voiced his misgivings about the risks of garrisoning India with so few troops,[60] and when news of the Mutiny reached Britain he succumbed immediately to bond-holder's panic and directed his bankers to sell his Indian railway shares at a loss. Once clear of ruin, Murchison congratulated himself on his escape, piously lamented the loss of Sir Henry Lawrence, the hero of the siege of Lucknow, subscribed for the suffering Anglo-Indians and, in talks with the Prussian King and Tsar Alexander II, blamed the British ministers for indecision and timidity in suppressing the uprising.[61] Murchison never directly reinvested in India, though he retained some Indian bonds inherited after the Mutiny because of his conviction that the authority of his friends Denison and the Earl of Elgin, Canning's successor as Viceroy, constituted sufficient guarantee of their security.[62]

Murchison supported the restoration of British supremacy by a strengthened military presence buttressed with applied science and technology. As during the Crimean War, he passed on to official connections such as the Earl of Ellenborough, President of the Indian Board of Control, the strategic suggestions of Major-General Shaw-Kennedy.[63] Even the defeated enemies of empire could be turned to its account, he argued: rebellious sepoys should be transported to found a new penal colony in northern Australia.[64] Visiting the Tsar and his ministers during a German tour in 1857, Murchison and his Russian friends compared their empires' subject races, likening the atrocities suffered in the India Mutiny with those endured in the Caucasus campaign and agreeing that both areas remained dangerous to their conquerors.[65] Despite this severity, Murchison's views were not as extreme as those of Denison, with whom he argued about the racial inferiority of Indians. The Mutiny nevertheless lingered in Murchison's mind, for nine years later his strongest argument in defence of Edward Eyre's controversial actions in putting down a black insurrection in Jamaica was that the governor had prevented the repetition of the horrors perpetrated in India.[66]

Kashmir and Tibet

Murchison also encouraged exploration beyond India's northern frontier for scientific, economic, and political motives. In 1852 he awarded an RGS gold medal to Captain Henry Strachey for his meticulous mapping of the Himalayas in Kumaon, Garwhal, and parts of Kashmir and western Tibet. Assisted by his brother Richard, Strachey provided valuable information on the geography, geology, botany, climate, trade routes, ethnology, and tribal politics of this remote region.[67] Richard Strachey's discovery of Silurian fossils near the axis of the Himalayas particularly recommended him to Murchison, whose pride in empire was inextricably linked to the extension of British stratigraphic nomenclature.[68] During the 1840s Murchison had considered visiting these mountains to link his Russian work with Indian research, but in the event he undertook their conquest by proxy.[69] As a Political Agent and amateur geologist in the Northwest Provinces remarked to Sedgwick in complaining that his local stratigraphic classifications had been superseded by the Silurian gospel: 'the Himalayan Mountains are British; *both* in their physical character & their political position, they bow their head to the British flag'.[70]

Murchison helped arrange the presentation of several memoirs by the Stracheys at the RGS and the British Association, and he chaired a Council meeting of the latter organisation which resolved to petition the Indian government to publish their results in full.[71] Although this request was unsuccessful, Richard Strachey's natural history collections were transmitted to the India Museum in London, and Murchison incorporated his results in *Siluria*.[72] Joseph Hooker, who had recently returned from a botanical expedition to Sikkim and Nepal, expressed his doubts to Murchison about the originality of the Stracheys' work after hearing the President's Anniversary Address to the geographers for 1852. In a remark testifying to the quasi-official influence which RGS presidents – and more especially Murchison – enjoyed throughout the empire, Hooker complained that the attention shown the Stracheys would raise jealousies in India because of the *'immense ... importance'* of Murchison's address, which would 'be reprinted in every Indian paper, within a few weeks, & duly canvassed.'[73]

To salve Hooker's ire, Murchison took care to extol the Himalayan explorations which Hooker and his fellow botanist Thomas Thomson had accomplished in the printed version of his address,[74] but he remained convinced of the value of the Stracheys' research. When Richard Strachey took home leave from 1850 to 1855, he served his first term as a Council member of the Geological Society, won election to the Royal Society, and attended the inaugural dinner of Murchison's Geographical Club. He was also recommended by De la Beche for the professorship of geology at Roorkee College in the Punjab, but the Indian authorities were unwilling to dispense with his valuable services as a military engineer.[75] During his subsequent years as an administrator, Strachey kept up his metropolitan contacts. In 1866, for example, he submitted soil samples from northern India for analysis by the Museum of Practical Geology, began his second term on the Geological Society's Council, and through Murchison's influence was elected to the Athenaeum Club as an eminently distinguished member.[76] After retirement in 1871, Strachey became a member of the Council of India as well as occupying several scientific posts such as the presidency of the RGS, from which he helped formalise the links between science and Indian administration.[77]

Murchison similarly supported the elaborate trans-Himalayan explorations conducted during the 1850s by the Prussian Schlagintweit brothers. As a Council member of the Royal Society, he helped Edward Sabine and W. H. Sykes secure the appointments by the East India Company of these three disciples of Humboldt. 'I find a feeling seeming to prevail against employing *Germans* in which I do *not* participate,' Murchison remarked to Joseph Hooker, 'indeed we have not better & fitter men ready'.[78] When British colleagues took umbrage at the Schlagintweits' claim to discoveries accomplished by officers of the Trigonometrical Survey and in consequence criticised Murchison for securing their admission as visitors to the Athenaeum, Murchison acknowledged their indiscretion. He defended the Prussians, however, against charges levelled by naturalists such as Hooker and Falconer, who were smarting under the Indian government's lack of recognition of their own researches, that they had swindled the East India Company of some £30,000 and published their results in German scientific journals.[79] Mingled with these personal grievances was resentment at Murchison's domination of scientific patronage and patriotic anger at the export of scientific results from British imperial turf. The King of Siluria, however, was willing to risk the displeasure of British savants to court the favour of Humboldt and accomplish explorations for which no equally qualified Britons presented themselves. He continued to support the value of the Prussians' work and used the murder of Adolph von Schlagintweit as a *cause célèbre* with which to promote further exploration of this politically sensitive and scientifically fascinating region.[80]

H. H. Godwin-Austen of the Great Trigonometrical Survey, the son of Robert Godwin-Austen, a geologist distinguished for Palaeozoic research in Devonshire, also won Murchison's praise for his discovery and mapping of the Karakorum glaciers in 1860–61.[81] Godwin-Austen was likewise commended for subsequent geological researches in Kashmir, Bhutan, and Tibet.[82] His work contributed to the metropolitan controversy over glaciers, as it appeared before the Geological Society, the RGS, and the British Association, as well as in *Siluria*.[83] Murchison had been interested in the geology of Kashmir since the 1830s, when he had examined fossils collected there by French explorers. During his first RGS presidency in the following decade he had considered awarding the Society's gold medal to the traveller Godfrey Vigne, whose observations on volcanism and seismic activity in the region lent support to his own views on convulsive mountain uplift.[84] The assumption of British sovereignty in 1846 opened Kashmir to systematic scientific investigation. Andrew Fleming's researches further whetted Murchison's appetite for data, while the commencement of mapping by the Trigonometrical Survey in the late 1850s made possible the consideration of a geological survey of the province.[85]

In 1861 the military commander in the Punjab approached Murchison to recommend a geologist to survey the vassal state of Kashmir on behalf of its Maharajah.[86] The Director-General agreed that '*mines & ores* are of course the main object',[87] but though the salary offered was £1,000 per annum, several candidates including James Hector, Archibald Geikie, and George Wall, formerly of the West Indian Geological Survey, declined the post.[88] It was finally accepted by Frederick Drew of the British Survey, a former Royal Exhibitioner at the School of Mines.[89] In his *Annual Report* Murchison commented on the rationale behind the installation of geologists in client states: 'Although Cashmire is not within the jurisdiction of our Sovereign, an acquaintance with the mineral structure of a region coterminous with our great Indian empire, and with which we have such intimate relations, is necessarily of public importance.'[90] Drew served ten years in Kashmir as a geologist and provincial governor for the Maharajah. He also acted as a Political Agent for the Indian government, supporting the forward school of foreign policy.[91] His research won the praise of British geologists, and he remained in touch with his old chief, transmitting geographical news and specimens for the Museum of Practical Geology.[92] After retiring, Drew published a geographical account of Kashmir,[93] served on the Council of the Geological Society, and for sixteen years held the appointment of science master at Eton.

Central Asia

Himalayan exploration led naturally to penetration of the Central Asian khanates. This region had been visited by occasional military explorers since the early nineteenth century, for officials of the East India Company and diplomats such as Palmerston who were suspicious of Russian activities wished to monitor developments there. Murchison used the observations of these intelligence agents to speculate upon the geology of Central Asia from the basis of his Russian research.[94] Because of his special relationship with Russia, Murchison was able to promote the southward expansion of Tsarist authority into this sphere, as well as British probings northward from India, without any sense of contradiction. Towards the end of his life, however, as the rivalry between the two empires intensified, the buffer zone shrank, and fears of a Russian invasion of India sharpened, he found himself increasingly isolated in his opinions.[95]

Murchison's tour of southern Russia in 1841 had established Central Asia as an area of his special interest. Deeply impressed by the drama of Russian imperial conquest, he noted the interplay between mineral discoveries, territorial annexations, and opportunities for scientific research. Observing that discoveries of alluvial gold near the Altai Mountains were prompting a military extension of Russia's frontier with China, he planned a visit to the region in order to add a new province to his Silurian empire, but he was forestalled by the naturalist Chikhachëv.[96] A scientific expedition under the geographer Nicholas Khanikov also left Orenburg during his visit with the purpose of surveying gold deposits believed to exist in the mountainous tract between Bukhara and Samarkand. The results of this exploration, published in 1852 and transmitted to the RGS through Murchison, probably influenced the Russian annexation of Bukhara in 1868, for one of the world's most productive gold mines now exists in precisely this area.[97]

Murchison's own research and subsequent Russian work extended the known range of Palaeozoic strata from the Urals to the Pacific. The British artist–explorer T. W. Atkinson, for example, traced these formations across the Kirghis steppe, Mongolia, and Chinese Tartary while travelling under the patronage of the Tsar during the early 1850s.[98] After furnishing Clarendon with political intelligence about 'Russian proceedings in Central Asia',[99] and the metropolitan audience with a mass of scientific and commercial information,[100] Atkinson was rewarded by Murchison with fellowships of the Geographical and Geological societies, public praise of his illustrated travel account, schooling expenses for his son, and a government pension for his family.[101] Rising Siberian gold production, coupled with the meridional axis of many Asian ranges and the apparent presence of the 'golden constants' identified by Murchison in the Urals, suggested that more treasure would be forthcoming in China and Central Asia.[102] The lure of precious metals in the khanates of Turkestan thus became an issue in the growing rivalry between Britain and Russia during the late 1860s. But Murchison, torn between his desires for further Silurian gold discoveries and the preservation of amity between the nations, hoped that, should the region prove sterile of such riches, it could be maintained as a buffer zone between the two powers.[103]

Murchison's promotion of Central Asian exploration brought dividends to the RGS, for the region's awesome topography and hostile inhabitants fascinated the public. The President lionised visiting Russian explorers like Khanikov, corresponded with Russian scientists as well as Britons in India to keep abreast of discoveries, and advised British travellers such as Lord William Hay on new routes to follow.[104] Foreign explorers were also encouraged. The Hungarian Armin Vámbéry, for example, introduced to Murchison in 1864 by the retired British Minister to Persia, was awarded a cash prize by the RGS, displayed at its meetings, and given favourable notice in the Anniversary Address to promote the sale of his narrative.[105] William Denison admitted to the President that this book had excited his cupidity to annex Central Asia, and the Russophobic Vámbéry sought further employment in Central Asia as an agent of the Indian government through Murchison's influence.[106] Murchison's private influence with Indian statesmen increased his voice in

Central Asian affairs. Lord Ellenborough, FRGS, who had requested Murchison's opinion about quarry development on his estate, turned to him again in 1865 for advice and maps concerning Russian movements in Central Asia.[107] Clements Markham's dual position after 1867 as Geographer to the India Office and Honorary Secretary of the RGS also augmented Murchison's leverage over Eastern matters. Markham owed Murchison much for advancing his career, and his assistant at the India Office, Trelawny Saunders, was similarly indebted for his appointment.[108]

Running through all of this activity was the leitmotiv of preventing rivalry between the two powers to whom Murchison owed his principal honours. Following the Crimean War, he hoped the old allies could support each other's imperial and commercial ambitions and that the 'ultimate definition of the boundaries' between them would 'set aside all jealousy & bring [the] two civilized powers to divide between them the barbarous East'.[109] At the British Association he favoured memoirs urging co-operative development of Central Asia,[110] and innumerable comments to the RGS dwelt on the same theme. He inveighed against the Indian press and Englishmen 'who only look at maps of Asia on the scale of perhaps 100 miles to the inch' for creating alarms 'at the bugbear of a Russian overland invasion of British India'.[111] Such remarks were noted with approval in St Petersburg, and Murchison took care to pass along to British statesmen the messages of gratitude received from his highly placed Russian friends.[112] After 1865, however, he found himself arguing against the sometimes measured, often strident opinion of Sir Henry Rawlinson that the Russian threat was real and must be aggressively countered.

Rawlinson was an RGS gold medallist, a frequent member of the Society's Council, Vice-President from 1864 to 1868, and eventually Murchison's chosen successor as President. He had been suspicious of the Russian threat to India since his association with the deep-dyed Russophobe Sir John McNeill in Persia during the 1830s. After serving as a Political Agent and Consul in Afghanistan, Arabia, and Baghdad, the scholarly Rawlinson retired to London in 1855, where as an MP and member of the Council of India he carried on a tireless propaganda campaign to alert the public to Russia's Asian designs. The House of Commons and the RGS, which because of its popularity and prestige he considered 'a power in the state', were his main platforms.[113] Rawlinson was supported within the Society by several other influential advocates of an Indian forward policy, including McNeill, Sir Henry Bartle Frere, Sir Justin Sheil, Sir John Kaye, and Sir Robert Montgomery.[114]

In a paper read before the British Association in 1865, Rawlinson expressed his fear that the power vacuum resulting from the withdrawal of Chinese authority from Eastern Turkestan in 1864 would be filled by the inexorably advancing Russians, who besides swallowing the markets, minerals, and potential cotton-producing tracts of the region, would be brought dangerously close to the British frontier in Kashmir.[115] Rawlinson went even further in an article published the same year in the *Quarterly Review*. Here he taxed

Murchison for having inadvertently aided Russia's designs by refuting the threat of an invasion of India. Claiming that no accommodation with Russia was possible in Turkestan, he also called for the occupation of Kandahar and Herat in Afghanistan should Russia annex any more of the region.[116] He then made a milder proposal to the RGS for the co-operative exploration and exploitation of Central Asia by both powers and the installation of joint consuls in its cities.[117] To this Murchison agreed, but he voiced his doubts that the buffer zone could ever be absorbed by Russia and rebuked Rawlinson in spirit if not in name for fostering suspicion of Russia's motives at the Society.[118] At a meeting in 1868 Rawlinson advocated British alliances with Persia and Afghanistan, which he saw as more likely invasion routes than Turkestan, and argued that the RGS should urge the government to develop Central Asia in the interests of Anglo-Russian amity.[119]

Rawlinson had by now incurred Murchison's direct ire for using the Society to foster distrust of Russia. Viscount Strangford, a Turkophile diplomat on the RGS Council, condemned Rawlinson as alarmist, but thought the President too sanguine of Russian motives. A collection of Strangford's geographical and political essays, largely concerned with Russian policy in Asia, appeared posthumously in 1869. The book had much to say on the role of the RGS in imperial affairs and on Murchison's style of leadership. It included an article from the *Pall Mall Gazette* of 1866 in which Strangford criticised the President for allowing his sympathy for Russia to lead him to focus public attention on Africa rather than 'the trans-Himalayan California of the future, the spot likely to turn out the one specially auriferous district of the old world'. Murchison, he claimed, as 'the one indispensable test and index of geographical novelty or importance', was more interested in generating discussions that enhanced popularity than in encouraging exploration in Central Asia. He cited the duel with Rawlinson as an example, accusing Murchison of distrusting him as 'a noxious reptile, fanged with poisonous rear-thoughts against inoffensive and growing Russia'. Believing that political discussions had no place in a scientific society, Strangford also censured Rawlinson's transgressions.[120]

In truth, Murchison had unsuccessfully attempted to goad Strangford into opposing Rawlinson at a recent RGS meeting.[121] Other influential Fellows who desired the RGS to adopt a more aggressive stance in imperial affairs supported Rawlinson's bellicosity. Captain G. H. Richards, the Navy's Hydrographer, also perceived a clear threat of Russian 'encroachment'.[122] Sir Samuel Baker, warning that the Russian advances required careful surveillance, pledged that the RGS was not only 'forewarned of the encroachments of neighbouring powers, should their expeditions be pushed beyond the limits of necessity, but we form a nucleus for all geographical information, should the Government resort to us in an emergency.'[123] Here was a clear bid for a larger role for the Society as an imperial intelligence bureau capable of monitoring and perhaps even countering Russian moves.

The importance of the issue of Central Asian policy caused the internal dispute at the RGS to spill over into the public journals, for the debate focused on the larger questions of defining national goals, choosing methods and instruments for achieving them, and establishing the disciplinary limits of geography. The geographers were thus in a sense mapping their own cultural frontiers by laying down the appropriate political role of their science's representative institution. While Murchison was willing to use the RGS to support British commercial and imperial expansion, and even on occasion to air suspicions of the expansionist motives of France or the United States, he was not prepared to allow it to be used by others as a pulpit from which to hector or threaten Russia.

Rawlinson, however, because of his conviction that India was at risk, was willing to challenge Murchison's authority on this issue. His attempt to redefine the role of the RGS caused him to reverse the President's priorities as to the geographical and political aims of scientific expeditions. He told the RGS in 1865 that 'the good such enterprizes did was two-fold: in the first place they furthered the political interests of the British Government; and in the second place contributed to the advancement of geographical science'. [124] In 1866 Rawlinson urged the government of India to relax its policy of restricting exploration north of Kashmir in order to avoid political complications. He also called upon Britain to exploit every opportunity to push its feelers into Central Asia, and suggested that the RGS could perform a real service in facilitating expeditions into this zone. [125] In 1868 he argued that the exploration of Central Asia should be reserved for Britain and that the region's trading potential should be developed 'under the auspices of British geographers' to advance the interests of British India and spread the benefits of civilisation into backward native states. [126]

The same year Rawlinson had an opportunity to test these convictions when the RGS sponsored George W. Hayward, a former Lieutenant in the Indian Army, to explore routes from Kashmir to Yarkand and Kashgar, which were now independent of Chinese rule, and to reconnoitre the Pamir plateau. [127] Hayward had been introduced to the Society by Rawlinson, who also wrote his instructions for this expedition. [128] In view of Rawlinson's imperial ambitions, it is revealing that Hayward's goals, as contrasted with those of the 'Russian armies of exploration', were given forth at the meeting of the British Association as 'the pursuit of new geographical encroachments'. [129] In his Anniversary Address Murchison admitted his anxiety for fresh discoveries, but dissented from the views of those who, 'overlooking all obstacles where British *prestige* and power are to be extended', blamed Viceroy Sir John Lawrence for stopping trans-Himalayan explorations. This remark, as well as others praising Russia's extending influence in Western Turkestan and denouncing the Indian invasion threat as 'a mere chimera and a political bugbear', were certainly aimed at Rawlinson. Still, Murchison hoped that Hayward would be allowed to proceed on his journey as an unauthorised private traveller. [130] Rawlinson's machinations were thus forcing Murchison to walk a fine diplomatic line. Striving to preserve both his Tsarist ties and good relations with the Indian government, he nevertheless needed to maintain the momentum of exploration in order to sustain the popularity and power of the RGS.

In 1869 Lawrence was succeeded as Viceroy of India by the Earl of Mayo, who continued the foreign policy of nonintervention and, like Clarendon, the Foreign Secretary, hoped that a buffer zone of semi-neutral states would continue to separate the two European empires. [131] Mayo, however, took a more active interest than his predecessor in Central Asian exploration as a means of gathering political, military, and economic intelligence. He began sending his old friend Murchison 'semi-official information' concerning the movements of Hayward and Robert Shaw, an English planter who had set out to explore the possibility of opening a commerce in tea between India and the khanates of Turkestan. [132] Murchison used Mayo's information in composing his Anniversary Address of 1869, a year in which two meetings of the RGS devoted to the subject of trade routes between India and Turkestan had been transformed by Rawlinson into seminars on Russian aggression and Central Asia's probable mineral wealth. [133]

In this cautious document Murchison balanced praise of Russia's Bukharan explorations with pride in the British surveying of south-western Tibet, including its 'extensive gold fields'. He expressed his hope that the two nations' traders might compete peacefully in the intervening area. He noted further that Shaw's crossing of the Karakorums to Yarkand promised the development of a valuable tea trade, since the khanates had lost their traditional supply by revolting against their Chinese overlords. Shaw's success was ascribed to the diplomatic skill of Thomas Douglas Forsyth, FRGS, Commissioner of the Punjab, another advocate of British expansion in Central Asia who had inspired Shaw's journey, propitiated the xenophobic Amir of Yarkand, and, just prior to the Anniversary meeting, provided Murchison with a report for the RGS on the subject of this tea trade. Murchison also advertised newly available maps of Turkestan and cited historical sources on the fertility and population of the region, as well as mentioning reports of an ancient walled city and paved road suitable to wheeled traffic from the Himalayas. [134]

All of this bait was well calculated to excite interest, and the drumbeat of publicity continued while Hayward and Shaw remained in the field. At the meeting of the British Association in August, Frere's address to the Geographical Section congratulated Britain and Russia for opening Central Asia and praised Lord Mayo for fostering peace with the independent khanates. [135] In the audience was Forsyth, lately returned from an inconclusive mission to St Petersburg on behalf of Mayo to settle the Russian–Afghan border. Murchison had used his friendships with Clarendon, Brunnow, and Argyll, the Secretary of State for India, to facilitate this diplomatic initiative. He had provided Forsyth with letters of introduction to his Russian friends and helped to reconcile the opposing views of the Foreign and India Offices in order to counter what he considered the baneful influence of Rawlinson in official circles and further his own plan 'to bring about a right understanding & good will between Russian interests & our own in the *Far East.*'[136] Forsyth and Saunders of the India Office each delivered memoirs on trade routes between India and Central Asia to the Geographical Section,[137] and two Russian papers were also presented to preserve the façade of scientific internationalism in what had become an extremely politicised debate. Khanikov's described Samarkand, and Chikhachëv's memoir echoed his friend Murchison in dispelling fears of Russian invasion and promoting co-operation in Central Asia.[138]

Soon after this meeting M. E. Grant Duff – Russophobe, Under-Secretary of State for India, and RGS Council member – introduced a visiting Yarkandi to Murchison in the hope that his information would prove interesting to the RGS.[139] At the same time the President was receiving intelligence from Mayo and Clarendon concerning the progress of Hayward and Shaw, who had joined forces and reached Yarkand. Largely because of the recent conclusion of a treaty between the Amir and the Russians, they believed, they had been imprisoned for six months, frustrated of their commercial and exploratory aims, and escorted back to the Kashmiri frontier.[140] Planning a second journey to the Pamir steppe via Gilgit and Chitral, Hayward transmitted to Murchison a report and maps of his route to Yarkand, including the intelligence that one pass from Kashmir could be made passable for military convoys and artillery.[141]

Opening the 1869–70 RGS session in November, Murchison promised that Central Asia would be much discussed in the coming year. He reported that Hayward was preparing to depart again with the aid of RGS funds and, in order to counter suspicions about the nature of the mission, carefully explained that 'Mr. Hayward is solely our agent for purposes of pure geography'. He mentioned once more the commercial potential of Yarkand, in which Indian tea and British factory goods could be exchanged for precious metals, silks, and fine wools.[142] Hayward next informed Murchison that he had won the support of the Maharajah of Kashmir for his journey and had been lent surveying instruments by the Indian government, though he had agreed to travel without official status.[143] Murchison now wrote to the Imperial Geographical Society in St Petersburg requesting aid for the explorer if and when he made his way north to the Russian frontier.[144]

Hayward's memoir on his first journey, which Forsyth considered 'splendid and valuable', was presented at the RGS meeting of 13 December 1869.[145] It contained a good deal of geological observation designed to please Murchison as well as much reference to mineral wealth of interest to members of the forward school eager to cast the prize of Central Asia as worth contention. Indeed, British merchants were already responding to these advertised opportunities, for a Yarkund Trading Company had been formed in anticipation of the opening of the lucrative new field of enterprise.[146] The meeting witnessed reiterations of position from Murchison and Rawlinson – the former dwelling upon the opportunities for co-operation between the two great powers in Central Asia, the latter upon the proof Hayward's work offered that forward pickets must be kept in the field to monitor the locations of Russian forces. [147] Rawlinson too had been receiving highly political dispatches from Hayward, many of them discussing the possibility of siphoning Central Asian trade southward to India. [148] Mayo was already working along these lines by promoting new trade fairs in northern India to attract the Central Asian caravans. [149] Once again, the two geographers' views of the priorities of exploration were starkly contrasted.

Mayo meanwhile wrote privately to Murchison relaying Hayward's request for official aid for his second expedition, which the Viceroy refused in order to protect his own plans for a gradual extension of trade and communications. He admitted being unable to prevent Hayward's travelling in a private capacity, though he clearly but indirectly appealed to Murchison to forestall the expedition. [150] The President in fact tried to do so, sympathising with Mayo's fear of the consequences 'if a recognized agent of the British Govt. were to go up there acting under our direct orders and get knocked on the head', but Hayward's ambition and Rawlinson's determination to pursue a forward policy impelled the explorer onward. [151]

At this point Forsyth advised Murchison that Shaw was on his way to Britain and asked that his protégé be elected to the RGS. He also passed along at Shaw's request a letter the explorer had written after being released from Yarkand, but asked that '*all political* allusions' be excised should Murchison use it. [152] This extraordinary letter provides evidence that Shaw and Forsyth, like Rawlinson and his supporters, were attempting to use the RGS to advance their political views regarding Central Asia. Shaw noted the Russian military and commercial threat to Yarkand, the Amir's consequent desire for commercial relations with India, the fact that Kashmir's exorbitant caravan levies constituted the only hindrance to this trade, and the absurdity of English noninterventionists believing that the price of friendship with the fiercely independent Amir might ever be the despatch of an army to defend him. [153]

On 11 February 1870 a memoir on Tibetan explorations accomplished by the Trigonometrical Survey prompted Shaw, now duly elected to the RGS, to comment that south-eastern Turkestan also 'abounded in gold'. In the same discussion Rawlinson, probably thinking more of the movement of troops than of goods, urged construction of the road which Hayward had proposed earlier. [154] Murchison was at this time also lionising Shaw at the Geographical Club and arranging with John Murray for the publication of his narrative. [155] Mayo meanwhile confided to Murchison that he was considering using Shaw for government service, that he would continue to help Hayward as much as his own objectives permitted, and would soon receive an envoy from Yarkand whom he hoped to awe with the power of the Raj. [156] Two weeks later Murchison had a heavily edited version of Shaw's alarming document read to the RGS. Despite being deprived of its lurid detail, Rawlinson spoke up strongly for a forward policy in the ensuing discussion. He demanded that the Indian government fortify Hayward's pass and compel the Maharajah of Kashmir to renounce his transit duties in order to throw open to British traders the markets of Central Asia. [157]

Hayward, turned back by snow in the passes, consoled himself by denouncing to his correspondents the recent depredations committed by Kashmiri troops in the district of Gilgit. He recommended that the Maharajah be confined south of the Indus River to prevent frontier instability leading to an encounter with Russia. Hayward now admitted that the resources of Yarkand had been exaggerated, but still believed the khanate could absorb large quantities of Manchester cotton goods. These reports were eventually published by the RGS in bowdlerised form, though Hayward, determined that the home public should know of the Maraharajah's aggressions, had specifically requested Murchison not to publish them unless nothing was omitted. [158]

Meanwhile the Society's promotional machine rolled on. In April, two papers were presented on Russian progress in Central Asia, with Rawlinson urging Britain and Russia to continue their rival explorations and publications. [159] At the Anniversary meeting in May, Murchison awarded a gold medal to Hayward for his first journey in Turkestan. That he was no longer denoted as a mere explorer, but called by the grandiose title of RGS 'Envoy to Central Asia', reveals the Society's – and especially Rawlinson's – pretensions to official status. Murchison emphasised that the medal was awarded only in recognition of geographical services, but pointed out that Hayward's travels had been 'at least as valuable in a public as in a scientific point of view' by having quietened invasion alarms in India, opened 'a new field for British trade and enterprize', and laid the foundation for a potentially valuable political alliance with Yarkand. Rawlinson, fittingly, accepted the medal on Hayward's behalf, describing how he himself had planned the expedition to gather geographical, commercial, and political intelligence, and berating the Indian government for withholding official recognition from the explorer. [160]

The Anniversary Address likewise demonstrated that Murchison, while compelled by the Russian annexations to accept a good many of Rawlinson's arguments, still hoped to defuse the rivalry between the two powers. The President here all but announced a British sphere of influence in Eastern Turkestan 'in contradistinction to Western, or what is really at present "Russian Turkestan."' An alliance between British India and Yarkand, he felt, based on trade but likely to include defensive agreements, would almost certainly result from Forsyth's mission to Russia, the explorations of Shaw and Hayward, Mayo's reception of the Yarkandi envoy, and the Viceroy's consequent despatch of Forsyth and Shaw on a return embassy to the Amir. He also described recent Russian surveys in Western Turkestan in terms almost suggesting that their results justified the occupation of the territory, and announced the election of the Central Asian explorer Baron Osten Sacken, Secretary of the Imperial Geographical Society, as an Honorary Corresponding Member of the RGS. This was another pseudo-diplomatic gesture of concili- ation intended to ease the tension between the two governments which the rival geographical societies had done so much to foment.[161]

An indiscretion on Hayward's part now upset the calculations of both factions at the RGS. His letters describing the Maharajah of Kashmir's atrocities leaked to the Indian press, infuriating Mayo because of the disgrace reflected on the Raj and the furore engendered between the rival schools of Indian foreign policy. Since the forward faction wished to use the Maharajah's territorial ambitions as a stalking-horse for a northward extension of British authority, Hayward's revelations provided the forwards' opponents with a golden opportunity to discredit their aims and methods. Yet such a debate could only hinder Mayo's moderate policy.[162] Hayward wrote immediately to the Indian papers in complaint and apologised privately to Murchison, offering to sever his connection with the RGS in order to limit damage to the Society's reputation and its relations with the India Office.[163] Mayo was somewhat mollified, but Hayward's reputation was ruined with the Indian authorities, who now focused their attention on the official Forsyth–Shaw mission to Yarkand.[164]

In September 1870 Mayo informed Murchison that Hayward had probably been murdered, but the letter did not arrive before the meeting of the British Association, at which Murchison again played down the threat of Russian invasion and dilated upon the market opportunities in Central Asia.[165] Papers by Forsyth and Rawlinson followed the latter theme.[166] By November Hayward's death was confirmed, and Murchison, after serving as the official conduit by which the news was transmitted to Argyll at the India Office, announced it at his last appearance before the Geographical Society.[167] Rawlinson, whose perception of the Russian threat seemed to shift with each geographical discovery, now returned to a theme which he had first put forward in 1865 to argue, on the basis of Hayward's recent evidence of the Chitral valley as the best route to Central Asia, that a chain of forts should be built across this region.[168]

Having lost its principal player on the board, the RGS found its active role in Central Asian affairs henceforth diminished, though it remained important as a publicity medium. Frederick Drew was detailed by the contrite Maharajah of Kashmir to recover Hayward's body, and he also collected the explorer's papers and erected a monument over his grave inscribed with an epitaph composed by Rawlinson.[169] Mayo continued to post Murchison on the murder investigation and relations with Yarkand. The Forsyth–Shaw mission was unsuccessful, and the celebrated envoy to Calcutta had proved 'a rascal' but the Viceroy was monitoring the situation with disguised native agents.[170] Shaw also forwarded several reports on the Yarkand mission to Murchison, including geological observations and specimens.[171] In the spring of 1871

Rawlinson, filling in for the ailing Murchison, kept up interest in the subject by discussing reports from the native Indian explorers sent into Turkestan by the Indian Trigonometrical Survey. Probably anticipating his succession to the presidency, his fulminations against Russia were temporarily muted, but his sense of immediate threat was still palpable.[172] Murchison's last Anniversary Address to the RGS in May reiterated his old theme of Anglo-Russian co-operation in Central Asia, but in designating Rawlinson as his successor he was perhaps acknowledging that the younger man's more aggressive style was better fitted to lead the RGS forward into an era of perceptibly quickening imperial rivalry.[173] Certainly Rawlinson maintained his aggressive stance during his tenure as President.[174]

Stimulated by Forsyth's discovery of abandoned jade mines in Yarkand, Mayo continued his schemes to channel the trade of Central Asia towards India. Like Rawlinson, he now suspected the Russians of orchestrating an annexationist plot against the Amir of Yarkand.[175] As a result, Forsyth and Shaw were sent back to Yarkand and on a second mission in 1873, this time accompanied by a complete scientific staff which included the chief palae-ontologist of the Geological Survey of India. Murchison was now dead, but evidence in the resulting geological reports proved the existence of Silurian and other Palaeozoic rocks as far north as the Russian border.[176] Though Forsyth succeeded in negotiating a commercial treaty with Yarkand more advantageous than a similar one secured by Russia, the Chinese reoccupation of Easter Turkestan during 1878 put paid to the intricate manoeuvrings of both European powers in this remote zone of contention.[177]

The focus of imperial rivalry shifted westward once more as the Russian occupation of Khiva in 1873 seemed to expose the British flank in Afghanistan and Persia. Rawlinson's skill at the 'Great Game' therefore failed to win the territorial annexations he coveted in Central Asia, but Murchison had at least the posthumous satisfaction of expanding the domain of Siluria across the debatable land. There was another echo of Murchison's policy at the RGS in the late 1880s, when the Indian government temporarily stopped transmitting information to the Society about official explorations beyond the northern frontier on the grounds that such data had become strategic secrets. With nice irony, however, the geographers were able to regain their privilege by demonstrating that they had been able to secure this information from sources in Russia.[178]

Because of the efforts of Rawlinson and other advocates of a forward policy, the Geographical Society narrowly missed achieving a significant mediating role in Central Asian policy making. Hayward had been put into the field against the wishes of the Indian government and his reports, together with those of Shaw and Forsyth, contributed to a deepening British involvement in the native states beyond the frontier. Geographical exploration was a social activity with multiple layers of meaning variously interpreted by its participants. As the crucible for resolving these conflicting interpretations the RGS never achieved true consistency of purpose, yet the explorations it initiated had profound, if often unforeseen, consequences. The Hayward affair illustrates the complexity of shifting factors affecting the equation of the influence of Murchison and the RGS upon imperial affairs. Not only were the physical geography and inhabitants of a particular area of activity critical to this equation, but so too were the motives and goals of the explorers themselves, factions within the Society, British officials both on the spot and in London, and the rival great powers in the region.

Murchison had always been interested in India because of its significance as the cornerstone of empire and his family's historic connection with the Raj. His earliest activities involving India had been geological – promoting research to extend the reach of metropolitan nomenclature and inform theoretical arguments, while at the same time providing administrators with precise information for efficient resource development. Very early, as in other regions, his related interest in geography found expression in the encouragement of both the systematic mapping of Britain's possessions in the subcontinent and the exploration of neighbouring territories from this regional base of operations. Once he had helped establish Oldham's Survey firmly, geological research, as in Victoria and Canada, could be left to the local institution modelled on his own establishment. But as opportunities to found new surveys in outlying dependencies such as Kashmir or the Straits Settlements presented themselves, largely because of the success of his own propaganda, they too were promptly seized.

Murchison consistently supported the expansion of British India because of his military outlook, imperial patriotism, and realisation that campaigns of conquest added new provinces to the domain of British science. For the same reasons he advocated the attachment of scientific personnel to all military or diplomatic missions penetrating unknown lands beyond the frontiers. His reaction to the Mutiny represented a brief faltering, due to self-interest, in a lifelong belief in the permanence of British rule. As geography came to dominate his attention during the 1860s, so too in India he turned increasingly from the monitoring of mundane geological research to the excitement of speculating on entirely unknown topographical features and mineral deposits beyond the pale of existing maps. This led to an enthusiastic promotion of trans-Himalayan geographical exploration which, in turn, brought up the issue of Russian rivalry – an added stimulus to those intrigued by the Great Game, but a topic that caused Murchison to suffer once more the tortures he had endured during the Crimean War. As a consequence of his crisis of conscience, Central Asia remained the single region of the world where Murchison's zeal for exploration failed to march in step with his pride in empire.

6

The Far East

Murchison's capacity for Far Eastern exploration was circumscribed by two factors. First, the region's highly organised civilisations proved remarkably successful in resisting foreign penetration. Second, vast areas of the Far East were under the control of other European empires – a consideration which, if not forbidding British exploration, nevertheless lessened the chances that reconnaissance could lead to the establishment of dominant commercial or political influence. Still, the possession of the Indian empire as a base of operation and the Royal Navy as an instrument of policy gave Britain peculiar advantages over local states and a freedom of action that her great-power rivals could scarcely contemplate. From the inception of public interest in the Far East during the Opium War, Murchison exploited every opportunity created by military expeditions, missionary endeavours, diplomatic missions, and commercial enterprises to promote research which would contribute simultaneously to the advancement of British science and the expansion of Britain's power influence, and commercial reach.

China

From an early stage in modern relations between the chief Eastern and Western powers, British scientific initiatives in China were linked with her economic policies. In 1793 Lord McCartney's embassy had sought to impress the Chinese with the superiority of British science and technology and gather information on Chinese manufacturing industries in order to open the country to European commerce and improve British production methods.[1] The Geological Society of London received a few Chinese specimens early in the nineteenth century collected by visiting naval officers and amateur naturalists serving at the East India Company's establishment in Canton, but the structure of China remained essentially an enigma until the confrontation with the West in the 1840s.

China's mining industry meanwhile languished under government discouragement, her stocks of silver were drained by the expanding opium trade with India, and her legitimate import and export commerce in metals was similarly revolutionised to the benefit of the British trading network as cheaper supplies of English copper, iron, tin, zinc, and lead flooded the Oriental market.[2] The Anglo–Chinese war of 1839–42 – the Opium War – witnessed the first extended deployment of steam warships and revealed a shortage of coal in Chinese waters. Depots were established for English fuel, but as requirements outran supply, coal exploration was stimulated throughout the Indian

Ocean and attention was turned to the possible exploitation of Chinese deposits.[3] The geologists were quick to offer advice on this strategic subject in order to display the usefulness of their science.

In his presidential address to the Geological Society in 1842, Murchison commented that, though British scientists had 'not yet penetrated to Pekin', fossil evidence from Siberia coupled with Russian reports of an extensive coalfield near the Chinese capital encouraged his anticipation of 'planting the Silurian standard on the wall of China'.[4] Shortly after the war ended with the Treaty of Nanking in the following year, he noted that the Society could expect an 'accession of natural knowledge' from the vast region 'laid open to British enterprise'. The large stocks of coal which the British naval force found piled on the quays of Nanking, 'as if stationed there to supply our invading steam-vessels', now proved the existence of Palaeozoic deposits in China, but Murchison was anxious to determine their location, extent, and stratigraphy. Arguing that this knowledge was 'indispensable for a people like ourselves, whose commercial and maritime advancement depends so essentially upon the application of steam power', he suggested that geologically trained agents 'should be attached to those permanent stations which are to be occupied by our forces; whence, if a friendly spirit of intercourse is continued, excursions could be made into the interior'.[5] The access provided by British gunboats was reflected almost immediately in the Society's Museum. In 1842 China was still the most poorly represented region of its size in the Society's cabinets, but the following year fossils collected by naval surgeons began to swell the collection, and by the 1860s the number of cabinets devoted to the Far East had increased dramatically.[6]

While China remained fixed in the public eye, Murchison made a similar bid for official support on behalf of the geographers to exploit the breach created by British arms and diplomacy. In 1844, extrapolating from recent gold discoveries in Siberia, he called for the government to finance exploration in China since it would probably soon prove to have an 'abundance of gold and other precious metals to exchange for our manufactures'.[7] The following year he pointed out that explorers could also ascertain details of demand for cotton cloth so that British manufacturers might systematically eliminate their rivals. Secure in his belief that Britain would lead the development of China, Murchison called for international scientific co-operation in assessing the results of Admiralty surveys, Russian explorations, and German historical researches.[8]

During the Opium War the RGS mobilised to gather information bearing on the campaign, which had exposed Whitehall's humiliating ignorance of China. President William R. Hamilton, for example, provided Peel's Foreign Secretary Aberdeen with a Jesuit map of Peking procured in Italy by Sir Woodbine Parish, FRGS, for use in the event of an attack on the capital.[9] Sir John Barrow of the Admiralty, an RGS Council member who had accompanied McCartney's mission to China, emerged as the only official with the slightest knowledge of Chinese geography. Barrow used this authority to convince John Backhouse of the Foreign Office – FGS, FRGS – to urge upon Palmerston the policy of retributive annexation that led to the acquisition of Hong Kong.[10] The hostilities also stimulated the Royal Navy to begin charting China's coast and the RGS to construct a new map of the entire country. Various China experts were asked for information and the Council, chaired by Murchison, discussed seeking permission to copy old maps in the Vatican.[11] Historical geography could thus provide strategic as well as commercial and resource intelligence.

While memoirs on China and Hong Kong appeared in the Society's *Journal*, Murchison commented on the possible relationship between the geology and climate of the new colony.[12] He also advised the publicist Robert Montgomery Martin on this subject for a pamphlet describing China as 'the next great arena for the development of British civilization'.[13] During the 1840s and 1850s most of the 'old China hands' not already affiliated with the RGS joined it in order to promote exploration, lobby for government support of British trading interests, and keep Chinese issues before the public. Their influence within the Society increased during these years for the same reason it did with British policy makers: the government and the public were dependent on their expertise.

Murchison worked closely with these influential men. The merchant and financier James Matheson, for instance, who returned from China in 1842 to manage his firm's London operation, was elected to the RGS on Murchison's recommendation in 1845.[14] He also proposed Matheson to the Royal Society – a rather hypocritical act, considering Murchison's prominence in the attempt to reform the Society in 1830 and his disparging comment that 'any wealthy or well known person, any M.P. or Bank Director or East India Nabob who wished to have FRS added to his name, was sure to gain admittance by canvassing'.[15] Matheson contributed information to the RGS garnered from his far-flung Pacific trading network and, because of his renown, was invited to dine by the Raleigh Club during Murchison's presidency.[16] The mutual Highland ancestry of the King of Siluria and the prince of opium traders formed another link between the two men. During the 1860s Murchison often made the Hebridean castle of Sir James, whom he enjoyed addressing as 'Lord of the Isles', his headquarters for geological tours in western Scotland.[17]

Another Highlander prominent at the RGS, Murchison's friend John Crawfurd, acted as parliamentary agent and publicity manager for the Calcutta agency houses during their campaign to abolish the East India Company's monopoly of the China trade.[18] Sir George Staunton, MP, former Chinese diplomat, and fellow of several scientific societies, served as one of Murchison's vice-presidents at the 1846 meeting of the British Association. Staunton donated Chinese specimens to the Geological Society, and his calculations of the erosional capacity of the Yellow River provided data for Charles Lyell's theories of denudation.[19] Karl Gutzlaff, missionary, Foreign Office interpreter, and propagandist for an aggressive policy throughout the Far East, likewise provided the RGS with information from his arrival in China during 1832. Visiting Britain in 1848, he offered his maps to the Society as well as several memoirs urging the exploitation of Indo-China's resources.[20] Murchison used Gutzlaff's comments on the abundance of gold in Yunnan province to reiterate his prediction of an imminent increase in Chinese production of the precious metal.[21]

During the period of uneasy peace between the two Anglo-Chinese wars, De la Beche's Museum of Practical Geology profited from the donations of British officers, diplomats, and missionaries exploiting the limited access granted to foreigners by the Treaty of Nanking. These specimens permitted the first informed hypotheses about the Celestial Empire's geological structure and mineral endowments. In 1849, for example, John Bowring, Consul at Canton, forwarded specimens of silver ore to the Museum; in 1854, upon being promoted Governor of Hong Kong, he donated a collection of Chinese fossils.[22] Murchison received coal samples from the same source as Director-General, and Bowring also sent the Geological Society specimens of copper ore gathered on a commercial mission to the Philippines.[23] In 1856 Murchison helped elect Bowring to the Royal Society, partially in gratitude for his aid in returning Russian prisoners captured during the Crimean War.[24] Bowring clearly understood the interlocking relationship between science, trade, and diplomacy as instruments of British paramountcy, for he also forwarded Chinese cotton samples to London, collected plants for Kew Gardens, gathered information for the RGS, and promoted research through the Hong Kong chapter of the Asiatic Society of Bengal.[25]

De la Beche performed another service regarding investigation of Chinese resources which reveals the refinement with which scientists were applying their expertise to industrial espionage as well as to the discovery of new sources of coal and metals. In 1850 Harry Parkes, interpreter to the Amoy consulate, came home on leave, where he joined the RGS and received six weeks' instruction in geology and metallurgy at De la Beche's Museum to prepare him for acquiring 'knowledge of the mineral products of China and the mode of treating them in that country'. De la Beche, who considered that British manufacturers could profit from information regarding Chinese smelting and casting techniques, arranged this scheme with Palmerston in order partially to circumvent the difficulty of conducting actual geological research in China.[26] The plan echoed the goals of the McCartney mission as well as Murchison's entreaty that geologically trained diplomats be stationed in China. Throughout his subsequent career in the Orient, Parkes contributed to the Survey Museum and the Geographical Society. Making no distinction between Britain's political and commercial interests, he believed science served the same national goal.[27] In 1855, for example, after assisting Bowring to negotiate the first commercial treaty with Siam, he presented the British Association at Murchison's urging with a memoir on Siamese market and resource prospects, and the need for aggressive diplomacy to make them available.[28]

Murchison meanwhile continued to promote Chinese exploration at the RGS. His 1852 address stressed the need for an accurate map of the country and praised a new book by Robert Fortune, the botanist responsible for transferring the tea industry to India, for its account of the tea plant's cultivation and suggestion that Silurian formations might be found in the Chinese tea-growing districts.[29] The following year Murchison commented that the opportunities for Western penetration presaged by rebel successes against the Manchu imperial government were opening out 'a vista of unbounded interest' to merchants and scientists alike, and that the 'unfolding' of China's riches would affect the world economy as profoundly as the gold-rushes of California and Australia. Labour, he believed, might prove China's most valuable resource, for the country's 'redundant population' could be exported for the 'improvement' of tropical regions as well as organised by Europeans to develop China's own potential.[30]

Murchison's hopes began to be realised as the Geological Society published a Chinese coal report which had been passed on by Clarendon at the Foreign Office from a scientific missionary.[31] When Anglo-Chinese relations once more deteriorated to the point of conflict, Murchison capitalised on public interest by inducing Sir John Davis, FGS, FRGS, noted sinologist and former Governor of Hong Kong, to deliver a paper to the RGS in 1857 on the geographical aspects of the war. At the meeting of the British Association the same year, Davis presented a similar memoir advocating Britain's retention of Hong Kong and Canton.[32] As hostilities continued, the RGS published a memoir on the Yangtze and Yellow rivers by William Lockhart, a medical missionary who also collected Chinese fossils and minerals for the London museums.[33] Lockhart echoed Clarendon's statement that the time had come for China to be 'opened up to the industrial enterprise of foreign nations'[34] by dilating upon the need to secure access by steam navigation to the vast resources and markets of the interior. Murchison chaired the extremely political discussion of Lockhart's paper which was dominated by Crawfurd and Rutherford Alcock, a former Consul in China who would subsequently serve as British Minister in Peking from 1865 to 1871 and preside over the RGS during the period 1876–78.

Having taken the pulse of the audience, Murchison gave rein to his own aggressive instincts in his Anniversary Address by dwelling on the importance of the Yangtze as a commercial highway to Hankow and 'all the sites of mineral wealth'.[35] He also noted approvingly how the Russians, under cover of their religious mission in Peking, were filtering mining engineers, geologists, and other scientists into China to evaluate the country's resources. One of these savants had given him a fresh account of the Peking coalfields and evidence for the exisence near the capital of Silurian, Devonian, and Carboniferous fossils. From these data and a series of Devonian shells Lockhart had forwarded from

Szechwan – evidence incorporated in subsequent editions of *Siluria*[36] –
Murchison anticipated a vast oriental extension of Britain's stratigraphic
empire. Practical benefits, he pointed out, would necessarily follow from this
scientific conquest: 'Possessing these palaeozoic rocks, with many ores and
metals, and vast and rich coal fields, the empire of China, with its rich
products of the soil, lies before us as a wondrous mine of wealth and lucrative
commerce.'[37]

As events in China continued to claim attention, Murchison gleaned
information for the RGS from manifold sources. He boasted to the astronomer
John Herschel, 'we are up to our neck in the great river Yang-tse-keang',[38] and
Woodbine Parish warned that 'everything now connected with the route to
Pekin & the country beyond it, must be useful to the public – & to the officials
who will be sent out there.'[39] The campaign, in fact, had temporarily taken on
the aspect of a militarised RGS expedition during the Earl of Elgin's voyage up
the Yangtze to treat with the Chinese at Hankow. Sherard Osborn, the captain
of Elgin's frigate, produced a memoir on the coastal portion of the voyage as
well as forwarding a selection of decorative loot to enliven the Society's rooms
during the 'China night' designated for discussion of his paper.[40] Osborn's
letter from Shanghai accompanying his chart of the Yangtze enthused over
northern China as 'the first field for our extended Commerce in this country', as
well as extolling the beckoning opportunities for merchants and explorers in
Japan and Korea.[41]

Captain Richard Collinson, RN, an RGS gold medallist and Franklin
search veteran at that time surveying the Amur River, saw similar prospects in
Manchuria. He warned the Society, however, that the Foreign Office should
back China in its boundary negotiations regarding the Amur region, which
was annexed by Russia in 1858. 'The Russians are accumulating an empire
there which will be a worse thorn in our sides than Constantinople', Collinson
advised, and he predicted that unless Elgin received immediate instructions to
secure the right of free navigation regarding the Amur, 'the Russians will
absorb the country & shut us out from the trade'.[42] Recent gold discoveries had
been a factor in the Russian decision to acquire this region, which was later to
prove highly productive of the precious metal.[43] Murchison, who as usual was
unprepared to take a hard line regarding Russian expansion, publicly praised
Russian scientific explorations of the Amur basin and held them up as an
example to the British government of its duty to fund similar expeditions in its
own colonies.[44]

Elgin's private secretary Laurence Oliphant, future Honorary Secretary of
the RGS, also contributed a memoir on the Yangtze voyage. It was illustrated
by Osborn's chart, which was rushed into print to satisfy public demand
frustrated by the delayed production of Admiralty charts. Oliphant's paper
stressed that, while the trading capabilities of the interior had been overestimated, British goods could still be introduced into China on a large scale
once a depot had been established at Hankow.[45] Murchison arranged that Sir

John Davis attend the discussion of these memoirs so that his expertise, as well as his Chinese drawings and photographs, might add weight and interest to the proceedings.[46] Elgin himself had dined in Belgrave Square before departing for China, and when he returned Murchison arranged his election to the RGS and attempted to lionise him for publicity purposes.[47] The results of the naval survey of the Yangtze were also reported to the Society.[48] Murchison's Anniversary Address of 1859 praised Elgin's voyage for demonstrating the military might and technological superiority of Britain. The President had by this time modified his earlier belief that the Taiping rebels served the cause of Western penetration in weakening the authority of the Chinese central government. With entry now effected, he encouraged his audience to expect the development of a vast market in China's interior once the rebellion was crushed with foreign aid.[49]

At the end of 1859 the RGS received a memoir on Manchuria by Alexander Michie, an employee of the trader Hugh Hamilton Lindsay and later a publicist for British commercial expansion in China. Lindsay, also a persistent advocate of a forward policy in China, was elected to the Society at this time by John Crawfurd, who recommended to Murchison the immediate publication of Michie's paper because of the stimulating effect he believed it would have on British trade. The resulting presentation provoked intense political discussion between Lindsay, Crawfurd, Oliphant, Lockhart, Osborn, and Davis regarding the wisdom of occupying Peking.[50] In editing this discussion for the *Proceedings*, Crawfurd, whose free-trade expansionism did not extend to support of coercive military expeditions, retained the debate, arguing that the RGS 'cannot be better employed than in the discussion of a march which may lead to disaster and is quite sure to cost millions!!!'[51] Such caustic comments earned the conservative Crawfurd the title of 'Objector-General' at the RGS, for he opposed nearly every exploring, trading, or colonising scheme brought forward. The forward policy advocated by his opponents, however, was the one that went home as the Society's own. Aggressive British merchants, exploiting public interest created by the war and backed by diplomats, missionaries, military officers, and journalists, were thus able to impose their views on the Society and broadcast them under its influential aegis.

In 1860, as the signing of the Treaty in Peking ended the second Anglo-Chinese war, the spate of Chinese material presented at the RGS slackened. Murchison kept himself informed about 'doings in China' through contact with Palmerston and other politicians.[52] The following year, filling in for RGS President Lord Ashburton at the anniversary meeting, Murchison detailed the Admiralty's progress in charting the Chinese coast for, as he believed, 'the labours of the Surveyor have always been, and always must be, the precursor of Commerce'.[53] Crawfurd wrote the Oriental section of this document, commenting on the importance of establishing commercial routes from China to India and the indispensability of Chinese trade to the maintenance of the Indian empire.[54]

Crawfurd also presided over the Geographical Section of the British Association the same year. Its presentations highlighted the importance of siphoning the mineral resources of China's Yunnan and Kwangsi provinces through British Burma and described an unsuccessful British expedition to explore a route from the headwaters of the Yangtze to India.[55] Detailed accounts of this expedition, along with the first accurate chart of the upper Yangtze, were processed by the RGS soon afterwards.[56] Harry Parkes, now British Consul at Shanghai, composed the Chinese section of Ashburton's Anniversary Address for 1862 during a home leave. He noted that the opening of the Yangtze and treaty ports had already produced a valuable trade which further explorations could rapidly increase.[57] While helping Murchison referee Chinese papers, Parkes also secured the election to the Society of Thomas Wade, Secretary of the Peking legation, interpreter for Elgin's mission, and future British Minister to China.[58]

Late in 1862 the RGS received a memoir describing a journey from China to Siberia across the Gobi desert. Its author, C. Mitchell Grant, next travelled to St Petersburg on commission for Baring Brothers, Jardine, Matheson and Company, and the Oriental Bank. His goal was to induce the Russian government to construct a telegraph line to Peking and, if possible, Russian North America, from where it might easily be extended to the British colonies on the Pacific coast. Great hopes were entertained of Grant's success, for the reading of his memoir was attended by Gladstone, the Chancellor of the Exchequer, as well as Horatio Nelson Lay, a former British consul serving as Inspector-General of Customs for the Chinese government. Lay was in London to oversee the fitting out of a fleet of gunboats purchased by the Chinese authorities for the suppression of piracy on their coast, and he took care to advertise the merits of this scheme to facilitate commerce in the discussion of Grant's paper.[59] Though Sherard Osborn accepted command of the squadron and delivered it to China in 1863, a dispute over the chain of command would soon cause the flotilla to return to England.

Murchison having provided Grant with introductions to the Russian court, the traveller promised in turn to inform the President, who had suggested the telegraphic link across the Behring Sea, of any interesting geographical discoveries.[60] The London banks with large commitments in China — several of whose principals or former partners, including James Matheson, Thomas Baring, and Ashburton, were Fellows of the RGS — thus attempted a communications breakthrough to provide advance intelligence on Chinese and possibly western American market conditions by means of an agent hired on the basis of a reputation acquired through the Society. Barings was simultaneously lobbying for a trans-Canadian telegraphic link to monitor the Oriental market.[61] Murchison, who stood to reap considerable credit for the commercial and strategic advantages this scheme offered the empire, and who hoped that such projects might reconcile Britain and Russia to co-operation in China as well as Central Asia, for several years encouraged its completion.[62]

Throughout the 1860s Murchison continued to promote the exploration
and commercial development of China at the RGS and the British Association.[63] He also used his influence with John Delane, editor of *The Times*, to
ensure that Chinese information processed by the RGS which he considered of
particular mercantile interest was given adequate publicity in exchange for
exclusive 'tit bits' of geographical news. One such memoir won the President's
approval for its geological observations and discussion of laws obstructing
foreign mining and railway development in China.[64]

A sub-theme in this decade remained the exploration of trade routes
between the Indian empire and the temporarily independent provinces of
south-western China, towards which the French were pressing from Saigon.
General Sir Arthur Cotton, a retired Indian engineer, characterised this issue
of secure overland access to China as 'an imperial question of the very first
importance', for besides the commercial opportunities offered by such a route,
Chinese labour could be imported to serve the expanding tea industry of
Assam.[65] During the period 1860–62, Captain Richard Sprye, RN, enlisted
Murchison's support for a scheme to develop the commercial capabilities of
south-east Asia by establishing a trade route and telegraph line between Burma
and China. To attract the RGS President's crucial influence, Sprye requested
him to bring the slave-trading propensities of the King of Siam to the notice of
Palmerston as a pretext for intervention in that country. Sprye also informed
Murchison that his prediction of gold in the mountains of south-east Asia had
been confirmed, and warned that while the British authorities remained
'inactive', the French were maturing designs to annex or monopolise the
mineral wealth of south-western China and upper Burma.[66]

Sprye was consequently asked to contribute a memoir on his project to the
RGS. It was read in December 1860, the President deeming it 'too important
to be postponed'.[67] During the ensuing discussion between Crawfurd, Davis,
Lockhart, and Oliphant, the view was expressed that, in the light of the
possible takeover of a dying Chinese empire by France and Russia, Britain
should implant a settlement in south-western China and stimulate a gold-rush
to finance it. Sprye's alarmism having activated Murchison's Francophobia,
the President agreed with these views as a stimulus to further exploration.[68]
By 1862 Sprye had produced a pamphlet and warned Murchison that since *The
Times* had voiced alarm over French and German explorations towards Yunnan,
'the Royal Geographical Society should look to it'.[69] He himself, backed by
the British governor of the Burmese province of Pegue, still believed that the
passes of northern Burma offered Britain the best opportunity of reserving to
herself the development of Yunnan.[70]

The exploration of Indo-China was thus involved in the debate over
China–India trade routes. Sir Robert Schomburgk, FRS, British Consul-
General at Bangkok and former Guianese explorer, supplied the RGS with
information on Siam and an abortive scheme to shorten the route to China by
cutting a ship canal through the Isthmus of Kra.[71] Murchison and Crawfurd

encouraged Schomburgk's researches because Siam had by the 1850s 'become an object of interest and importance to all the European nations, and more especially to ourselves'.[72] Similarly, Schomburgk provided the Geological Society, via Lord Clarendon, with a report on Siam's coal and ore deposits, their access to water transport, and the ease of winning mining concessions from the Siamese court.[73] He also won telegraph concessions, collected specimens for Kew Gardens, and countered the anti-British machinations of the scientific French Consul in Siam.[74]

When a private explorer wrote to Murchison in 1867 of his plan to cross the mountains of Yunnan into Burma, Thomas Wade of the Peking Embassy, who had hoped that a practicable pass for a telegraph line to Calcutta might be found, asked the President to keep the plan to himself so that the Chinese authorities would not interfere.[75] An RGS special committee drafted instructions which Murchison transmitted to China, but the explorer was turned back by the Chinese and subsequently failed in a second attempt from the western side financed by Calcutta merchants.[76] The same committee, consisting of retired generals and administrators from the Indian service, then lobbied the government of India on behalf of the British Association to explore this route in order to forestall the perceived French challenge.[77] By 1871 the authorities had responded by sponsoring an expedition. Its results, however, proved the region too politically disrupted for safe transit.[78]

The mineral potential of China received considerable attention during these years, since the travel privileges specified by Elgin's treaty had at last unleashed the scientists to examine the country's endowments. Much of this investigative work was conducted by British consuls who, primarily actuated by the economic motives which explained their presence in China, were nevertheless often stimulated to such research by scientific curiosity. A memoir offered to the RGS in 1863, for example, accompanied by specimens for the Geological Society from Robert Swinhoe, the Vice-Consul in Formosa, drew attention to the coal deposits and commercial potential of that unexplored island.[79] Swinhoe, who combined enthusiasm for ornithology with approval of a forward policy in China, also advocated Britain's annexation of Taiwan and suggested a local bird which might serve as the envisioned colony's symbol.[80] The low-grade Formosan lignites he mentioned had already been analysed for the Admiralty by De la Beche during the 1840s and discussed at the scientific societies.[81] In 1868 the Geological Society published a definitive comparison between these and other Far Eastern fuels.[82]

In 1870 Swinhoe also provided the RGS at Murchison's request with a report on a commercial mission up the Yangtze which Alcock, Britain's Minister in Peking, had organised in response to pressure from the foreign merchants represented by the Shanghai Chamber of Commerce. This memoir advocated the development of riverine steam navigation so that Europeans might dominate China's internal trade. At its reading the missionary Lockhart and Admiral Sir William Hall, who had commanded the celebrated gunboat

Nemesis during the first Opium War, urged the Foreign Office to secure permission for Europeans to develop China's coal mines in order to supply British shipping.[83] A missionary's memoir of the previous year relied on Murchison's theories of Palaeozoic mineral occurrence to advertise the coal, iron, and gold resources of Manchuria and the ease with which they could be developed with railways in harmony with the divine laws of progress and 'the new order of things' decreed in China by the Western powers. At the readings of this report Murchison, Lockhart, a diplomat from the British embassy in Peking, and General Sir Hope Grant, who had commanded the army during the second Anglo-Chinese war, united in calling for a programme of railway, telegraph, and mining development of the region. Lockhart stressed that China could supply India and all of Europe with coal once Britain's supplies were exhausted.[84]

In 1870 a similar report from the British Consul at Chefoo on the metal-rich province of Shantung was published with a colour-coded map of the province's products, complete with details of road, river, and coastal access – in short, a blueprint for foreign exploitation. At its discussion Lockhart, Murchison, and Hall again demanded development of China's coal resources by steam technology.[85] The Geological Society meanwhile published the first attempt at a comprehensive classification of Chinese strata. This memoir was both scientific and economic in content, and its discussion once more suggested Murchison's meridional theory to explain the distribution of China's mineral wealth.[86] Murchison also took up the theme of Chinese economic development in his last address to the RGS in 1871. Here he argued that railways and extended riverine navigation were required to increase the sale of British manufactures in the interior and to encourage the exploitation of the region's vast coal and iron deposits, recently reconnoitred for the Shanghai Chamber of Commerce by the Prussian geologist Ferdinand von Richtofen.[87] He thus envisioned science, commerce, and diplomacy co-operating in the development of China as a partial solution to Europe's market and resource depletion problems.

In the era between Murchison's death and the First World War, European involvement in China largely followed this pattern. The Foreign Office encouraged the Chinese government to relax commercial obstructions and grant mining and railway concessions, and foreign investment focused on the mineral regions which eventually formed the basis for the European spheres of influence.[88] Richtofen's appreciation of the strategic position of the Shantung Peninsula in relation to the coal and iron deposits of Shansi and the markets of northern China, for example, was largely responsible for Germany's acquisition of the port of Tsingtao in 1898.[89] The railways providing the structure of these foreign spheres were likewise routed on the advice of geologists and mining engineers to tap mineral deposits whose exploitation might justify the costs of construction. British mining concessions in this period, while utilising the expertise of Foreign Office and colonial geological survey veterans, never paid dividends because the Chinese government feared such developments might be used as a political wedge for territorial cessions. Nevertheless British companies carried out mineral explorations and founded mining colleges, while at the same time Chinese engineers were trained at Western institutions such as the Royal School of Mines, where over a dozen Chinese students graduated before the First World War.[90] These activities helped to create the corpus of information on mineral deposits and the cadre of experts capable of exploiting them which subsequent governments required to initiate the development of China's modern mining industry.

Japan and Korea

Because Britain had no merchants' pressure group oriented towards Japan comparable to that which lay behind the forward policy in China, the Foreign Office long regarded the opening of the Japanese islands as a subsidiary aspect of Chinese affairs.[91] British scientists were therefore denied access to information about Japan until Admiral Sir James Stirling, FRGS, negotiated a preliminary commercial treaty in 1854. Stirling, who envisioned Britain organising the Far East on the Indian model,[92] also sent the Museum of Practical Geology a collection of Japanese rock specimens in the hope of establishing the stratigraphic potential for coal. As Governor of the new colony of Western Australia during the 1830s, Stirling had likewise displayed a keen interest in geological research with a utilitarian bent.[93] In Japan, local fuel supplies would not only facilitate naval operations, but would be useful for the carrying trade. Murchison, however, could make very little of Stirling's poorly chosen specimens: though lignite coal could be identified, no light was thrown on Japan's stratigraphy.[94]

From the very opening of British relations with Japan, therefore, Murchison took part in the scientific evaluation of the country's resources. His fascination with Japan grew as a function of public interest in the termination of its isolation. As usual, his curiosity represented an amalgam of personal, professional, and patriotic motives. Even before Stirling's treaty, Murchison had praised American and Russian efforts to open the islands.[95] He also armed himself with what little scientific and political information was available about Japan in order to prepare to arbitrate discoveries there. He consulted, for example, Baron Philipp van Siebold, a Bavarian naturalist who had amassed unrivalled collections of art, literature, maps, and natural history specimens while serving as a military engineer with the Dutch trading establishment in Japan. A few years later Siebold would act as a mediator between Japan and the nations of Europe, so that his opinion caried great weight. In consequence, Murchison was careful to record his views on Japanese hostility to the American presence, though he could not draw Siebold on the details of a recent secret treaty with the Russians. 'Siebold believes', Murchison noted, 'that the Japanese can be made a great deal of, and that they can be treated with.'[96]

In 1857 Murchison helped plan a scheme to steal a march on Britain's commercial rivals by clandestinely investigating Japan's market and resource potential. Stirling had promised the Japanese Emperor the gift of a steamboat in order to display Western technology and secure advantages for Britain. As the yacht prepared to sail from Britain, the privy councillor Lord Hatherton, who had recently had the benefit of the Director-General's advice on the chances of discovering coal beneath land he was considering enclosing,[97] warned Murchison that the government was throwing away a fine opportunity 'of acquiring information of various sorts as regards the natural appearance of the Country, the People, their habits & wants & possibly many useful things they may possess'. At the same time Hatherton urged Captain Edward Inglefield, RN, an RGS gold medalist seeking an assignment in the Far East, to approach Murchison on the subject. Inglefield, who was also a Fellow of the Royal Society, had been involved in the search for Sir John Franklin, and he stood high in Murchison's favour for having named a Greenland sound in his honour and contributed evidence of the presence of Silurian rocks above the Arctic Circle.[98] Hatherton believed the presentation of the vessel and its trial voyages might be arranged so as to give her officers – especially one with scientific training such as Inglefield – 'access to what they might otherwise not see', and he urged Murchison to press the issue upon Clarendon and Prince Albert.

Murchison encouraged the Foreign Secretary in consequence to adopt the plan, arguing that 'the visit of Englishmen "Dona ferentes" would give us advantages which would be denied to a threatening & hostile mission like that of the U. States.' He cautioned Clarendon, though, against mentioning his name at the Admiralty, where his opinion was at a discount because of his repeated attempts to force a continued search for Franklin.[99] In the event, the yacht was sent out under a different officer with orders that it be simply handed over to the Japanese. Lord Elgin, however, who was instructed to secure a new commecial treaty with Japan before concluding his mission to the Far East, intercepted the vessel at Shanghai and used its delivery to travel to the Japanese capital for a ceremonial reception rather than humiliating negotiations at an outlying port.[100] Though no scientific or economic data resulted, the incident demonstrates that the attitude of the British governing elite towards overseas development was largely an extension of their traditional approach to exploiting domestic resources. It also illustrates Murchison's adroitness at exploiting opportunities to conduct reconnaissances, his use of friendship with cabinet ministers to bypass obstructionist departmental officials, and how his influence could be mobilised by ambitious explorers whose projects meshed with the vision of science as a servant of national expansion which Murchison himself had done more than anyone to create.

Murchison's address to the RGS in 1857 encouraged Western penetration of the 'unbroken unit' of Japan and recommended the narrative of the American Matthew Perry's pioneering voyage as 'replete with valuable geographical and ethnological notices'.[101] The excitement which Japanese prospects awakened in geographical circles may also be gauged from Murchison's receipt of an application for a passage with Elgin's mission from Edward Cullen, a bankrupt emigration promoter who now hoped to recoup his Latin American losses in Japan.[102] The Japanese comments in Murchison's 1859 address constituted a virtual paean to the imperialism of free trade and the links between commerce, diplomacy, military power, and science. He praised the Admiralty's port surveys for mapping the approaches to the Japanese market and, referring to Lord Elgin's negotiation of a second set of Japanese capitulations, rejoiced that British subjects connected with the new consulate could now explore throughout Japan and that British products could be imported 'into every corner of the empire'. 'Doubtless', he pointed out, 'it may require time to create wants in a population hitherto so independent of the rest of the world, but the acquisitive and imitative instinct of the native of Japan is so remarkable that he will rapidly discover the merits of Western arts and manufactures, and apply them to his own uses.' Murchison was confident that British woollens would replace native cotton garments and that other articles such as sugar could be forced on the Japanese. Similar prospects beckoned in the export sector: cochineal had already 'proved a most profitable investment' among the products and manufacturing processes available from 'this new field for the energy and enterprise of Europe'. As a sop to the evangelical element in the RGS, the President concluded his aggressive speech with the hope that foreign intercourse should not bring upon the 'docile' Japanese 'those curses of demoralization which too often attend upon the influx of a higher civilization'.[103]

In 1861 the Society received a Japanese memoir from Clarendon by Vice-Consul Christopher Hodgson, FRGS. Its discussion provoked Murchison's statement that 'It was to the public interest that such papers should come from the Secretary of State, being of real use to the country in pointing out new channels of commerce, and ... the Society in discussing and publishing them became an important auxiliary to the State.'[104] At the same time Murchison asked Laurence Oliphant, who was returning to Japan as Secretary of the British Legation, to supply more information, and this was soon received by the Society.[105] In Murchison's Anniversary Address the same year Crawfurd continued to celebrate the opening of Japan as a splendid commercial opportunity,[106] but diplomatic Fellows with first-hand knowledge were no longer so sanguine. Murchison published their views as well in order to maintain good relations between the Society and the Foreign Office and to rein in aggressive merchants whose blunders might precipitate a disaster to British interests in the Orient.

Two such memoirs were submitted by Rutherford Alcock, now British Consul-General in Japan. Murchison arranged for the more geographical of the two to be presented at both an RGS meeting for which he had engaged Lord Elgin, and the autumn meeting of the British Association.[107] The other – and extremely political – paper was read to the RGS in Alcock's presence the following spring. It criticised Japanese feudalism as obstructive to Western interests, but appealed to Western governments to resist the demands of their mercantile classes for an immediate removal of all barriers hindering free intercourse with Japan. A breakdown of authority and a bloody war would inevitably result from such a forward policy, Alcock argued, and peace and patience would yield more fruitful results.[108] Murchison took Alcock's quasi-official warnings on board, for he abated his aggressiveness regarding Japan in his address to the RGS and the Geographical Section of the British Association in 1863.[109]

Murchison was simultaneously active in encouraging Japanese geological investigations. In 1864 he communicated to the Geological Society a memoir and collection of fossils sent to him by a naval surveyor engaged in charting the islands' coasts.[110] Japan's coal reserves were of particular concern, for Murchison considered them to be naturally situated to service a trade between India, the Far East, and the British colonies in north-western America.[111] Harry Parkes sent the Museum of Practical Geology a report on Japanese coal mines in 1867 which Murchison communicated to the Geological Society because of its commercial interest.[112] Similar reports on the availability and performance of Japanese coals were processed in subsequent years.[113]

Throughout the period between the opening of Japan and his own death, Murchison promoted the exploration of the country as beneficial to both British science and British trade. His initial enthusiasm gradually waned, however, as the reluctance of the Japanese to allow foreign powers to reorganise their society and resources became apparent. His activities nevertheless raised the British public's awareness of Japan as a sophisticated culture, geometrically increased the amount of Japanese data available to British scientists, and contributed to the growth of Anglo-Japanese trade.

Murchison's interest in Korea was awakened by the second Anglo-Chinese war of 1856–60, but his only actual involvement with this Chinese tributary state occurred in 1865. In that year Captain Allen Young, another veteran of the Franklin search, presented a memoir to the RGS demanding that a direct embassy open the Hermit Kingdom to British exploration and commerce in order to forestall French and Russian threats to its sovereignty. Sherard Osborn likewise supported the policy of using military force, if necessary, to coerce the Koreans into signing a commercial treaty similar to those which Elgin had extorted from the Chinese and Japanese.[114] Having learned the Japanese lesson of too-hasty Western penetration, however, the President cautioned the geographers that such bellicose methods were inappropriate and advised Britain to wait until a less scrupulous great power penetrated the country.[115]

Despite his distrust of French foreign policy, Murchison seems in this instance to have taken a mild line because of his unwillingness to think ill of Russia, into whose sphere of influence he felt all of central and north-eastern Asia fell. Since he deemed Korea vital to neither Britain's commercial nor strategic interests, he implied that Britain could afford to allow others to surpass her in the competition for influence so long as her goods won access to the market. The ambiguous tone of his public language, however, as well as a private letter he wrote to William Denison commending Young's proposal, suggest that the President was not entirely happy in warning his countrymen off Korea. His repeated references in remarks on Korea to the Elizabethan era, which he considered the great age of British exploratory enterprise, support this interpretation.[116] Because the country was not opened until after Murchison's death, he received no evidence with which to prognosticate upon its geological structure, but others did not hesitate to apply his theories to the prediction of the kingdom's mineral wealth.[117]

The Philippines and the East Indian Archipelago

Throughout his geographical career Murchison encouraged the exploration of the islands of the East Indies and the Philippines. His concern was partly scientific, for the archipelago harboured probable clues to the geological transformations which had isolated Australia. But the East Indies also possessed potential for resource and market development and fell within the sphere of Murchison's plans for linking Australia with India and the markets of China. Nearly all of Murchison's addresses to the RGS reiterated the mutual advantages to science, commerce, and empire which exploration of this region would provide. His earliest pronouncements, for example, which owed much to the professional aspirations of the naval surveyor Owen Stanley, called for a hydrographic survey of the archipelago to facilitate the establishment of a steamship line between Australia and India, provide geographers and merchants with a vast accession of useful information, and aid military planners in preparing to counter any attempt by a hostile power to conquer Britain's Antipodean colonies or interdict her Oriental commerce.[118]

Borneo held the most fascination for Murchison because of its size, known endowment with economic minerals, and, owing to Sir James Brooke, Rajah of Sarawak, its close association with British interests. Brooke had sailed for Borneo in 1838 as a Fellow of the RGS, armed with a list of exploring desiderata which included the collecting of '*all particulars*' concerning precious metals, and he had corresponded with the Society about his progress in Sarawak.[119] The leaders of the Geological Society also interested themselves in Borneo's minerals. In 1840, three years after an American trader had first reported the existence of coal in Brunei, the Society published a memoir on Borneo which mentioned a deposit of coal on the nearby island of Labuan and anticipated a large supply being obtained there. This was written by George Tradescant Lay, a naturalist in the service of the East India Company who would later serve as a British consular interpreter in China and was the father of H. N. Lay.[120] The following year President William Buckland predicted to the Society that if Lay's report proved true, Labuan might 'become a station of inestimable value for effecting intercourse by steam between China, India, and Australia, and the great islands of the Malay Archipelago.'[121] The island was in fact soon to be acquired as a colony by Britain precisely because of its mineral resources.

In 1842, once the voracious consumption of steamers deployed during the Opium War had revealed the need for fuel supplies in the Far East, the Admiralty began expressing interest in Labuan's coal. Brooke, who had been granted the privilege of mining coal in Labuan by the Sultan of Brunei, meanwhile opened negotiations with Peel's government through Henry Wise, FRGS, his London agent. Brooke advertised Labuan's potential as a coal source and sought naval protection for Sarawak in exchange for exclusive mining rights to these deposits. In consequence, the hydrographer Captain Sir Edward Belcher was ordered to reconnoitre the coal outcrops of Labuan, Brunei, and Sarawak in 1844. Belcher was a member of several London scientific societies and had made extensive natural history collections on previous surveying voyages.[122] He was unimpressed by the coal seams he found, but tests of the resulting samples carried out by the navy in England produced promising results. After John Crawfurd urged the Admiralty to acquire Labuan for strategic and commercial reasons, a second investigation was commissioned in the same year. It was conducted by another naval officer and a mining engineer.[123] This time De la Beche's laboratory analysed the specimens of what was identified as bituminous lignite.[124] Their favourable performance as steam fuel, coupled with Labuan's potential as a commercial entrepôt and the threat that Brooke might grant his mining rights to the United States in return for a guarantee of Sarawak's independence, led to Britain's purchase of the island from the Sultan of Brunei in 1847 and the appointment of Brooke as the new colony's governor.

A chartered company was then formed by Wise to work Labuan's coal seams on the basis of Brooke's mining rights. The arrangements worked out between the Eastern Archipelago Company and the Colonial Office, by which Wise was granted a lease of the seams in exchange for an obligation to produce a stipulated quantity of coal for naval use and pay a small royalty on any excess mined for commercial sale, provided the precedent for the Admiralty's subsequent negotiations with the Hudson's Bay Company to develop the coal resources of Vancouver Island.[125] Before its eventual failure due to a falling out between Wise and Brooke, the Company transmitted further samples to De la Beche to convince the Admiralty of the fuel's reliability, which had proved to be less than anticipated. The superintendent of the Company's colliery also sent the Director-General a detailed description of the stratigraphy of the coal measures of Labuan and Borneo which was passed on to the Geological Society.[126]

Brooke's regime in Sarawak also depended on the exploitation of minerals. The White Rajah drew his principal revenue from a royalty on production of the rare alloying metal antimony, of which Borneo was then Europe's chief supplier. His second largest source of income remained the royalty on coal, but he also encouraged the exploration and development of nickel, copper, and gold deposits in order to broaden his financial base.[127] Brooke was awarded an RGS gold medal upon his first return to Britain in 1847. In 1851 Murchison presented a memoir from the Rajah to the British Association which mingled description of northern Borneo's topography, geology, and botany with advertisement of the Dutch threat to British interests in that quarter.[128] When Brooke visited England again in 1857–58, Murchison attempted to make much of him for the Society's ends, and his Anniversary Addresses continued to praise Brooke for laying the foundation of an important settlement which he believed must 'be ever held as a British dependency'.[129]

Murchison promoted the exploration of Borneo and the surrounding region by other means as well. In 1852 he praised a memoir by John Crawfurd for his discussion of gold, diamond, antimony, iron, and coal deposits in the island.[130] Murchison was aware of the alluvial gold deposits of the East Indian Archipelago and the Philippines and was interested to test whether his theory comparing the auriferous mountain ranges of Asia and Australia could be extended to include these linking islands. This concept had been suggested by G. W. Earl, who dreamt of a great British trading empire in the archipelago based on the axis of Singapore and a permanent settlement to be established on the north Australian coast.[131] Travelling in Germany in 1857, for example, Murchison spent several days interviewing retired officials of the Batavian government about Dutch exploring and mining projects in Borneo and the neighbouring tin island of Beliton. He communicated this information to the RGS the following year.[132] In 1859 his Survey laboratory assayed gold specimens from Borneo, and an enthusiast who was convinced that New Guinea contained extensive deposits of gold and other metals approached him with a scheme for Britain to colonise the island with 'an organized corps' of miners.[133] Intrigued, Murchison queried Crawfurd on the project, but was informed that the hostile climate and inhabitants and the uncertainty about the reports of gold rendered the idea worthless.[134]

The naturalist Alfred Russel Wallace, who had come to Murchison's attention as a result of his investigations on the Brazilian Rio Negro, also received support for his researches in the Malay Archipelago. In 1853 the President arranged Wallace's passage to Singapore, as well as introductions to Dutch and Spanish colonial authorities in the region, by emphasising to Clarendon that his investigations would provide insight into the natural productions of the islands.[135] In a situation demonstrating how wars could hinder as well as help the progress of field sciences dependent on naval and military expeditions, the vessel which was to carry Wallace east was ordered to the Crimea on transport duty instead. Murchison nevertheless managed to procure a complimentary passage on a P & O steamer to send Wallace to his destiny as the independent co-discoverer of the theory of natural selection.[136]

When the naturalist's first memoir concerning the geology of north-western

Borneo was discussed by the RGS in 1857, Crawfurd reiterated the island's mineral wealth. Murchison announced that Lieutenant de Crespigny of the Royal Navy was on his way to investigate the interior of the island, which he himself described as 'largely productive of some of the choicest desiderata for the advancement of civilization', and especially worthy of development because of its position astride the China–Australia trade route.[137] Crespigny was soon sending the Society information which, along with Wallace's, permitted Murchison to expatiate in subsequent addresses on the commercial and resource potential of the region, the example of British capitalists opening the mines of Labuan, and the value of scientific investigation in creating such opportunities.[138] Similarly, when the British Consul-General in Borneo provided the RGS with a highly geological memoir in 1860, Murchison took special care to bring it forward and supplied its author with scientific instruments to encourage further explorations.[139]

The same year, as Director-General, Murchison identified rock specimens forwarded from a naval surveyor in the Bangka Strait by Captain John Washington, Admiralty hydrographer, former RGS Secretary, and a like-minded advocate of overseas expansion.[140] From his knowledge of the vast tin deposits of Bangka Island and Beliton, which had been described to the Geological Society in 1846 by the East India Company naturalist J. Forbes Royle, and the lithological resemblance of this assemblage to the tin-bearing formations of Cornwall, he advised that these geological characteristics of the unexplored region be engraved on the Admiralty's charts.[141] Murchison was thus able on this occasion to extend his geological mapping operations into territory controlled by another European power, and to encourage development in both the Dutch and neighbouring British possessions of new sources of a crucial industrial metal threatened with domestic depletion. As Royle's memoir and contemporary attempts by the government of India to correlate regional knowledge of the metal's distribution demonstrate, exploitation of tin deposits in the Straits Settlements and the Burmese province of Tenasserim had long been a priority of economic development policy in the Indian empire.[142]

Murchison was supported in these publicising activities by the Dutch authorities, who in 1858 presented his Museum with specimens of tin from the East Indies, and for many years afterwards co-operated with the British Survey in an exchange of colonial mining reports and geological maps.[143] In 1863 the Chief of the Mining Department of the Dutch East Indies presented Murchison with a large collection of fossils in exchange for their identification. Since the Survey Museum displayed only specimens from British possessions, however, the Director-General forwarded the collection to the Geological Society. Officers of the Society arranged for the classification of the specimens, advised the Batavian mining chief on further researches they wished to see conducted, and published his memoir describing the archipelago's economic minerals.[144]

Murchison's 1865 Anniversary Address, a review of geographical progress since the founding of the RGS, summed up explorations in the Malay Archipelago in a passage implying dissatisfaction with the comparatively 'small portion' of the region under Britain's 'political influence'. He linked the growth of trade with scientific exploration, and in praising recent Dutch research for having 'fully redeemed the short-comings of the preceding centuries', suggested that the production of scientific results for the international community was, if not sufficient justification for imperial rule, at least the foundation of efficient administration.[145] Murchison thus signified that the competition between British and Dutch scientists was an important aspect of the two powers' commercial and imperial rivalry in the region. His call for an exploration of New Guinea on behalf of expansionistic Australians made the same point: the failed attempts of the Dutch to found settlements suggested an opening to British explorers and traders. Spanish inactivity in the Philippines presented a like opportunity.[146] In 1870 Murchison even went so far, under the influence of the Australian politician Sir Charles Nicholson and his own anxiety to guarantee the inviolability of that continent's northern coast, as to modify the pessimistic views he had imbibed from Crawfurd and suggest that Britain should colonise New Guinea because her future interests 'would be greatly damaged if any other Power were to possess itself of the south–eastern shores of New Guinea'.[147] The hostility of the natives, however, soon compelled him to moderate these sanguine expectations.[148]

Murchison never had the same opportunities for prosecuting geographical and resource surveys in the Far East that he enjoyed in other regions, but Britain's possessions offered strategic bases for exploratory endeavour, and diplomatic and military initiatives provided invaluable chances for access to China, Japan, Korea, and Siam. In the Malay Archipelago, Dutch co-operation, the activities of James Brooke, the surveying operations of the Royal Navy, the researches of naturalists, and the probings of Australian traders provided a flow of information for the monitoring of this marginal region as a potential theatre for intensifying trade, resource development and, possibly, colonisation. The Philippines, of all the counties in the region, received the least attention, though Murchison periodically encouraged their reconnaissance.

Given these constraints and advantages, Murchison devoted considerable attention to the Far East. His activities closely reflected the ebb and flow of public interest as the various countries of the region were brought into prominence by events. Only in the East Indian Archipelago was this pattern broken: there he remained considerably in advance of public opinion because of his sympathy with Brooke, expansionistic Australians, and members of the forward school of naval strategy. As elsewhere, he strove to co-ordinate the research of natural scientists with the activities of other agents of British expansion, demonstrating particular concern for mineral exploration because it served his own speciality and neatly fitted in with the disparate practical, theoretical, and evangelical concerns of these interest groups. Because of the distances involved, the initiative in Britain's Far Eastern policy lay with the men on the spot, nearly all of whom were Fellows of the RGS, active both in forwarding information and dominating discussions at the Society's meetings. RGS policy towards the Far East was thus very much that of the 'old China hands'. The Society became, as it did for 'forwards' concerned with other regions, a valuable metropolitan platform for the dissemination of their particular peripheral views. At the RGS, which constituted his chief institutional instrument for activity in the Far East, Murchison is thus once more revealed as reconciling the aims of various interest groups concerned in exploiting the region with public opinion and government policy.

As his views on the opening of China and Japan indicate, Murchison was not averse to the use of force when other means of gaining ingress failed. Yet as his attitude towards the subsequent development of Japan and the opening of Korea suggest, his zeal for Western penetration was tempered by appreciation of the uncertain commitment of Britain's commercial community and the policy limitations of her government. Like contemporary cabinet ministers, Murchison preferred the process of informal imperialism as a means of extending British influence and opportunities, rather than the ultimate obligation of annexation. Where such moves were deemed necessary, however, as at Hong Kong, expedient, as at Labuan, or took place as *faits accomplis* outside official channels, for instance Brooke's adventure in Sarawak, Murchison backed them with his usual staunch patriotism.

7

Africa

Europeans knew little about the interior of Africa in the first half of the nineteenth century. A challenge to explorers, the continent also came to represent for Murchison a testing ground for the power of natural science to help organise and develop alien environments. As President of the RGS he spearheaded Britain's geographical assault on Africa, while as Director-General of the Geological Survey he encouraged reconnaissance of the continent's mineral resources. Throughout the 1850s and 1860s Murchison worked closely with Palmerston and Clarendon to harness science to the Prime Minister's policy of reforming Africa by the introduction of legitimate commerce. At times, he even advocated colonising ventures.

Cape Colony and Natal: the theory of Africa's structure

Murchison's concern with Africa, as with other overseas regions, began with his desire to extend his Silurian System. In 1836 his friend Sir John Herschel sent him some fossil trilobites from the Cape Colony which were cited in *The Silurian System*.[1] Two years later, Murchison's prominent role in organising a public dinner in honour of Herschel's return was to earn him the influential astronomer's enduring gratitude.[2] In 1842 Murchison's presidential address to the Geological Society praised research accomplished at Cape Town by W. B. Clarke during his voyage to Australia as confirming his own inference that Silurian rocks were common in southern Africa.[3] He also illustrated the connection between scientific and commercial expansion by extolling the efforts of Dr William Stanger, FGS, geologist to the ill-fated Niger exploring expedition of 1841.[4] This enterprise had been sponsored by the British government with the backing of Thomas Fowell Buxton's Society for the Extinction of the Slave Trade, but though it was equipped with a light-draught steamer to avoid lingering in the unhealthy interior, most European members of the expedition perished.

In 1844 Murchison's first presidency of the Geographical Society provided a new venue for encouraging African exploration. As successor to the African Association, and because of the influence of Sir John Barrow,[5] the Society had long been interested in the continent, but as its financial difficulties worsened during the 1840s it fell back on stimulating research by secondary means. Still, the RGS *Journal* reflected a growing concern with Africa sparked by the agitation of philanthropic groups and the interventionist policy of Palmerston. The great Whig Foreign Secretary had launched a diplomatic initiative to create stable native polities which could serve as strategic fulcra for introducing legitimate commerce to extinguish the slave trade. Palmerston also hoped that these treaty states might, under the influence of British advisers, prevent rival powers such as France from gaining control of key stretches of the African coast and blocking access to the trade of the interior.[6]

Motivated as much by desire to associate the ailing Society with topical issues as by interest in scientific exploration, Murchison in his first Anniversary Address hewed closely to the Palmerstonian programme of fostering British influence in Africa through the extension of trade. In discussing Abyssinian exploration he commented that the recent British occupation of Aden, the opening of steam navigation in the Red Sea, the explorations of Dr Charles Beke, and the secret British political mission to Shoa in 1841 had focused European attention on this remote kingdom. He dwelt on the escalating French presence, implied that France's failure to publish detailed maps and reports of their explorations was motivated by exclusionism, and cast the British mission in contrast as an effort to open general European intercourse with the Abyssinian interior. Murchison also emphasised the need for exploration of Africa's sub-Saharan interior to solve the mystery of the continent's general structure.[7]

Changing tack the following year, he compared the lacklustre reports of British scientific agents in Abyssinia with the newly published array of French scientific and commercial data. In an attempt to enforce the authority of the Society as Britain's national repository of geographical data and enhance its control over the global flow of such information, Murchison also chastised the geographer James McQueen for communicating details of the German missionary Krapf's exploration of the Juba River in Somaliland – an area advertised as a potentially lucrative market in the previous address – to the Geographical Society of Paris. He went on to mention that a British army officer had volunteered to cross southern Africa and determine the continent's geographical axis.[8] The attempt failed despite official support organised by the RGS, but it demonstrated Murchison's ability to motivate explorers and the Society's capacity to mobilise aid for projects cast as related to imperial interests.

Murchison's interest in Africa found no real outlet in the period between his terms as RGS President in 1844–45 and 1852–53. His mid-century pronouncements on gold mentioned African sources and speculated that the location of King Solomon's Ophir might lie in central Africa.[9] However, when he resumed the chair of the RGS in 1852, his Anniversary Address contained a bold prediction of the overall structure of the continent (Figure 2). He based this prognostication on four factors. First, Andrew Bain, an amateur geologist in Cape Colony, had discovered a series of fossils which led him to speculate upon palaeoclimatic conditions in southern Africa.[10] Second, the discoveries by David Livingstone of Lake Ngami and by Francis Galton of vast salt pans suggested progressive desiccation in this region's interior.[11] Third, Adolph Overweg, companion of the Prussian explorer Heinrich Barth then employed

Figure 2 Murchison's postulated rim of African uplift in relation to topographical features

by the British Foreign Office, had at Murchison's request collected fossils in the central Sahara which included Devonian specimens.[12] Finally, the geologist William Hopkins had postulated a theory of crustal uplift and catastrophic river gorge formation which might be employed to correlate these phenomena.[13]

In 1844 Bain had sent the first specimens of a new order of extinct saurians, the Permian Dicynodonts, to the Geological Society of London.[14] He also reported Silurian fossils which promised to extend Herschel's evidence. Murchison had urged him to provide more organic remains[15] while the Geological Society secured a grant from Peel's government to encourage further research. Bain also received help from the Cape government, which was anxious that he complete a geological map of the colony because of its use in mineral development. He subsequently began tracing the shoreline of a huge lake he believed to have existed during the era of the Dicynodonts. He cautioned Murchison, however, that the Silurian System, while apparently well developed in the southern Cape, did not seem to extend to the colony's northern frontier.[16] In 1851 Bain transmitted his completed map and report to the Geological Society. Murchison then wrote to his friend Sir John Pakington, Secretary of State for the Colonies, with whom he was in communication regarding his Australian prediction, to press for Bain's appointment as geological surveyor of Cape Colony. Murchison also helped to arrange for Bain's fossil collection to be purchased by the British Museum.[17] De la Beche likewise encouraged Bain's appointment, but the matter was dropped when the necessary funding was not forthcoming from the British Survey.[18]

Synthesising the data from these sources, Murchison theorised that Africa comprised an elevated central trough ringed with Palaeozoic uplands. He predicted the discovery of Carboniferous coal on the inland slopes of southern Africa's flanking ranges, and though he did not announce as much at this time, he clearly believed gold to exist in this locale. Reasoning that the Dicynodonts had once occupied an environmental niche comparable to that of the hippopotamus in the interior depression of which the present lakes and marshes were remnants, Murchison concluded that southern Africa had not changed its main features for aeons, and predicted that this structure would be found true of northern Africa. He believed that the centre of Africa had not been submerged since the remote period when the continent had been uplifted and chasms had been rifted in the subtending uplands through which the major rivers found escape. He called for exploration in east and west Africa to determine if his postulated 'girdle' of primeval rocks could be verified by the discovery of formations analogous to those in Cape Colony and the Atlas Mountains. In conclusion, Murchison begged all African travellers to collect fossils and record stratigraphic observations to help determine the continent's geological history.[19]

Soon after his address Murchison displayed Bain's geological map at the

Colonial Office, the Geological Society, and the RGS in order to promote colonial science, illustrate his theory, and advertise the addition of a new province to his Silurian domain.[20] Bain's results were also published by the Geological Society, but his supposedly Silurian fossils now proved to be Devonian.[21] Murchison nevertheless continued to cite these remains as indicative of both Silurian and Devonian strata in order to avoid relinquishing territory, and Bain accepted the master's view.[22]

In 1854, as a result of gold reports from the Transvaal and the Orange River Sovereignty and a copper rush in Namaqualand, Cape Colony at last offered to fund a temporary geological survey. The geologist's salary of £1,000 per annum was to be met from mineral leases. De la Beche recommended Andrew Wyley of the Irish Survey for the post and plied him with instructions on the evaluation of ores before he sailed.[23] At the same time specialists at the Museum of Practical Geology were analysing coal, iron, and copper specimens from Namaqualand,[24] and Bain was pressed into service to investigate the copper-mining district. He ascertained which formations contained the richest ores and recommended the construction of a tramway linking the mines to coastal shipping points. By dramatically reducing the cost of haulage, Bain argued, such a system would stimulate exploitation and prevent the waste of ores of less than the 45 per cent metal content which under prevailing circumstances could alone bear the cost of freight to the Swansea smelters.[25] At the destination of these ores, too, science in the service of government was promoting the improved flow of peripheral raw materials to the world's workshop as naval surveyors charted Swansea Bay and advised the city's importers on the construction of a breakwater for the copper ships.[26]

While Bain continued his amateur pursuits until his death in 1864, a year in which he was warmly received by Murchison during a visit to London,[27] Wyley served as the official geologist for Cape Colony from 1855 to 1859. He proved that much of the strata Bain had classified as Silurian was actually Devonian, sent De la Beche's Museum a collection of ore specimens illustrating the wealth of Namaqualand, and provided Murchison with another precedent for urging the founding of colonial surveys.[28] Wyley wrote six reports on the minerals of Cape Colony. His second is the most interesting in terms of Murchison's indirect influence on his work. When rumours arose that Australian miners had salted the gold diggings in the Orange River Sovereignty with imported nuggets, Wyley was ordered to investigate. He found no gold but, rejecting the idea that the nuggets had been planted, reported on the basis of Murchison's theory of gold occurrence that the metal probably existed at levels too deep for economic recovery. Wyley's paradoxical stance may in fact have been motivated by a desire to persuade the British government to alter its decision to abandon the Orange Free State to the Boers.[29]

Further reports transmitted to Murchison by a geologist assessing these districts for a mining company were published by the Geological Society.[30] When Wyley's appointment came to an end, Murchison encouraged amateur geologists by donating a set of British fossils to Cape Town's South African Institution for purposes of correlative dating; his Survey laboratories also analysed ore specimens for the Cape government.[31] The inception of geological inquiry in Natal followed a similar pattern. Gold discoveries in 1852 prompted Murchison and others to suggest on the basis of Bain's identification of the rocks of the Great Karroo as Silurian that this district would also repay exploration.[32] Since coal also existed in Natal, a coincidence of sufficient incentives – gold strikes, favourable opinions from metropolitan experts of the probability of larger deposits being found, and the presence of other valuable minerals – once more prompted the foundation of a colonial geological survey.

In 1853 Dr William Stanger, Surveyor-General of Natal and former geologist to the 1841 Niger expedition, was accordingly authorised by the Secretary of State for the Colonies to begin investigations.[33] Dr Peter Sutherland, a Scottish naturalist distinguished for geological work conducted during the search for Sir John Franklin, also sailed for Durban on the advice of Sir George Barrow (son of Sir John Barrow) of the Colonial Office, in the hope of winning the appointment.[34] Sutherland was furnished with suggestions for general research by the RGS, and on his outbound voyage he collected meteorological and oceanographic data for a memoir to the Society which was refereed by Murchison.[35] In 1854, through the influence of Murchison and the Barrows, Sutherland was appointed Government Geologist of Natal.[36] He at once began surveying the colony's coal resources, specimens of which De la Beche had been analysing for the Admiralty. Murchison took great interest in these tests because Natal's coal was of strategic significance to the empire and suggested the existence of a palaeozoic spine paralleling the south-east African coast as specified by his structural prediction.[37]

Murchison received reports from Sutherland during 1854 on Natal's coal and copper which he passed on to the Geological Society.[38] In 1855 Stanger died, and with the aid of Murchison and the Barrows, Sutherland was promoted Surveyor-General.[39] The following year new coal seams were discovered. While proved of slight use as steam fuel, these deposits, when correlated with other finds in Portuguese East Africa, seemed to demonstrate further the accuracy of Murchison's prediction.[40] He accordingly demanded an acceleration of the Admiralty's coastal survey in Natal and Sutherland's topographical mapping. The want of such surveys not only prevented geological mapping, he argued, but 'bars the progress of the settlers, hinders the development of the resources of the district, and is attended with loss to the colonial exchequer.'[41]

Murchison favourably noticed the fossil and mineral collection which Sutherland sent to the International Exhibition of 1862, and, in 1866, after Muchison gave his opinion of the commercial value of a marble deposit discovered in the newly annexed province of Alfred, Sutherland named the magistracy in which he made the find after the Director-General.[42] The following year, as part of a general inventory of imperial coal resources,

Sutherland issued a report on the extensive formations of north-west Natal which he now believed to be of more recent deposition than the Carboniferous period. Despite their abundance and accessibility, he concluded that these coals would never prove profitable because the negative effect on demand caused by diversion of shipping to the Suez route and competition from other Indian Ocean fuels would prevent the development of a sufficiently large market to justify construction of the railway necessary to open production.[43]

The Director-General did not receive this report in time for inclusion in his own overview of colonial coal resources, but he nevertheless urged the immediate survey of Natal's seams.[44] The publication of Murchison's comments in *The Times* provoked several applications for the anticipated geological post.[45] The Legislative Council of Natal, stimulated by a railway proposal, considered commissioning a survey that would satisfy the colonists as to their fuel resources 'and at the same time be accepted as conclusive proof by the English capitalists, that there exists here a safe field for the investment of capital'.[46] Despite a recommendation that Murchison be asked to name a geologist, Sutherland's offer to accomplish the survey won over the fiscally conservative legislators.

Having been provided with Sutherland's documents, Murchison was then requested by Lieutenant-Governor Robert Keate, a Fellow of the RGS who had previously co-operated with him in a proconsular capacity regarding the Geological Survey of Trinidad,[47] to give his opinion as to whether Sutherland's scheme for forwarding field data to the home Survey would suffice to determine Natal's coal potential. Keate was confident that the Legislative Council would reverse itself and accept Murchison's recommendation should he feel a special survey was necessary, and he asked in advance that the Director-General select an appointee. Murchison replied that only a mining engineer could estimate the costs of extraction and railway transport to coastal ports, but he himself believed 'that the mere opening out of the coal seams . . . would unquestionably be the means of enriching the colony, increasing its resources, and creating centres of manufacture'. From Sutherland's report, articles in the Natal press, and private letters from Dr Robert Mann, FRGS, Special Commissioner of the Natal Government, the Director-General judged that Natal contained more good coal than 'any part of Eastern Africa yet explored' and that its formations represented a southern extension of the seams on the Zambezi which Livingstone's expedition had found to contain true Carboniferous fossils. He recommended in consequence that a mining geologist be despatched at once.[48] The Lords of the Admiralty were similarly advised when they too expressed interest in Natal's coal.[49]

The Colonial Office agreed to Murchison's proposal, but when the Director-General demanded a salary of £1,000 per annum and a two-year contract for his candidate, the government of Natal balked.[50] Despite Murchison's attempt to influence them through his address to the geographers in 1868 and Mann's similar encouragement via the British Association, the expenditure proved too great for the colony to meet without aid from the imperial government. When this was not forthcoming, all action was suspended.[51] Thus in Natal as in several other colonies, stubborn legislators unconvinced of the ability of science to pay economic dividends prevented the installation of a geological survey, though Sutherland had held such an appointment before Natal was granted representative government.

The simultaneous discovery of gold between the Limpopo and Zambezi rivers by the Prussian geologist Karl Mauch and the local prospecting in Natal stimulated by this news made the colonists regret their parsimony. Linking Mauch's description of the ancient mining sites where he made this discovery with his own theories of Africa's structure and the occurrence of gold, Murchison erroneously concluded this highland tract to be Silurian and, as he notified the geographers, 'precisely in that position in which, as a geologist, I should have expected to find gold'. He asked the society to fund Mauch's further explorations in the interests of science and the economic progress of Natal, and speculated that these goldfields were the lost source of King Solomon's wealth – the mouth of the Limpopo being put forth as the probable site of the biblical Ophir.[52]

Mauch's discoveries and smaller strikes in Natal itself were also discussed at the RGS, advertised by Murchison in *The Times*, and publicised through the British Association by Mann, who had come to London to float the South African Goldfields Exploration Company.[53] A report from Sutherland, however, which relied on Murchison's Silurian theory of the metal's occurrence, soon demonstrated that Natal exhibited the wrong characteristics to be a rich gold-producing country. Murchison published this flattering, if disappointing, letter through the Geological Society, and its information also appeared in the book on the colony's geology which Sutherland brought out the same year.[54] Early in 1869 Murchison received two further reports from the Colonial Office on the gold of Natal and Matabeleland – one of them by a former student of the School of Mines temporarily hired as a gold surveyor by the mortified colonial government. These provided similar evidence that the Murchisonian 'constants' were lacking for significant gold production.[55] Murchison refereed these notices for the RGS but, as he concluded that neither held out 'any prospect of *productive* operations being conducted in the tracts examined', they were never published.[56]

In his Anniversary Address for 1869 Murchison reiterated that, while the Matabeleland goldfields were probably the source of Ophir's wealth, alluvial deposits seemed exhausted and held out little 'incitement to speculators'. Crushing operations might one day pay, he cautiously suggested – the lesson of Victoria having gone home by this time – but the region was too 'wild, distant, and uncivilized' to permit such operations at present.[57] Shortly thereafter, an exploration of the mouth of the Limpopo convinced Murchison that the region could not be the site of Ophir,[58] but the following year brought reports of further gold discoveries south of the Limpopo in the eastern Transvaal.

In 1871, a few months before the President's death, the RGS again discussed the gold potential of the Limpopo–Zambezi region as reported by Thomas Baines, the former artist of Livingstone's Zambesi [*sic*] Expedition. Baines had been engaged by Mann as a prospector and concession negotiator for the Goldfields Exploration Company.[59] Baines' memoir, compiled by Mann, advised Britain to establish a protectorate over Matabeleland in order to provide the settled government necessary for systematic mineral extraction. Two years earlier Baines had urged a similar course of action with regard to the rich copper reserves of south-west Africa, in a memoir presented before the British Association and based on the results of an abortive transcontinental expedition mounted in 1861.[60]

Rawlinson, chairing the Matabeleland discussion, followed the absent President's theory in predicting that 'a line of gold deposits' probably extended northward from this region to Lake Tanganyika. During the same year Mauch was rewarded by the Society for the discovery of further gold deposits and the ruins of Zimbabwe, for as Rawlinson had also commented in discussing Baines' memoir, 'one of the great objects of the Royal Geographical Society always had been and always would be to combine, as far as possible, geographical science with practical economical results'.[61] Despite his extensive explorations and seminal geological research on the structure of the Transvaal, Mauch was largely ignored and died penniless in 1875.[62] Baines died the same year, the company which employed him having gone bankrupt and failed to exploit the valuable mining concessions he had won. Gold-mining activity in the Transvaal and Matabeleland was to remain insignificant until the diamond rush to Kimberley in 1871 triggered a general wave of mineral exploitation throughout the region, but the initial excitement sparked by Mauch's discoveries had introduced the prospectors who made these subsequent finds.[63]

Sutherland continued as Surveyor-General in Natal until retirement in 1887. Just before Murchison's death he provided the Geological Society with a memoir on the colony's glacial detritus which, together with evidence of southern hemispheric glaciation simultaneously being received from New Zealand, helped transform the metropolitan debate regarding the erosive power and former wide extent of glaciers.[64] In a multidisciplinary role analogous to that played by Logan in Canada and Hector in New Zealand, he also collected plants for Kew Gardens, produced the first accurate topographical map of Natal, and generally promoted the scientific investigation and development of the colony's resources. In 1870 Sutherland played host to newly arrived Cecil Rhodes, whom he may well have encouraged to invest in the diamond boom.[65] The coal resources which Sutherland first surveyed were opened for exploitation by the construction of railways in the 1880s following further investigations, but a permanent geological survey was not founded in

Natal until 1899, shortly after Cape Colony and the Transvaal began their own.

The Niger

When Murchison proclaimed his African structural theory in May of 1852, he announced the first of many exploration schemes which he would promote to test its validity in the course of serving more general scientific and commercial goals. Lieutenant J. Lyons McLeod, RN, FRGS, proposed to ascend the Niger to its source using a steamer which the Scottish merchant MacGregor Laird was obliged to provide for exploring purposes by the terms of his west African mail contract with the British government. Once the RGS Council resolved to support this plan, McLeod secured additional backing from the Admiralty and the Manchester cotton manufacturers, who wished to develop new sources of supply for their staple.[66] Murchison then brought the proposal to the attention of the Prime Minister and it was again aired before the British Association.[67] After McLeod's plan was limited on the advice of naval members of the RGS Expedition Committee to a charting expedition of the lower Niger, Murchison convinced the Association's President, Sir Edward Sabine, to make another recommendation to Lord Derby. Despite these efforts the dissolution of Derby's government 'put an end to all hopes of McLeod's Expedition' receiving sanction that year.[68]

Early in 1853 it was learned that Heinrich Barth had charted the upper reaches of the Benue, or Tchadda River, the main tributary of the Niger. Murchison's advice coinciding with the judgement of Clarendon, Foreign Secretary in Aberdeen's new government, the expedition was transformed into a relief mission to rendezvous with Barth and verify the navigability of the Benue. Murchison's Anniversary Address linked McLeod's modified project with an account of the explorations of Barth and Overweg received from August Petermann, a Prussian cartographer employed at the London observatory. With the aid of Charles Bunsen, FRGS, Prussia's Ambassador in London, Petermann had originally recruited Barth for the Saharan expedition. Murchison had at that time attempted against his better judgement to persuade Bunsen to contribute funds for the Foreign Office mission,[69] and he now implied his resentment at the efforts of this champion of Prussian nationalism to help funnel Barth's results to Berlin.[70]

Laird remained the driving force behind the new Niger expedition, but Murchison arranged the appointment as naturalist of William Balfour Baikie, a Scottish naval surgeon who had recently planned a South American exploration.[71] By December of 1853 Murchison learned that McLeod had fled the country for financial reasons.[72] The steamboat being provided by the Laird shipyard had also overrun its contracted cost, and Sir Francis Beaufort, the Navy's Hydrographer, had turned against the project. To stop the rot, Murchison pointed out to Clarendon that the government as well as the RGS would lose popularity by abandoning its commitment and advised him to take the project out of naval hands and 'name *your own Consul Beecroft* of those parts to work the paddler'.[73] Murchison had previously engineered Beecroft's election to the RGS, and the Foreign Secretary now appointed him leader of the Niger venture. Before the expedition sailed in 1854, Murchison helped Beaufort compose its scientific instructions. He arranged with Clarendon to prevent what seems to have been envisioned as an attempt to 'Prussianise' the mission by unequivocally subordinating the research of Wilhelm Bleek, a philologist appointed as interpreter through the influence of Bunsen, to Beecroft's authority.[74] Murchison also exposed the attempt by Bunsen and Petermann to commandeer the results of Barth and Vogel, Overweg's replacement. Furthermore, he secured Clarendon's promise of prior access for the RGS to future despatches received from the Prussian explorers.[75]

Command of the expedition fell to Baikie upon Beecroft's death. Though Baikie missed meeting Barth, he demonstrated the navigability of the Niger and the effectiveness of quinine as a malaria prophylactic.[76] When the expedition returned to Britain in February 1855, Baikie and Laird began a campaign at the RGS for a second voyage. Palmerston was now Prime Minister, with Clarendon retaining the Foreign Office. Both were keen to encourage African cotton production as a means of undermining the slave trade in Africa while reducing Britain's dependence on slave-grown American cotton. Clarendon worked closely with Murchison to convince the public of the necessity for the renewed expedition. Murchison now considered the question to be 'one of policy & the improvement of our Foreign Affairs', rather than mere exploration, but he urged that scientists continue to accompany such annual voyages in order to justify the risk of 'a row' with the natives'.[77] He likewise cautioned Clarendon to weigh the political advantages of the undertaking against the opposition such an expenditure might arouse in the midst of the Crimean War.[78] The same day Murchison requested the Foreign Secretary to support his appointment to the Geological Survey. His simultaneous work for the war effort and Clarendon's confused minute – 'To be President of Geological Society' – on this note confirm that Murchison, rather than the institutions which were his instruments, constituted the essential element in the Foreign Secretary's plans to conduct the Niger mission and other scientific reconnaissances.[79]

Barth's arrival in London in October 1855 heightened interest in the new Niger expedition. As the Crimean War came to a close at the beginning of 1856, Baikie requested Murchison to bring the proposal before the RGS Council.[80] As Vice-President, Murchison did so, disingenuously expressing his hope that Clarendon would support the plan should the RGS recommend it.[81] The proprieties required that sanction for a project which had already been agreed upon in private should appear to have arisen organically from public demand. Murchison soon secured approval from the RGS Council, having chaired the evaluating committee and forewarned Clarendon of a formal appeal. In anticipation of opposition from the Admiralty, whose leaders had wearied of his demands for Franklin search expeditions, Murchison also explained that he had been pushed into promoting it by Baikie and the geographers.[82]

There could be no hope of a Niger expedition during 1856 since the flood season which would permit a rapid ascent of the river had passed. Murchison therefore adjusted the pace of his promotional campaign to culminate at the meeting of the British Association in August so that Clarendon might derive maximum credit from his patronage. Meanwhile he arranged that Barth, in accepting an RGS gold medal, should demonstrate the commercial importance of following up his own discoveries. President Beechey, likewise influenced by Murchison, also urged the renewal of Niger exploration, but he berated Baikie for omitting in his published narrative any mention of the parts played in his first expedition by McLeod, the RGS Council, 'and particularly by Sir Roderick Murchison'.[83]

Baikie and Barth now requested Murchison to urge Clarendon to act before the season passed away again. Murchison had delayed the public promotion of Baikie's proposal at the RGS until the Foreign Secretary had made a decision, but he now asked for authorisation to announce his approval and plan 'to have some expedition next year & then begin a regular intercourse'.[84] Like Baikie, he was thinking of some form of permanent involvement in the Niger valley. In July Baikie again brought Murchison up to date on the efforts of philanthropic and commercial interests to win approval for his expedition. Murchison expressed his hope to Clarendon that the project might lead to 'a bond of union between the *Sultan* of Sokatu [*sic*] & our Sovereign!'[85] As a deep-dyed patriot whose watchword was 'forward, ever forward',[86] Murchison found his enthusiasm was leading him into the grey area of intensifying political and economic activity in which Palmerston's diplomacy often spawned spheres of influence or protectorates.[87]

Besides apologising for his former omission,[88] Baikie now wrote a new proposal for the British Association which conspicuously cited Murchison's role in the 1854 expedition. This document emphasised the natural products of the region, including known deposits of copper and lead and the potential for gold discoveries, and stressed the key role the geologist would play in opening the country to commerce.[89] Murchison's notes on Baikie's manuscript reveal that this memoir awakened associations with his theory of Africa's structure.[90] These ideas found expression in his introduction to Livingstone's description of the crossing of southern Africa, which largely confirmed his 1852 prediction. Murchison implied that Baikie's proposed Niger voyage would prove north-central Africa also to consist of a marshy interior flanked by salubrious Silurian highlands replete with agricultural and mineral resources and probably containing alluvial gold to attract colonising emigrants.[91]

After this, Murchison reported to the RGS that Clarendon was 'very favourably disposed' towards the Niger project.[92] While arranging a deputation from the British Association to wait upon the Foreign Secretary, he expressed his hope 'that your fiat will not issue until the savants have thrown in their appeal, as we are naturally desirous of having some credit for our exertions.'[93] The delivered petition detailed how scientific research as well as anti-slavery activities could be widely conducted from commercial centres in the hills along the lower Niger.[94] Within a week, Clarendon secured the backing of other cabinet members including George Cornewall Lewis, the Chancellor of the Exchequer and an old friend of Murchison's.[95] Clarendon delayed his decision in order to consider a new offer by Laird to send two steamers annually up the Niger. With his suspicions aroused by Laird's apparent desire to establish a dominant trading position on the river, Murchison cautioned that 'grasping at too much' might delay the expedition itself.[96] At the end of November, Clarendon finally gave him permission to announce that it would go forward as planned. Murchison then encouraged the re-employment of Daniel May, FRGS, a naval pilot who had guided Baikie's previous expedition, while at the same time striving against Admiralty interference in the preparations.[97]

Early in 1857 Clarendon requested Murchison's evaluation of another plan spawned by interest in the Niger as a source of cotton. This visionary scheme proposed diverting the river to irrigate the Sahara. Murchison solicited Barth's opinion before denouncing the project as insane, but he noted that, since it might succeed in the City of London, 'until our citizens become better geographers they *deserve* to be *done* by schemes like this'.[98] Clearly, Murchison had made himself indispensable as a scientific adviser, and with geographical discoveries beginning to rival mineral strikes in stimulating bogus development schemes, his roles at the RGS and the Geological Survey as an authority capable of differentiating between facts and lies, probabilities and absurdities, became increasingly complementary.

During April Murchison protested that Baikie's draft instructions contained little regarding geological researches, and that the Admiralty had only solicited his opinion on the eve of the expedition's departure. Arguing that his own science was more important than zoology or botany, both of which were represented by a specialist, Murchison noted that if Clarendon's main object was to promote legitimate commerce, 'the ores of copper, lead, & iron; & then all the beds of coal which may be found, to say nothing of the sources of the gold sands of the rivers; are at the very head of the desiderata in *your* estimate'. He therefore recommended that his own detailed geological instructions be sent after Baikie and that the scientific officers be directed to 'attend equally to the structure of the subsoil'.[99]

These instructions directed Baikie to establish the region's stratigraphy by examining the natural sections the Niger had cut through the coastal hill ranges. Between the Niger–Benue confluence and the Bussa rapids further upstream, he was to search the river banks for coal, 'since it is at about the same distance from the seaboard on the opposite coast of Africa that Dr. Livingstone has discovered the Carboniferous junction'. The overland party marching north from the rapids to Sokoto was to make further geological observations and visit any gold, copper or iron mines revealed by the natives. During the downstream return to the rapids, in a stretch where the Niger ran perpendicular to the strike of the strata and the regional axis of elevation, Murchison expected the explorers to find metamorphosed Silurian rocks indicative of gold and other metal ores. Other overland journeys would provide additional lengthy transverse sections. The research programme was thus designed to confirm Livingstone's verification of Murchison's prediction regarding the structure of central Africa and to demonstrate that the Niger, like the flanking ranges of the south, would prove 'a field of great mineral wealth'.[100] This was truly geology on the grand scale, the sort of broad-brush, exploratory research in which Murchison increasingly specialised in the final phase of his career.

Once Baikie departed, Murchison advertised the Niger expedition in his Anniversary Address and spoke of the explorations of the murdered Vogel, for whose family he later secured a government pension.[101] The President duly apportioned credit for the new Niger initiative among the RGS, the British Association, and the government, intent as always on presenting a public image of seamless co-operation between science and officialdom in the accomplishment of national goals. Clarendon, especially, came in for much praise, and Murchison sent him a copy of the address afterwards with the personal references marked. In describing Baikie's project, Murchison complained that a geologist had not been appointed since the region probably contained minerals valuable as 'sources for future trade'. He also pointed out that Britain had no 'desire to establish colonies or settlements which might give umbrage and provoke quarrels'.[102] In the ensuing months Baikie steamed up the Niger to suffer shipwreck at the Bussa rapids and later found the settlement of Lokoja at the Benue confluence. Murchison meanwhile defended himself from the jibes of Sabine about 'we geologists always taking care of *ourselves*' – the terrestrial magnetician having been frustrated of this chance to appoint a specialist in his own discipline – by adducing the want of such a companion for Baikie. He also continued to express to fellow scientists his resentment of Admiralty interference.[103]

In January 1858 Murchison began receiving despatches on Baikie's progress from Clarendon. He used this favour to publicise the liaison between the RGS and the government and to provide Clarendon with an early return on his investment in science. Malmesbury, the succeeding Foreign Secretary, continued to provide Niger reports which Murchison employed to maintain public confidence in the stranded expedition.[104] In his Anniversary Address of 1859, the President reiterated that worthwhile results were accruing since he had received an account from May describing the cotton-producing potential of the Yoruba country and interviewed the returned officer about the future prospects of the expedition. Murchison had concluded from May's geological specimens that the hills near the Niger's rapids were composed of Silurian rocks, and in implying that his 'golden constants' were present he once more regretted that no geologist had accompanied the expedition.[105]

Baikie heard of this comment and wrote in protest that he had been making geological observations himself. 'I cannot venture for a moment to compete with a Livingstone or a Barth, & my duties are much more political than theirs & therefore less interesting', he explained.[106] Baikie was afraid he had forfeited the crucial support of Murchison and the geographers, as he believed he had already lost the backing of philanthropic and commercial circles. There was some substance to these fears, for the President was compelled by the Society's dependence on popularity – a dependence which he had done more than anyone else to create – to seek constantly for sensational discoveries. Proto-imperial activities appeared less thrilling than the explorations they succeeded, though Murchison himself remained fascinated with the process of consolidating British overseas interests.

During the same year, Palmerston's return to office, the outbreak of the American Civil War, and the publicity given May's memoir describing the Niger vallley as capable of an immense production of cotton revived flagging interest. The geographer James McQueen attempted to persuade the RGS to re-examine a plan he had projected in the 1840s for a chartered company to develop cotton cultivation at the Niger–Benue confluence. Francis Galton, refereeing the scheme, commented that Baikie's settlement at Lokoja might be rendered productive if it were taken in hand by the government. 'The growing uneasiness about our cotton supply, might induce a strong feeling in Manchester & elsewhere in favour of a new African plantation', Galton wrote. 'If such a feeling existed or appeared likely to exist, than I think the Geographical Society would be quite in the right to encourage and direct it.'[107] Though the RGS did not adopt Galton's proposal, the discussion in its inner councils of the formation of a new African colony for commercial reasons, and the proposal of the Society as the appropriate body to organise the enterprise, indicates the imperial sentiment which pervaded the geopolitical thinking of many key Fellows.

Murchison knew that openly projecting the RGS as a colonising agency would be disastrous, but there were subtler methods by which the Society might exploit the public fixation with cotton to enhance its image of national utility and further British penetration of west Africa. Aside from the RGS associating itself with cotton schemes to win support for expeditions elsewhere during 1860, Murchison had May's Yoruba memoir re-presented before the British Association. He also advertised the new Admiralty charts of the Niger drawn by Baikie's naval surveyor, Lieutenant John Glover, FRGS, and reminded the public of the continuity of British enterprise on the river.[108] Baikie and Glover were meanwhile advising the cabinet that annexations and extensions of consular authority were necessary to secure Britain's interests, and these views were publicised through the RGS with Murchison's concurrence.[109]

As hopes for the rapid development of a lucrative Niger trade faded and Baikie was recalled on grounds of health, Murchison was swept on to more exciting African prospects. Still, he took care to praise Baikie's demonstration that 'a British settlement can be made a centre of civilization'.[110] Baikie produced no mineral reports, but a company seeking a subsidy to operate steamships on the Niger nevertheless claimed that 'immense quantities' of the richest copper ores existed there.[111] Ongoing optimism about mineral resources and vegetable products which might be exchanged for manufactured goods probably influenced the decisions by Palmerston and his successor Russell to continue the attempt to develop the Niger region as a British economic satellite. In 1865 Murchison delivered an obituary eulogy of Baikie that left no doubt of his own belief in the long-term value of official involvement in the Niger interior.[112] In 1866, following the death of Palmerston, Murchison urged the government to maintain the establishment at Lokoja,[113] and when the discussion of a posthumous memoir of Baikie's the following year led Trelawny Saunders to urge the suitability of the Niger uplands for European settlement and press for the foundation of a chartered company 'to take a strong hold of Africa, and deal with the natives as we had done in India', Murchison's countenancing of this blatantly imperialistic remark probably indicated agreement with its views.[114]

The King of Siluria continued to tout the Niger system as a potentially prolific source of gold, and he remained convinced that the maintenance of a British presence would one day 'prove remunerative'.[115] As late as 1900 mineral exports were still negligible, but in the 1880s and 1890s agents of the Royal Niger Company had prospected and obtained mining rights over large tracts of what would become the colony of Nigeria. Geologists from the Imperial Institute, which was founded in 1887 to co-ordinate development-oriented scientific research throughout the empire, surveyed both Southern and Northern Nigeria for the Colonial Office during the period 1903–13. In the course of their work they helped ensure that no valuable mineral deposits fell to the French when the northern border was demarcated in 1905–06.[116]

The Central Lakes and the Nile

Murchison played a less prominent role than he had in the Baikie missions in organising the contemporary expedition to east Africa of Richard Burton and John Hanning Speke, both officers of the Indian Army. Though W. H. Sykes of the East India Company's Court of Directors bore most of the responsibility for launching this exploration via the RGS, President Beechey of the Society closely followed Murchison's recommendation about approaching the government for aid. Beechey's comment that 'I am sure your experience in these matters is such that your advice will always guide me not only in this but in all others connected with the Geographical Soc.'[117] demonstrates Murchison's ascendancy over those who briefly held the chair of the RGS in the midst of his own long tenure. Murchison took part in the Society's deputation to the Foreign Office and, having convinced Clarendon of the expedition's utility, requested the Treasury's support. [118] Burton's instructions, like Baikie's, reflected the interlocking concerns of Clarendon and Murchison: while attending to the usual range of investigations he was to prospect for economic minerals. The explorations of Burton and Speke had more far-reaching geological consequences, however, than their location of native iron mines. Their discovery of Lakes Tanganyika and Victoria suggested an east African corroboration of Murchison's structural theory which might be proved if it could be demonstrated that the Nile issued from this lacustrine basin through a chasm in the continental rim. Murchison's anxiety to complete the verification of his prediction thus represented one causal strand in the Nile quest, itself a factor in drawing Britain deeper into political involvement in Africa. [119]

Almost inevitably, Murchison was forced to mediate in the dispute between Burton and Speke over the honour of discovering the source of the Nile. He personally favoured Speke, but strove to remain impartial in public and also attempted to win Burton a new exploring or diplomatic appointment. As British Consul at Fernando Po, Burton continued to provide the RGS with information, though he annoyed Murchison by publishing his narratives before they had appeared in the Society's *Journal* and exciting Joseph Hooker's ire for his caddish cribbing of data from a Kew plant collector. While encouraging Hooker to chastise Burton for this transgression, the President continued to support Burton in order to maintain the reputation of the RGS, and even considered proposing him for the Royal Society. [120]

When Speke and James Grant returned to settle the Nile question Murchsion co-ordinated their government aid, instructed them in their scientific duties, and gave them a public send-off. [121] He also organised a relief expedition under John Petherick, British Vice-Consul at Khartoum, to meet the two explorers as they made their way down the Nile. Petherick had formerly worked as a prospector in Egyptian service and as a trader in the Sudan. [122] In 1859 he presented the Museum of Practical Geology with specimens of gold from the upper White Nile, analysis of which suggested further verification of the Director-General's African theory. [123] In consequence, Murchison was eager to put Petherick back in the field to help Speke confirm his own prescience. An approach to the Foreign Office to fund the venture failed, so Murchison fell back on a public subscription as he had when similarly rebuffed during the Franklin search. As the RGS coffers quickly filled, the government was shamed into a small donation, and Petherick was returned to Khartoum in 1861. [124] He produced no interesting geological data, however, and his failure to rendezvous with Speke precipitated another embarrassing quarrel.

Murchison temporarily despaired of the main expedition when, in vetoing the reading at the RGS of one despatch from Grant, he remarked that it showed 'so much of *mismanagement* & disaster as to turn African explorations into ridicule and regret'. [125] But Speke and Grant soon earned the President's highest admiration for their discovery of the Ripon Falls, which solved the fundamental mystery of the Nile's source and offered another proof of the accuracy of his structural prediction. By badgering his friends Russell and Layard at the Foreign Office, as well as Gladstone at the Treasury, Murchison persuaded the government to send funds to Egypt to bring the explorers home. [126] A series of letters to *The Times* raised anticipation of their arrival to fever pitch. Speke and Grant were given a triumphal reception at a special meeting of the RGS so crowded that men scaled the walls of Burlington House to glimpse the event through the windows and the delighted Murchison barely managed to force his way to the dais and take the President's chair. Here, flanked by the two heroes of the hour with a huge map of Africa forming a dramatic backdrop, he struggled to direct the proceedings (Plate 5). [127]

Murchison now said nothing about his earlier doubts. Advertising this huge publicity coup in advance in his Anniversary Address of 1863, he admitted having organised it 'in order that we may gratify the public, and do honour to ourselves'. The President also boasted of the confirmation of his theory, but having perused Speke's geological notes, he held out scant hope of the region around Lake Victoria proving auriferous. He implied, however, that the mountains of southern Abyssinia might contain gold. [128] Having employed Petherick's gold to create expectation when funding for the Nile expeditions was required, he was now shifting public interest to a region still requiring exploration.

Grant, like Burton and Speke, received an RGS medal from Murchison. The President remained on close terms with this fellow Highlander for the rest of his life. He introduced Grant to various statesmen, secured him a military staff appointment in India when he declined to accept the consulship of Fernando Po, arranged that he be awarded a C.B., and eventually installed him on the RGS Council. During the 1870s Grant was to become an enthusiastic proponent of African exploration; in the 1880s, he would use his influence at the RGS to generate public support for a policy of expanding Britain's commitments in Africa. [129] Speke was elected to the Athenaeum by special ballot and proposed for a baronetcy, [130] but he rapidly fell out of Murchison's favour because of his refusal to publish through the RGS and his unwillingness to lead a follow-up expedition proposed by the President as a means of establishing a commercial route to the central lakes via the Nile.

Murchison's plan, as approved by the RGS Council, formally proposed to Palmerston, and publicly advocated at both the Society and the British Association, envisioned a southward extension of the authority of the Khedive of Egypt. By this means he hoped to eradicate the Sudanese slave trade and render the upper Nile basin safe for the explorers and merchants who would begin the task of civilising the region. [131] In order to demonstrate the importance of the recent discoveries and the usefulness of the RGS, he thus

Plate 5 Reception of Speke and Grant at the RGS, 22 June 1863: Murchison at the speakers' platform flanked by the explorers

directly encouraged a subsidiary form of imperialism on the part of a local power anxious to expand its borders and legitimise itself in European eyes. The Khedive Said had recently assured Murchison, upon being questioned about his capability of helping Speke and Grant, that 'my frontiers are very elastic',[132] and soon after the proposal of Murchison's new scheme, his successor Ismail was elected an Honorary Member of the RGS. But as Samuel Baker and Charles Gordon were to find in governing the new province of Equatoria – whose annexation, like that of the other Sudanese and Somali territories acquired by Egypt in the early 1870s, may have been influenced by Murchison's encouragement – abolishing the slave trade was almost impossible since it was a central element of the regional economy and was connived at by the Egyptian administrators.

Speke, too, believed that the solution to the problems of east Africa was the imposition of a strong paternal government, but his model was India, not Egypt.[133] In consequence, Speke opposed Murchison's scheme, which he described to Layard of the Foreign Office as 'forcing my way with Egyptian troops'. Instead, he offered to return as a British envoy to the equatorial kingdoms to open legitimate commerce,[134] but Layard broke off negotiations with the explorer on Murchison's advice in order to maintain good relations with the RGS for joint work in Central Africa.[135] Alarmed by Speke's 'aberrations', which included a scheme to establish religious missions under the protection of a drilled black regiment and another plan concocted with Louis Napoleon to penetrate central Africa from French Gabon, Murchison was anxious to distinguish them from the Egyptian project proposed by the RGS.[136] Still, Speke possessed publicity value, and Murchison invited him to the British Association meeting of 1864 to debate with Burton and Livingstone about the Nile and the opening of Africa (Plate 6). Speke's death deprived the geographers of this long-awaited confrontation, but Murchison's announcement of the tragedy produced a like sensation. Speke's funeral and an appeal for a monument were similarly transformed into publicity exercises for the glorification of the RGS and Britain's mission in Africa.[137]

The Zambezi

The third African expedition which Murchison initiated during 1856 was David Livingstone's Zambezi mission. Murchison enjoyed a closer friendship with Livingstone, also a fellow Scot, than with any other explorer: their relationship was 'that of a Highlander to his chief'.[138] Livingstone provided Murchison with a series of breathtaking discoveries that catapulted the RGS to a pinnacle of fame and influence, while the President transformed the obscure missionary into one of Victorian Britain's archetypal heroes. By creating Livingstone's renown, Murchison made possible the government aid, public subscriptions, and profits from books sales which freed him to prosecute further African explorations. The ultimate goal of these researches was the

Plate 6 Murchison, in white shooting jacket, on a geological excursion at the 1864 meeting of the British Association held in Bath. Livingstone, in characteristic cap, stands behind him

encouragement of Western-guided economic development as a means of stopping the slave trade and improving physical and moral conditions for Africans.

Livingstone had been transmitting geological data to the London savants from the beginning of his African explorations,[139] but it was the missionary's discovery of Lake Ngami in 1849 that first alerted Murchison to his value as a scientific observer. When Livingstone made his epic trek across south-central Africa during the years 1853–56, he sent Murchison a series of reports and a transcontinental geological section which confirmed his structural prediction. Murchison publicised this information through the RGS[140] and intervened with Gladstone, the Chancellor of the Exchequer, to attempt to secure Livingstone a grant-in-aid.[141] He also arranged that Livingstone be awarded an RGS gold medal, convinced John Murray to publish a narrative he urged the explorer to compose, and promised him a hero's welcome upon his return to Britain.[142] Judging that the influential geographer might prove a better patron than the London Missionary Society, Livingstone had then sent Murchison a request for financial aid which implied his willingness to accept some other form of employment connected with developing the region. His tendency to overrate commercial opportunities to further his plans for regenerating Africa was evident in this letter: its emphasis on minerals, and more especially gold, was well calculated to stir Murchison's interest.[143] The latter immediately forwarded this report to Clarendon, suggesting that the missionary be given a diplomatic appointment to carry out the Foreign Secretary's plans for developing intercourse between Britain and Africa. Similar dispatches which Clarendon had received direct from Livingstone were now turned over to Murchison and used by him to demonstrate Clarendon's interest in the missionary and the success of Palmerston's anti-slave trade policy.[144]

Livingstone arrived in Britain in December 1856. It is indicative of the importance he attached to his link with Murchison that, after sixteen years in Africa, the portmanteau he carried down the ship's gangplank was filled with rock specimens from the Zambezi highlands. These were used along with Bain's geological map to illustrate the commercial, mining, and settlement potential of south-central Africa at the RGS reception which Murchison arranged to publicise Livingstone's discoveries.[145] Murchison then interrupted his Christmas holiday with Palmerston to initiate a nationwide subscription for Livingstone, again urging that the missionary be employed in a diplomatic capacity.[146] By the end of January 1857 he had arranged a meeting between Livingstone and Clarendon so that the explorer might explain his plans for introducing British commerce and establishing 'a great cotton-growing operation in the heart of Negro-land, with a free transit by the navigable branch of the Zambesi'.[147] Carried away once more with thoughts of healthy, accessible highlands endowed with iron, coal, and alluvial gold probably derived from Silurian rocks, Murchison suggested the acquisition of the upper Zambezi valley from the Portuguese. 'The establishment of the good British

Name in such a region would be a noble beginning', he wrote Clarendon. 'It might follow, that Portugal would be too happy to cede her 3/4 ruined settlement of Tete for a very small consideration.'[148]

A few weeks later, while Livingstone began composing his best-selling *Misionary Travels*, Murchison returned to the usefulness of gaining possession of the decaying Zambezi outposts as a source of cotton and other industrial raw materials which might render Britain independent of foreign suppliers. 'Either England and her ally Portugal may be made *one* for this great object', he explained, 'or the latter country might readily part with her colony of Quilimane and Tete, so useless to her, but which in our hands might be rendered a paradise of wealth.' Clarendon was apparently startled by this proposal, for the same day Murchison explained that his 'unguarded hint' was 'a mere crochet of my own' and apologised for his '*irregular* foray into your Department' in the interests of not prejudicing Livingstone's scheme.[149]

Livingstone's own proposal to Clarendon merely suggested the formation of a chain of commercial stations along the Zambezi, but to his closest supporters such as Adam Sedgwick and the Duke of Argyll – the latter a member of the cabinet – he was soon speaking of founding a British colony.[150] Murchison was almost certainly party to these clandestine plans. Though public men were aware of Livingstone's motives and the cabinet certainly had no intention of embarking on a new colonial adventure in Africa, the explorer's status as a national hero made it virtually impossible for the government to deny him support. The evidence suggests that, besides buying an opportunity to partake of Livingstone's popularity, Palmerston and Clarendon seem also to have envisioned his expedition as a lever to force the Portuguese to abandon their connivance in the slave trade and the prohibitive tariffs levied on foreign goods entering their colonies.[151] Here again Palmerston demonstrated his adroitness at using scientific expeditions to achieve political ends – not only for gathering intelligence, but for adjusting the diplomatic equation. This technique would be brought to a fine art during the imperial scramble at the end of the century, but Murchison established many of the precedents for the use of geographical expeditions as quasi-official instruments of policy.

By mid-April Murchison's negotiations with cabinet ministers had secured tentative approval that Livingstone was to receive a consulship and passage back to the Zambezi. Livingstone's obligation to complete his book prevented a departure during 1857, but Murchison advised that the delay be employed in organising his African enterprise, conducting a fund-raising speaking tour, and visiting Lisbon to secure Portuguese support.[152] In his Anniversary Address he noted that Livingstone's confirmation of his own structural prediction implied easy water communications between the headwaters of the Zambezi and Congo rivers. If true, this in turn implied rapid, low-cost access to the whole of central Africa by alternate routes, provided means could be found to circumvent the falls known to block navigation on both rivers. After outlining Livingstone's plan for developing the Zambezi's vegetable and mineral resources as a preliminary step to the civilisation of the region's inhabitants, Murchison again called on Clarendon to return the explorer to Africa with an official commission. No mention was made of a government expedition or the pending consular appointment, but the parallel drawn between the work to be done on the Zambezi and Baikie's efforts on the Niger left little doubt that another major government commitment was in the offing.[153]

While Livingstone awaited the healthy season for African travel, Murchison arranged the production of *Missionary Travels* with John Murray, publisher of Murchison's own geological works as well as the RGS *Journal*. With the aid of T. H. Huxley, his subordinate at the School of Mines, Murchison prepared Livingstone's sectional diagram of Africa for the volume, asking Murray in return to suggest that Livingstone dedicate the book to him. The remarks printed in consequence gave Murchison a direct share in Livingstone's fame in addition to what he received indirectly as RGS President.[154] Reciprocally, Murchison's unswerving support remained the explorer's greatest asset throughout his later career. Livingstone returned to London from a triumphal speaking tour in October 1857. Coming back himself from Germany, Murchison learned that a deputation from the British Association, co-ordinated with appeals from provincial chambers of commerce, would soon wait upon the Foreign Secretary to request aid for the explorer to chart the Zambezi and reconnoitre its resources. Murchison took over these nego-tiations, realising that the new proposal focused Livingstone's amorphous plans into a concrete scientific venture.

Palmerston approved the Zambezi survey, but the Admiralty pointed out that no river steamers were available.[155] Murchison, sensing another naval row, begged Clarendon to intercede in support of science and 'our stifled manufacturers', and the following day MacGregor Laird informed the Presi-dent that he was willing to provide a steamship for Livingstone as he had for Baikie.[156] Holding this trump card, Murchison could not well be refused. By mid-December Lewis, the Chancellor of the Exchequer who had supported the Niger expedition in the belief that Britain should 'attempt the establishment of factories on some of the great African rivers',[157] had announced to a cheering House of Commons that the government would vote £5,000 for the Zambezi venture. Thanking Clarendon, Murchison echoed the British Association in requesting that the expedition include 'at least two or three scientific men – the one of them combining a knowledge of geology with geography.'[158]

John Washington, the Admiralty Hydrographer, was now deputed by Clarendon to draw up plans for the expedition after soliciting Murchison's advice. Remembering the problems created for Baikie, Murchison panicked when he discovered that arrangements for Livingstone's project had been transferred from the Foreign Office to the Admiralty. A conference with Livingstone, Laird, and Washington, however, convinced him that the Hydrographer could be trusted to consult the appropriate authorities about the scientific research to be undertaken. Because of his hopes for major discoveries of Silurian strata and economic minerals on the Zambezi, Murchison was taking no chances that his own science would again be cut out, and he confidentially instructed Clarendon that Washington was to 'officially apply to me to recommend a good *mining* geological surveyor'. He also advised the Foreign Secretary to obtain suggestions from the Royal Society. Recalling his frustrations regarding the Palliser Expedition, Murchison requested on behalf of the RGS that as a reward for his own efforts the Foreign Secretary should 'sustain the dignity of the public scientific body which is so much consulted with by Government'.[159]

While personnel were being considered Sir Samuel Baker applied to join the party, but Murchison reluctantly rejected him because Livingstone wished to be accompanied only by thoroughly subordinate specialists.[160] Baker then concocted a new plan to march north from Natal to meet Livingstone on the Zambezi, but though Murchison and the Manchester Cotton Supply Association supported this project, the change of ministry destroyed all chance of pressing it upon the government.[161] Livingstone was already exhibiting the jealousy of command which would contribute to the mission's failure, but the mesmerised Murchison unquestioningly supported his demands for a free hand. Similarly, Murchison cautioned Clarendon to reject Washington's proposals for a large expedition: expensive ships meant naval officers who might challenge Livingstone's authority as well as unmanageable impedimenta. He had even come to disapprove of Laird's costly steamer, believing that it would founder like Baikie's and could only serve as a Trojan horse for Admiralty meddling. Washington riposted by informing Laird of these objections, after which the merchant convinced the government to buy his ship.[162]

In late December Clarendon requested scientific advice from the RGS and the Royal Society. As President of the Geographical Society, Murchison had the largest influence upon its reply. Once again, he was officially approached for advice which he had privately asked Clarendon to seek from him. Soon afterwards, he attended the meeting of the Royal Society's Zambesi [sic] Committee together with Livingstone and reported its resolutions in advance to Clarendon. These botanical and zoological instructions, drawn up respectively by Joseph Hooker and Richard Owen, also demonstrated an awareness of the research opportunities inherent in projecting science as a pioneer of empire, for they emphasised reconnaissance of agricultural potential and eradication of the tsetse fly as a prelude to colonisation.[163]

At the same time, Murchison approved the choice of Commander Norman Bedingfeld, R.N., FRGS, as Livingstone's second in command and advised that, if the troubles with the Admiralty continued, the Foreign Secretary should 'take the whole responsibility on *your own department*' and let the explorer spend the government grant as he thought best.[164] He also persuaded Livingstone to employ Thomas Baines, recently returned from an RGS commission in Australia, as the expedition's artist, and influenced Sir William Hooker's choice of John Kirk, a Scottish civilian unlikely to challenge Livingstone's leadership, as the economic botanist. Similarly, Livingstone's suspicion of rivals, the Admiralty's jealousy of army officers, and Murchison's entreaties that research be limited to immediately practical considerations, again thwarted Sabine of appointing a specialist in terrestrial magnetism, though other members of the party were trained to make these observations.[165]

As arrangements were being completed in the new year and Livingstone burst out at an RGS meeting with 'rather *too bluff*' an attack on Portuguese complicity in the slave trade, Murchison apologised to Clarendon to prevent the expedition being jeopardised.[166] On 13 February 1858 Murchison staged an elaborate 'Farewell Livingstone Festival' at the RGS. Attended by some 350 eminent men, this was another publicity exercise to demonstrate widespread support for the Zambezi expedition and raise the prestige of the Society. Clarendon declined to attend, confessing that he had 'for some time past thought that Dr. L. was being too much honoured for his own good and that the public was being led to expect more from his future labours than will probably be realized', but he allowed Murchison to announce his 'heartfelt wishes' for Livingstone's success.[167] Murchison promised that '*no gasconades*' would occur, but in recompense for Clarendon's absence he exacted a pledge that the Duke of Argyll would be allowed to attend as a representative of both the cabinet and the western Scottish Highlands which had nurtured Livingstone. A good deal of imperialistic sentiment was nevertheless expressed at the Festival. Argyll mildly countered this by pointing out that Livingstone would not be working for 'the just pride of national domination' and good-humouredly chiding Murchison for his Silurian territorial ambitions in Africa.[168] Here, once more, the government's delicate policy with regard to Portuguese sensibilities was revealed.

The Festival succeeded even beyond Murchison's expectations: he told Clarendon that 'I never had to do with anything of the sort that went off so well.' His contrition about his hopes for the establishment of a colony was plainly transparent, however. He assured the Foreign Secretary that his own comments had been:

directed against the too sanguine anticipations of immediate good results – for, if we fail in our warmest anticipations (I am pretty confident in the *mineral* line) we shall at all events gain a real cession of knowledge respecting a large part of Africa which it would be disgraceful to ignore & at all events we shall leave behind us as good a name among the natives as cannot but be productive of ultimate good results.[169]

The ambiguity of language in this statement reflects the same logical continuity between exploration and exploitation, the same desire that scientific endeavours should become agencies for creating political influence, as Murchison was simultaneously demonstrating in regard to Baikie's activities on the Niger.

While recuperating from the Livingstone extravaganza, Murchison drafted instructions for the geologist he had recommended for the expedition. Richard Thornton was a nineteen-year-old graduate of the School of Mines who had specialised in mining and metallurgy. Considering Thornton one of the most brilliant pupils ever to pass through the School, the Director-General overrode Livingstone's objections to his youth and advised him to turn down an offer from the prestigious Geological Survey of Victoria in favour of the virgin research field of Africa.[170] Thornton's instructions likewise demonstrate the applicability of Murchison's theories to the search for economic minerals which might underpin commercial development or colonial settlement in Africa.

In the highlands beyond Tete Thornton was to determine the character and accessibility of the coal beds known to exist near the head of deep-water navigation and evaluate any deposits of iron, copper, or lead which might be discovered. Murchison's related injunction to search for limestones which 'are necessarily of prime importance for manufacturing the ores, and other uses, and are most likely to yield fossils', reveals not only an interest in palaeontology but a concern with smelting metals at source for export. As Livingstone stated, the objects of the expedition were exploration and inducing the natives to take up agricultural and industrial pursuits 'with a view to the production of raw material to be exported to England in return for British manufactures'.[171] Murchison was also anxious to discover more evidence verifying his structural prediction and to prove Silurian strata as the source of the Manica gold which had long been traded to the coast. He therefore cautioned Thornton to search for specimens of Bain's Dicynodont, the Permian reptile considered diagnostic of Africa's ancient lacustrine basin, and establish if its barrier hill range contained venous or alluvial gold.[172] The discovery of paying quantities of gold, Murchison knew, would render almost inevitable the new colony which he and Livingstone envisioned, for the forces unleashed by a gold-rush would sweep away the government's misgivings about bullying Britain's oldest ally into ceding her east African possessions.

Murchison kept alive public interest in the departed expedition by adverting to the commercial potential of the Zambezi, but he hedged the possibility of disappointing economic results by cautioning against over-optimistic commercial expectations.[173] Thornton meanwhile read *Siluria* on the outbound voyage, and by the end of 1858 sent home a report on the coal seams near Tete which was published by the Geological Society.[174] He also reported what Livingstone described as 'immense quantities of the finest iron ore' and produced an accurate chart of the Zambezi channel which Murchison publicised through the RGS and *The Times*.[175] Unbeknown to Murchison, however, Livingstone had already abandoned his optimism about the Zambezi because of the impassable Kebrabasa Rapids and begun to fall out with Thornton. His subsequent examination of coal seams on the Ruvuma River was accomplished with the aid of Kirk instead.

After navigating up the Shiré River, the first main tributary of the Zambezi, as far as the cataracts he named after Murchison, and hearing reports of a fertile country to the north, Livingstone's hopes revived. His new plan for a Scottish colony in the Shiré highlands was first sketched out in a letter to Murchison. [176] Those that followed described the geology of the region as proving the President's prediction and showing much evidence of usable iron ore and cotton-growing potential. [177] These letters were publicised in the autumn of 1859. John Crawfurd, however, true to form, attacked the colonisation suggestions by arguing that trading prospects on the Zambezi were poor. [178] Livingstone meanwhile continued to drive home his conviction that a British colony should supplant the Portuguese and informed Murchison that Thornton had been dismissed from his post for insubordination and laziness. [179] These charges were largely false, but Murchison sided with the sacrosanct Livingstone until Thornton eventually offered his own confidential explanation. [180]

Thornton then prospected up the Zambezi before joining an expedition to Mount Kilimanjaro led by the German Baron Karl von der Decken. He sent Murchison reports of these journeys, for he knew that the Director-General's good opinion was essential to his securing further government employment. Thornton examined coal deposits near Mombasa and attempted to make a topographical map of the Kilimanjaro region, but failed for lack of instruments. Through Murchison's intervention he was then reinstated with Livingstone, who was anxious to secure Thornton's charts and notes in order to have positive results to show for the nation's investment in his expedition. Thornton began a geological survey of the area selected for the Universities Mission settlement, but died during 1863. Murchison publicised Thornton's findings and planned to have all of his geological data published, but the young scientist's notes proved indecipherable and the project was abandoned. [181]

Thornton ascertained that the axes of uplift in the Zambezi valley conformed to Murchison's prognostications, but the Director-General was disappointed of the richly mineralised Silurian province he had envisoned because none of Livingstone's party reached the Manica goldfields. Several years later Karl Mauch would provide this information. Thornton, however, gained some idea of the region's mineral potential, and as he was the first trained geologist to work in east Africa his reports represented a valuable contribution to knowledge of the continent's structure. This information was also co-ordinated with data received from Joachim Monteiro, a Portuguese classmate of Thornton's at the School of Mines employed as a mineral explorer by a copper-mining company in Angola. Murchison enlisted Monteiro as a speaker on the subject of the mineral resources of the Congo basin at an RGS meeting discussing a memoir by Bedingfeld on that river's commercial potential. [182]

Until Livingstone's return to England in 1864, Murchison continued to support his efforts to establish a settlement colony in the Shiré highlands. But as the explorer disassociated himself from the Universities Mission, the RGS

President also turned against evangelical endeavour unsupported by British colonial sovereignty as an impractical method for initiating African development, and tried to free Livingstone for continued exploration.[183] During this period Livingstone supplied Murchison with further geological reports tailored to support the latter's African theory.[184] Together with Thornton's observations and fossils and pottery fragments brought home by Kirk, these data contributed to a memoir Murchison offered to the RGS in 1864. In it, he restated his structural theory and argued that the Negro race had made no progress towards civilisation even as measured on the geological time scale.[185]

When Livingstone returned to England, Murchison took him to Palmerston to discuss the slave trade. But the failure of the Universities Mission and the disappointing results of the Zambezi expedition itself had cooled interest in Murchison's greatest African lion, and Livingstone soon retired to the country to write his second book, which Murchison advised him to dedicate to the Prime Minister. Predictably, Murchison ensured that Murray included a lithograph of the Murchison Cataracts in this narrative. Flattering references to Murchison's theories and invaluable personal support were interspersed with pleas for the foundation of a British settlement in east Africa to demonstrate the ineffectiveness of Portugal's territorial claims.[186] Murchison continued to publish the explorer's letters through the RGS, excising 'all pungent political or clerical allusions' to defend his protégé's reputation.[187] During the remainder of Livingstone's stay in Britain, Murchison displayed him at the British Association, obtained for his brother Charles the consulship of Fernando Po which Grant had declined, and arranged a new consular appointment and joint RGS–government funding so that Livingstone himself might clear up the remaining mystery of the Nile source.[188] In his Anniversary Address of 1865, the RGS President advertised Livingstone's new opportunities and hinted that the troubles encountered by the Zambezi expedition could have been obviated by Britain's establishment of a protectorate over the region.[189]

Following Livingstone's departure, the President again used the explorer's letters to maintain interest in his activities, and once news ceased arriving, steadfastly countered all rumours that he had died. In 1866 Murchison secured John Kirk's appointment as Vice-Consul in Zanzibar to lend support to Livingstone and locate another scientific observer in Africa.[190] Two years later he used the Zanzibar connection in another attempt to further exploration and imperial interests when he helped arrange a royal reception for envoys of the Sultan of Zanzibar who were seeking the status as a British protectorate which Kirk negotiated in all but name in 1873. Murchison invoked Palmerston's memory to urge Clarendon to seize an opportunity he himself considered 'of the *highest importance to our country*', as well as entertaining the envoys at the RGS.[191]

In 1867 Murchison was also instrumental in despatching an expedition to find and resupply Livingstone. Thomas Baines, the artist unfairly dismissed by

Livingstone from the Zambezi expedition, was favoured by members of the RGS Council to lead the new mission. Murchison, however, angered by Baines' exposure of the national idol's less than heroic behaviour in this affair, chose instead another Zambezi veteran, Lieutenant Edward Young, RN. At the same time, he was displaying Baines' paintings at geographical soirées to which the artist was not invited.[192] Murchison's exploitation of Baines, like his initially uncompromising stance towards Thornton, demonstrates that he had a blind spot where Livingstone was concerned. He felt a stronger identification with Livingstone than with any of his other field agents, and treated as personal attacks all aspersions cast on the man whose reputation he himself had largely created. The government's willingness to fund Young's expedition and a second search was as much a tribute to Murchison's prestige as it was an expression of national sentiment for the lost hero: Murchison literally blackmailed Clarendon into financing the 1870 operation with threats of adverse publicity.[193]

In a sense, Livingstone was a projection of Murchison, for both men attempted in their own ways to bring order from what was seen by Europeans as primeval chaos. In Geikie's words:

Murchison's labours among the older rocks stood indeed to geology in a relation not unlike that which his friend Livingstone's work in Africa bore to geography. Round these rocks there had gathered some share of the mystery and fable which hung over the heart of Africa. And he dispelled it not by intuitive genius, but by plodding and conscientious toil, directed by no common sagacity, and sustained by an indomitable courage.[194]

Murchison's loyalty to Livingstone during the explorer's last quest, like his unyielding adherence to outmoded geological views, may also be seen as a symptom of his refusal to surrender to the forces of change and decay. He admitted in fact that the unfulfilled hope of welcoming Livingstone home a third time prolonged his own life.[195]

Abyssinia and the Sudan

The quest for the sources of the Nile also revived Murchison's interest in Abyssinia. Sir Samuel Baker's 1862 RGS memoir on his exploration of the Nile's Abyssinian tributaries, for example, contained geological observations and descriptions of copper and lead ores of interest to the President.[196] In 1864 Baker had shifted his attention south-westward while Murchison was approaching the Foreign Office with his own scheme to extend Egyptian authority to Lake Victoria. The following year news was received of Baker's discovery of Lake Albert and Murchison Falls, the cataract which broke the flow of the White Nile between Lakes Victoria and Albert. After accepting Lord John Russell's congratulations on behalf of the Foreign Office, the President announced Lake Albert to the public as a second feeder of the Nile in such a way as to dampen further acrimony over the river's ultimate sources.[197] Baker was given a triumphal reception at the RGS upon his return. Murchison was thrilled with the explorer's descriptions of the lake and falls, for they offered not only another confirmation of his African structural theory, but evidence to support his contention that glaciers were incapable of gouging out deep lake basins in solid rock. Contemporary discoveries by British scientists working in Canada, British Columbia, New Zealand, and Tibet were likewise providing data which bore upon this important geological question.[198] Revealing the mental yardstick he used to reckon the magnitude of discoveries, Murchison emphasised that Baker's Lake Albert, like Speke's Victoria, was 'estimated to be about as long as Scotland'.[199]

Baker remained the great geographical lion throughout 1866. He appeared before the British Association and, through Murchison's influence, was elected to the Athenaeum Club and granted a knighthood.[200] Baker was also introduced to A. H. Layard of the Foreign Office in order to provide advice about the plight of the British consul imprisoned by King Theodore of Abyssinia and the related problem of the vacuum of authority in the southern Sudan.[201] Baker longed to return to the upper Nile and begin developing the region according to a plan much like Murchison's. He explained to the President early in 1867 that central Africa could only be opened to European influence by 'annexing to Egypt the equatorial Nile Basin'. This was to be accomplished by an Anglo-Egyptian expedition led by Baker and consisting of twenty vessels and a thousand troops supplied by the Khedive. Baker believed 'the fact of an Englishman commanding would give us a powerful influence in Egypt hereafter', and he advised Murchison that Britain should support Egypt to counter Turkish and French influences in the region.[202]

Such plans were superseded by the decision to send a military expedition to Abyssinia under the command of General Sir Robert Napier. Murchison, remembering that his advice regarding the Crimean invasion had been ignored, immediately proposed that a scientific staff accompany this force. The Foreign Secretary, Lord Stanley, agreed to support this scheme.[203] At the meeting of the British Association in September, Baker presided over the Geographical Section with Murchison as his second. The President's address stressed the crucial role exploration played in national expansion, discussing the effects of the Suez Canal on the defence of India and urging the government to attach scientists to the Abyssinian expedition.[204] Murchison was 'sorely put out' by Baker's remarks that the long-absent Livingstone had probably perished and that Canada and Australia should achieve independence, but he delivered a stout defence of Livingstone which swayed public opinion in favour of his own view that the explorer remained alive.[205]

Murchison was soon informed by Sir Stafford Northcote, the Secretary of State for India, that the Governor of Bombay was appointing a botanist, a meteorologist, and a geologist to accompany Napier's force. Another geologist with Australian experience was also requested, but though the authorities seem to have been thinking in Murchisonian terms of possible gold discoveries, this appointment was never made.[206] Murchison did succeed in securing a geographical post for Clements Markham, Honorary Secretary of the RGS and Geographer to the India Office. Markham's unique appointment owed something to his connections with the India Office, but more essentially it recognised the status to which Murchison had raised geography as an instrument of imperial policy. This is illustrated by the private advice offered by George Hamilton of the Treasury. He suggested that Murchison formally propose Markham's appointment in order to 'give him what you may consider a better position, as sent out by the R.G. Society with the concurrence of the Imperial Govt., than if sent out under the Indian Dept., as an attaché to the Expedition'. Possibly recalling how even the scanty scientific findings of the Zambezi expedition had permitted Murchison and the government to claim some practical benefit from an expensive mission, Hamilton also confided that as much scientific information as possible should be gathered in order to 'perhaps attain some return for a huge outlay'.[207] The scientific community divided in its reaction to the expedition. Archaeologists of the British Museum requested Murchison's aid in appointing a collector of antiquities,[208] but Joseph Hooker of Kew Gardens considered the affair a 'fiasco' and refused to involve himself in the choice of a plant collector despite Murchison's encouragement.[209]

To take advantage of public interest, the RGS organised a meeting in November 1867 to examine the geographical aspects of Napier's expedition. Murchison spoke about the army's route and water supply, the Abyssinian explorer Charles Beke discussed the likelihood of disaffected tribes aiding the British advance, and Henry Rawlinson advocated the establishment of permanent sanitaria in the highlands for the recuperation of Britons serving in the East. The meeting ended in consideration of what should be done if King Theodore fled his capital of Magdala, Murchison having permitted the proceedings to stray far into the realm of politics.[210] Baker had meanwhile made another proposal to the Foreign Office to lead a flying column from the Sudanese port of Suakin to rescue Theodore's European captives, but when this was rejected he fell back on a modification of his Nile scheme which Murchison then presented to the RGS. The new plan involved persuading the Khedive to underwrite an expedition to launch a steamer on Lake Albert in order to explore its southern shores and search for Livingstone.[211]

While Napier's army approached Magdala in early 1868, Murchison chaired another 'Abyssinian night' at the RGS to consider two memoirs Markham had produced. Once again the discussion waxed political, with Baker and Rawlinson dominating the floor and Northcote representing the government. Baker advocated Britain's retention of all Abyssinia, while Rawlinson preferred a mere coastal enclave. The embarrassed Northcote was forced to counter these views with an assertion that, once the captives were rescued, 'our forces would be withdrawn, and no other consequences would follow.'[212] Markham would later claim that increasing feebleness made Murchison disinclined in these years to resist opposition within the Society. But Murchison was hardly senile, even in 1868 – Markham himself admitted finding him 'still flourishing' upon his own return from Abyssinia.[213] Judging from his countenancing of other proposals for British annexations in Africa, it is probable that he permitted this political disgression because he approved of its direction. He made little attempt to quell Baker's excesses because he was simultaneously promoting his scheme of Anglo-Egyptian conquest; he chose not to rein in Rawlinson because Abyssinia, unlike Central Asia, did not call into question his own special relationship with Russia.[214]

This extraordinary outburst of imperialistic rhetoric did not go unnoticed. Viscount Strangford, a representative of the Asiatic element at the RGS, criticised the politicisation of the Society's proceedings in a slashing article in the *Pall Mall Gazette*. Strangford had his own reasons for taking the leading players in this affair to task, however. He opposed Rawlinson for his excessive belligerence in imperial affairs, Baker for his condemnations of Ottoman rule, and Murchison for having so concentrated upon Africa that an explorer's only hope of winning attention remained to 'go off to the Nyanza with a Fletcher rifle and Sir Roderick's blessing'.[215] Strangford taxed the geographers for beguiling Northcote 'by making their great fountain of honour play liquid butter before him, over their absent secretary . . . insomuch that the minister was fain to imagine that butter rather than research was the final cause of the Presidential machinery of that learned body'. This quite accurate assessment of Murchison's promotional style was followed by criticism of Baker for advocating the annexation of Abyssinia in order to pre-empt an Egyptian take-over, of Rawlinson for predicting the retention of the country's Red Sea ports, and of Murchison for subjecting Northcote to such an ordeal. Strangford then described the encounter between Rawlinson and Baker in fine imagery:

It was not in nature for Sir Henry, who has a great deal of the wild elephant in him, to see Sir Samuel rushing and trumpeting about, and crashing down everything, with tusk and trunk, exalting in his strength, without eagerly accepting the challenge of the rival monarch of the forest. When these two huge male geographers were in the full shock of their collision it would not have been possible to call them to order; nor would it have been safe for even Sir Roderick, that skilful driver, to have handled them like tame elephants and dug his iron hook into the brawny corrugated napes of their great bull necks, trying to coerce them.

Fearing that the rhetoric of the two contestants would appeal 'to the mischief-making propensities of foreign quidnuncs on the *qui vive* for signs of British annexationism', Strangford once more advised that discussing politics was improper and dangerous for the geographers.[216]

Following the storming of Magdala, Murchison was approached by his Russian friends for a share of the scientific loot from the expedition,[217] and the Berlin Geographical Society telegraphed its congratulations 'for this new

success of British valour, benefiting geographical science'.[218] As Hamilton
had predicted, the scientific results of the campaign were winning international prestige and helping allay suspicions of British acquisitiveness.
Indeed, the campaign was viewed as another crusade in the cause of
civilisation, and Murchison, as the originator of the scientific corps, reaped
much of the credit. In his Anniversary Address for 1868, Murchison expressed
his astonishment at the public's apathy towards the expedition, which had
proved that Europe could march 'a scientifically-organized army into an
unknown intertropical region'. The scientific results of the expedition, he was
convinced, would ensure it a worthier place in history than 'many a campaign
in which greater political results have been obtained, after much bloodshed,
but without the smallest addition to human knowledge'.[219] To extract as much
publicity as possible from the episode, Murchison then had Markham's final
reports presented before the RGS and the British Association.[220] He also
organised the election of Sir Robert Napier to the RGS and invited one of the
Magdala captives to address the Society.[221] William Blanford, the School of
Mines graduate from the Geological Survey of India who had been the
expedition's official geologist, likewise produced memoirs for the Association
and the Geological Society before publishing his complete results in 1870.[222]

Baker had meanwhile been appointed Governor-General of the equatorial
Nile basin by the Khedive. He approached Clarendon to support his
expedition to occupy the upper Nile, but the Foreign Secretary declined
because of the cost and the diplomatic awkwardness of treating independently
with an Ottoman vassal.[223] Murchison's Anniversary Address of 1869,
however, left no doubt that Baker's 'grand project' was to be regarded as a
British enterprise.[224] Indeed, Sir Thomas Fowell Buxton, grandson of the
anti-slavery crusader, had to be corrected by Murchison for his belief that
Baker was directly employed by the British government.[225] Murchison also
helped elect Baker to the Royal Society,[226] and personally saw off his flotilla of
boats from London. In the spring of 1870, prevented by ill-health from
accepting the Khedive's invitation to attend the gala opening of the Suez
Canal, Murchison deputed the MP Lord Houghton as his ambassador to
express the thanks of the RGS to Ismail for aiding Baker.[227] Murchison
continued to publicise Baker's progress, but he died without the satisfaction of
announcing Baker's annexation to Egypt of Equatoria or his attempts to
acquire the kingdom of Bunyoro.

Murchison used the opportunities provided by British involvement in Africa
to expand the frontier of metropolitan science. He simultaneously schooled the
official mind in the belief that science could serve the drive for national
expansion by evaluating the commercial capabilities of new environments and
analysing their fitness for colonisation. He promoted many other British
expeditions organised on a smaller scale than those described here, as well as
providing valuable backing for several important foreign explorations. Since

he occupied a central position in the network of individuals and organisations which constituted informed opinion on Africa in Britain,[228] Murchison's views on the physical conservatism of the continent helped provide scientific validity for the concept of Africa as a 'lost world' whose floral, faunal, and human inhabitants had remained for ages in the same primeval state. These ideas, as expressed in Murchison's own writings and the narratives of explorers, meshed with the growing body of ethnological literature to explain African primitiveness to the progress-obsessed Victorians and justify their efforts at improvement. In the second half of the century the Dark Continent was to replace the Antipodes as the premier laboratory of archaic conditions.[229] Here, more intensely than anywhere else, the explorer acted out the European longing to be challenged by nature in a wild and exotic setting. In so doing, he simultaneously verified the superiority of European civilisation, opened new frontiers for expansive capitalism, and provided an outlet for emotional impulses stifled by industrialisation and urbanisation.

Murchison's African activities, like his efforts to promote Australian exploration and development, represented an intensification of the tradition established by Banks and Barrow. In directing Britain's mid-century assault on Africa, he frequently appeared to have succeeded in making the RGS an independent source of national policy towards the continent. His ability to realise the colonising schemes to which the geographers were often led by their enthusiasm to apply the fruits of discovery, however, was ultimately constrained by the government's unwillingness to incur the necessary costs and risks of such a forward policy. Murchison's scope of action was most extensive during Whig administrations because of his friendship with Clarendon and his ability to tailor his goals and methods to the policies of Palmerston.

Murchison played a key role in organising nearly every British expedition to enter Africa during the two decades between 1850 and 1870, and in evaluating their results. His geological theory regarding Africa's structure, informed by the desires to extend his Silurian empire, score another scientific prediction of continental scope, and lay the foundations of a series of new colonies around the rim of the continent, also influenced the routes of explorers. The rivers upon which they focused offered not only potential commercial highways, but natural geological sections which could provide clues to the location of exploitable minerals. In turn, the activities of the explorers and the promotional machine to which they were geared created the public anticipation, actual information about resources, and climate of scientific and commercial rivalry which set the stage for the *fin de siècle* partition of Africa, and likewise affected the pattern of British annexation in the continent.[230]

Murchison took advantage of the random probings of many explorers to add detail to his developing picture of Africa's physical configuration. Most of the major British expeditions mounted during the period of his ascendancy, however, were planned in part to test his structural theory and its implications for the discovery of economic minerals. As was the case with his Australian prophecy, Murchison's African prediction proceeded from several inaccurate premises, but it demonstrated high powers of inductive and deductive reasoning. He not only synthesised a wide variety of isolated data into an apprehensible theory of the configuration of a major portion of the earth's crust, but went on to derive important economic and political consequences from that explanation. Murchison's formulation proved substantially accurate for the same reason it caught the imagination of his contemporaries – the King of Siluria was the only geologist of his day with the breadth of vision to think in terms of entire continents. For all his shortcomings, Murchison had in this regard inherited the mantle of Humboldt.

8

The architect of imperial science

British natural scientists in the nineteenth century, led by a few farsighted and aggressive commanders such as Murchison, successfully exploited the burgeoning opportunities for research created by the nation's expansion overseas. As with other professional groups, the desire to establish secure career structures impelled scientists to seek their share of political, social, and economic power. At the same time, utilitarian logic encouraged them to experiment with deriving practical benefits from their work. In order to win public support, therefore, scientists pressed for recognition of the applicability of their disciplines to a broad range of technical and social problems.[1] The empire – conquered, but yet awaiting systematic exploration and exploitation – offered an obvious arena for such demonstrations of utility, and to a savant of Murchison's militaristic mind it must have seemed particularly inviting.

While the scientists won access to widening career options and new data, the imperial government gained accurate information for administering and developing its sprawling possessions. The commercial, industrial, and financial communities meanwhile received reliable reports of new products, markets, sources of supply, and investment opportunities both within and beyond the empire. The public at large gained for the first time an authentic, though culturally conditioned, vision of the diversity of environments and cultures into which their representatives were prying. Science also contributed to the emerging ideology of imperialism which justified British rule over distant lands.

In serving these functions, British scientists performed according to the role of intellectuals defined by Antonio Gramsci – that is, as agents and propagandists for the ruling class. Murchison's efforts to expand his personal power base, to increase the authority of the scientific organisations with which he was associated, and to serve Britain's national interests on a global scale exemplify, at different levels, Gramsci's doctrine of hegemony. This theory postulates that a ruling class maintains its supremacy not only by direct economic and political control, but through cultural hegemony which sanctions its social and moral authority. As an increasingly important aspect of culture, science represented a new resource commandeered by hegemonists. Once he had mastered the principles and programme of science as conducted within the traditional British ideal of the gentleman amateur, Murchison worked assiduously to integrate it with the larger cultural, political, and economic systems which likewise served to perpetuate the established order. As a scientist and an administrator of science, Murchison ranks in Gramscian terms as both a 'traditional' and an 'organic' intellectual.[2]

Murchison's diverse activities spanned the economic and cultural spheres. But for one who identified himself completely with Britain's governing elite, his advocacy of national expansion and the use of science in the service of imperialism was as natural as his earlier military career and conservative political efforts. Indeed, Murchison's evolution – from landowning army officer, to bondholding country gentleman, to metropolitan scientist, to bureaucrat and publicist of empire – typifies the alliance forged in the interests of maintaining hegemony between the traditionally dominant landed class and the new financial, commercial, industrial, and professional elites during the early nineteenth century. Significantly, too, Murchison had little religious faith beyond the natural religion he had imbibed as a geologist, though he defended the established Church 'as a great and essential moral engine'.[3] Science and patriotism seem to have largely supplanted traditional Christian belief in his personal creed. This secular theology was singularly appropriate to the era which witnessed the rise of modern capitalism and the nation-state representing its political embodiment.

Murchison's science thus contributed to the hegemony of his own class at several levels. Culturally, it helped perpetuate the belief system of the landed elite while simultaneously propagating a new action-oriented ideology tuned to the needs of the emerging capitalist elite. Economically, it directly supported the preponderance of both groups, aiding the search for industrial minerals in Britain while discovering new markets, sources of treasure, and supplies of raw materials abroad. Activists such as Murchison were crucial in integrating new fields of opportunity as well as new means of control into the existing hegemonic system. In this respect, too, empire represented an extension of Britain, for in the colonies as in the home islands scientists were concerned with uniting theory and practice to engage and transform the physical world in the interests of perceived social needs.

Murchison's scientific activities were in fact multifunctional, using knowledge, in Steven Shapin's phrase, 'to legitimate structures in the wider society' while simultaneously employing it 'for the prediction and control of phenomena'.[4] By purveying science to the imperial government on both levels, he helped to make science a state instrumentality. By ratifying this interdependence, Murchison transformed the meaning of science and also contributed a new dimension to the concept of empire as a powerful emotive symbol available to Britain's rulers.

Science, as an aspect of British culture not dissimilar in this respect to the missionary movement, was thus both a beneficiary of imperial and commercial expansion and an integral part of that process. Murchison played a crucial role in forging this symbiotic relationship. He found expression for his own ambitions and will in the nation's outward-looking *Zeitgeist*, helping to channel it into the organised activities and styles of thought which came to characterise imperialism in its culminating phase. In the larger view, Murchison's efforts to promote imperial resource investigation as well as reconnaissance outside the empire represented one facet of the general wave of scientific exploration that occurred during the nineteenth century. Specifically, his activities contributed to the cultural consolidation and economic development of Britain's self-governing white colonies; to the administration and improvement of dependencies believed to lack the capacity for political independence; and to the exploration of territories outside the area of formal British control whose commercial potential might be developed in coordination with the imperial economy.

Murchison's public remarks, private observations, and tolerance of annexationist political discussions at the RGS demonstrate that he remained throughout his career a staunch advocate of the extension of British influence along Palmerstonian lines. His competitive instincts and deep-seated suspicions of French and American aggression frequently overcame his professed scientific internationalism, but he was astute enough to realise that open espousal of territorial acquisitions would compromise both his own ostensibly apolitical stance and that of his institutional instruments. Given Britain's unrivalled ascendancy during Murchison's era and her consequent capacity to achieve most objectives abroad through informal means more compatible with the dominant economic philosophy of *laissez-faire* than were assumptions of formal political control, there was in fact little need to urge or adopt such a policy. At times, however, when he felt Britain's imperial interests could be guaranteed in no other way, as in the cases of the Zambezi highlands and the islands fronting northern Australia, Murchison passed beyond mere encouragement of the extension of informal influence to advocacy of outright annexation. In these cases it must be admitted that his opinion had no appreciable impact on the policy makers' calculus of the costs and benefits involved in extensions of sovereignty.

Murchison's greatest significance for the history of empire lies in his role as a promoter of cultural imperialism. While concepts of empire constituted an important influence upon his science, his personal identification with Britain's prestige as a world power caused him to encourage the exportation of British 'know-how' in the form of scientific expertise, organisational modes, and new technologies both in his private capacity and through his direct links with the state. Murchison's *modus operandi*, based on patronage, influence, and mediation of the global flow of information, worked most advantageously in existing colonies whose administrative and cultural traditions facilitated the installation of metropolitan institutions and personnel. But his impact was similarly felt outside the empire in both developing nations and unorganised regions. His career demonstrates that a broader definition of imperialism than that confined to the territorial extension of political control[5] is required to encompass such aspects of cultural expansion.

Murchison's ramified influence both in and outside government also demonstrates that rigid historiographical conceptualisations of the structure of authority in imperial policy making require modification. His capacity to

Plate 7 Murchison chairing the Geographical Section of the British Association, 1864

operate through the interstices of the civil service suggests that further research on the growth of the imperial bureaucracy, pursued either through area studies such as Roy MacLeod has initiated in regard to Indian science,[6] or more holistic descriptions of the evolving power and organisation of individual departments such as Oliver MacDonagh's pioneering account of the Colonial Land and Emigration Commissioners,[7] would prove worthwhile.

Geology

Murchison's position as Director-General of the Geological Survey gave him authority over most aspects of mineral reconnaissance and a direct access to the cabinet which he never fully relinquished to his actual superior, the Secretary of the Science and Art Department. Like the Hookers at Kew Gardens, Beaufort at the Hydrographical Department of the Admiralty, and Sabine through the facilities of the Kew Observatory,[8] he constructed a global arena for self-fulfilment in the vacuum created by the imperial government's failure to establish a central department to manage official scientific enterprise and provide advice on technical and policy issues. In a sense, these men were all partial inheritors of the unique position which Joseph Banks, the premier 'scientific imperialist', had created for himself in the last two decades of the eighteenth century.[9]

The loosely knit structure of the mid-Victorian civil service and the reluctance of both political parties to expand the sphere of government authority allowed Murchison to build on the foundations laid by De la Beche and elaborate a mutually advantageous relationship between the Geological Survey and the Colonial, Foreign, India, Admiralty, and War Offices. Murchison provided a range of services to these departments in exchange for research opportunities throughout the world, despite his own bailiwick's subordination to a domestic hierarchy with no authority in imperial affairs. The imperial focus which Murchison brought to the Survey during the high tide of British commercial expansion can be viewed as an historical parallel to De la Beche's meshing of scientific goals with his social and political interests during the preceding era of reform.[10]

Murchison managed to maintain the integrity of this bureaucratic edifice by sheer force of personality during his own lifetime, but it was dismantled afterwards as a result of shifting official priorities and corresponding changes in governmental organisation. In 1873 the Royal School of Mines was reorganised when the courses in chemistry, physics, and natural history were transferred to South Kensington as the nucleus of the new Science Schools. The entire faculty was finally incorporated into the Imperial College of Science and Technology in 1907. The old Survey establishment was further dismembered in 1883 when the Mining Record Office moved to the Department of Inspectors of Mines at the Home Office.[11] The Museum of Practical Geology continues to serve as the national geological reference centre. Rebuilt as part of the South Kensington complex in the 1930s and renamed the Geological Museum, it has now been amalgamated with the British Museum (Natural History). Outside government, Murchison simultaneously maintained overlapping power bases in several scientific institutions which he and other statesmen of science had upgraded into advising and co-ordinating agencies with considerable quasi-official authority. Men of apparently minor significance in the process of formulating imperial policy sometimes thus achieved the status of major players by manipulating official departments and private organisations whose functions complemented those of government.

Murchison offered overseas career opportunities to many young explorers and geologists who thereby became foot soldiers in the Director-General's worldwide campaigns. As his extraordinary physical vigour diminished, he similarly relied on younger colleagues to act as 'light infantry' during European field operations. One reviewer of *Siluria* remarked that 'Sir Roderick is as essentially a general in science as Napoleon or Wellington was of troops', and Edward Forbes observed that he was 'a true Highlander in this respect, holding that the glory of a chieftain lies in the number and power of his clan'.[12] Murchison's grasp of global geology, his ambition to score further scientific triumphs of continental scope, and his complementary offices perfectly qualified him to command this corps of investigators and marshal their results for the simultaneous benefit of Britain's imperial interests and his own reputation. In the process of extending the reach of British science, these men gave testimony to Murchison's unique role by applying his name to twenty-odd features throughout the world.

Within the service of the empire, Murchison created a shadow sovereignty of patronage and reciprocal obligation. He was particularly keen to promote the careers of fellow Scots. While his countrymen played prominent roles in imperial administration, defence, medicine, and science, Murchison acted as an important metropolitan connection in their network for mutual advancement.[13] The Murchison family exemplified the Scottish imperial diaspora, and Roderick himself remained keenly interested in both Highland emigration and the drama of Scots reclaiming ruined patrimonies with fortunes won in distant climes.[14] His influence at the Geological Survey also extended to the institutional level. Not only was the Royal School of Mines the prototype for a similar school in the colony of Victoria, but mining academies at Berlin, Vienna, and Harvard and Columbia Universities were set up, in Murchison's words, 'copying the very scheme and syllabus of our own establishment'.[15] The Italian Geological Survey was also founded with Murchison's advice on the model of the British Survey.[16]

The London scientific societies, as venues for the publication and validation of the work of colonial scientists, served as monitoring devices whereby British savants culled data to incorporate into their own synthetic theories. The original scheme of the Geological Society for co-operative research by metropolitan theorisers and provincial fact-gatherers[17] was thereby perpetuated on an imperial scale, the London scientific elite using the labours of colonial field workers to help maintain its ascendancy. As was demonstrated by the radical difference between the King of Siluria's treatment of W. B. Clarke, who openly opposed his views, and other colonial geologists such as Bain, Logan, Sutherland, Hector, Haast, and Selwyn who either supported his theories or deferentially suggested their modification, Murchison's central position in London's scientific establishment permitted him to make or mar the careers of many colonial scientists. His changes in attitude towards Grant and Thornton – from disillusionment before they made substantial discoveries to enthusiasm afterwards – his grudge against Thomas Baines, his distaste for individualists like Burton or explorers who strayed from the accepted norm of national service such as Speke in his final phase, suggest that in the field of geography, too, Murchison used his authority to enforce conformity. In a sense he was driven to do so, for the popularity and prestige of the RGS depended on the marketing of practical results and standardised heroes.

Though he generally displayed tact and discretion in exercising his omnipotence, it inevitably provoked resentment. The Astronomer Royal George Airy, for example, cited Murchison in complaining of 'the influence which one or two men of science can exert beyond all legitimate limits on the society of West London'.[18] J. B. Jukes, too, almost certainly had the Director-General in mind when he publicly protested against scientists who 'pandered too much to the utilitarian quackery of the age' and thereby attained 'far wider reputations than the real men of science', being 'named knights, or labelled with C.B.', and accepting 'other crumbs that might fall from the table of the politically great and powerful'.[19] Similar criticisms had been levelled at the imperial implications of the King of Siluria's eagerness to annex foreign territories to his stratigraphic realm.[20]

In the main, British scientists appreciated Murchison's invaluable services in furthering their claims for public recognition and patronage, and were willing to tolerate his vanity and elitism. But even in the colonies, whose scientists, excepting the prominent contingent of Germans, were predominantly British emigrants, practitioners of metropolitan stature who felt their reputations and research were being traduced by Murchison successfully revolted against his autocracy. In the case of the dispute over the long-term profitability of Victoria's gold-mines, where no transcendent criteria existed to judge the outcome of deep-level mining which was at once an experiment and a speculation, Murchison's climb-down, like the positions adopted by Selwyn and Barkly, was probably conditioned by economic and social interests – that is, desire to facilitate colonial resource development financed by British capital investment.

Despite such exceptions, Murchison's dominant influence on honours, publication privileges, fossil purchases by public museums, book sales, and reputations in general constituted an indirect system of control which, together with his official patronage of geological survey and academic appointments, effectively subordinated colonial science to the imperial metropolis. The lines of scientific authority in the empire thus paralleled those of political authority for much of the nineteenth century. In all the colonies of white settlement, the growth of indigenous scientific institutions became an important part of the nation-building process. As Roy MacLeod has emphasised, British science like British politics accommodated itself to colonial aspirations for self-determination by a policy of devolution and assimilation aimed at consolidating the dominance of metropolitan interests.[21] Murchison's efforts to link the colonies organically with the mother country illustrate this policy of envelopment at both the cultural and political levels. His patronage of colonial scientists and his sponsorship of colonial members in the Athenaeum Club complemented his support for colonial representation in Parliament and his attempts to throttle talk of colonial secession.

The overseas activities of the Geological Survey rose to a peak under Murchison. De la Beche had accomplished a great deal abroad, but his successor transformed the Survey into a clearing house, operating, in conjunction with the RGS and the Geological Society, to funnel scientific and economic data into Britain.[22] Many colonies anxious to attract immigrants and investment capital founded geological surveys during this period in order to discover and evaluate mineral resources and provide the data necessary to regulate their private exploitation. This significant, if limited, intervention by government in the economic process represented an attempt to systematise the pace of colonial development. Surveys were usually requested after initial discoveries, but the propaganda of metropolitan geologists went far towards preconditioning the official mind to the need for such undertakings. Colonial surveys served to bring under state control and organise according to the metropolitan formula the amateur research already initiated in many cases by Fellows or correspondents of the Geological Society.

Geoffrey Blainey has argued convincingly that the timing of mineral strikes in the nineteenth century was not random, as contemporaries believed, but varied with the short-term oscillations of local economies. The cases of Queensland, Cape Colony, Natal, and Jamaica demonstrate that the founding of geological surveys often followed this pattern, although intercolonial rivalry seems generally to have been a greater stimulus in the Antipodes while competition with the United States influenced the Canadian decision. The duration and scope of surveys varied with local circumstances; mineral priorities varied with the actual resources available and, again, with the rhythms of colonial prosperity. In both Australia and New Zealand, gold tended to be more eagerly sought for and discovered in slumps, while the cry was invariably for coal during booms.[23] Except for an obvious emphasis on precious metals, industrial raw materials took precedence as in Britain. Exploitable deposits of coal, iron, copper, tin, and lead were critical to the eventual development of transport systems and industry, but in the short run they provided valuable export earnings for nascent colonial economies.

Mineral resources offered the quickest means of fostering native capital formation, generating government revenues from the sale or lease of claims, and attracting foreign investment. Gold remained a priority because it offered an instant solution to these chronic deficiencies, especially after the bonanzas in California and Victoria transformed opinion about the undiscovered quantities available. The Victorian rush in particular captured the British imagination by proving a colony could mature almost overnight on its golden profits.

The instructions Murchison gave to Hector, Bauermann, Baikie, and Thornton, as well as his hopes concerning Tasmania, northern Australia, New Zealand, China, and Borneo, illustrate his fixation with discovering new El Dorados to match the one he had predicted in New South Wales. Further gold-rushes would triumphantly vindicate his geological theories, prove the indispensability of science for inaugurating the rational exploitation of new lands and, by attracting sufficient population, ensure the consolidation of existing colonies or the foundation of new ones. De la Beche, in contrast, did not provide the geologists he sent abroad with specific theoretical suggestions since he had no personal commitment to the extension of British stratigraphic nomenclature or theories of mineral occurrence to test.

As industrial technology evolved during the nineteenth century and metal production rose as a share of total manufacturing output, economic dependence on mineral resources and knowledge of the earth's structure intensified. Both the variety and quantity of minerals consumed by industry grew at an accelerating rate, especially after 1850. When the more easily accessible supplies of the scarcer ores were exhausted, production costs rose, stimulating exploration for new and cheaper supplies, and a number of peripheral production centres successively emerged to feed the voracious North Atlantic markets. The pattern of world trade and the international movements of labour and capital were significantly affected by these regional shifts in the centres of supply.[24]

Geological surveys formed part of this pattern, and their researches profoundly affected the pace and nature of colonial economic growth. Surveyors' itineraries and priorities were partially predetermined by previous discoveries, but their explorations affected the timing and location of subsequent mineral booms which, in turn, helped define permanent patterns of development. As surveying proceeded, theoretical problems such as the age of coal deposits and the distribution of gold were revealed, whose solutions governed future successful prospecting for economic minerals. As the work of Williams and Fleming in India, Dieffenbach in New Zealand, Bain in Namaqualand, Sutherland in Natal, Poole in Turkey, and Thornton in Portuguese East Africa demonstrates, geological surveyors also made important recommendations regarding the construction of roads, railways, and harbours required for efficient mineral exploitation as well as for the deployment of river steamers. The recommendations made at second hand by

Richard Taylor and De la Beche concerning mining leases in the Queen
Charlotte Islands also illustrate how geologists working in the context of
empire focused their skills on establishing the resource exporting capabilities
of new hinterlands being integrated into the world economy centred on
Britain.

The connection between mineral exploration and transport planning was
inescapable, for given the international mobility of labour, capital, and
entrepreneurial skills, the key factors governing the initiation of development
were an increased knowledge of natural resources and the availability of cheap
transport to provide access. Since the progressive and technology-intensive
mining industry was 'potentially . . . the most effective form of production for
diffusing technical skills and raising the level of labour productivity',[25] its
introduction into colonial economies could exercise a multiplier effect,
stimulating the growth of related sectors. Its impact, however, was rarely
realised except in colonies of white settlement because of the industry's reliance
on imported expertise and its failure to upgrade the skills and wages of
indigenous work forces. The impact of science, too, was diffused from the
surveys through the cultures of settlement colonies by the foundation of
museums and laboratories, the encouragement offered colonial researchers, the
training given local assistants, and interaction with educational institutions.

From the vantage point of Whitehall, Liberals and Conservatives alike
approved of geological surveys. By the application of knowledge at very slight
cost, they 'improved' dependencies by discovering new resources which
encouraged emigration, created markets for British goods, and provided the
mother country with new sources of supply. In the colonies, however, the view
was often very different. Since the economic, much less cultural, benefits of
science were less well understood by most colonial legislators, De la Beche,
Murchison, and their appointees were constantly forced to emphasise geology's
utility in order to secure appointments and maintain funding. In Canada and
New Zealand, Logan and Hector successfully managed to balance their
scientific ambitions against the economic demands of rapidly growing colonies
and use their surveys to carry out vast programmes of miscellaneous research.
In India, Oldham's survey flourished equally under the autocratic patronage of
the Raj, and in Natal Sutherland prospered on a smaller scale under the aegis of
his Surveyor-Generalship. But in the Australian colonies, Newfoundland, the
West Indies, and the Cape, the geological surveyors all came to a parting of the
ways with their official patrons because disciplinary goals could not be
reconciled with the financial imperatives of colonial societies.

Because of preoccupation with social and theoretical issues, several historians of science have argued that British geology developed during the early
nineteenth century independently of the needs of the mining and manufacturing industries.[26] Indeed, it has been established that a comprehensive
stratigraphic column – the central accomplishment of British geology during
its 'golden age' – was elaborated by the savants of the Geological Society as an

intellectual exercise with only slight input from the mining industry. Other research, however, has demonstrated that the larger pattern of interrelationships between the science and the economic sphere, in Britain as in continental Europe, was more extensive than this narrow and paradoxical interpretation admits. From the late eighteenth century, in fact, geology played an increasingly important role in exploration for the coals and metallic ores required by the industrial revolution as miners' traditional empirical rules proved inadequate to this task.[27]

Murchison had been closely concerned with the practical bearings of his own research upon mineral exploitation in both the home islands and the colonies from the early stages of his career, and he never ceased reiterating and demonstrating the usefulness of earth science. His acceptance of the post of Director-General of the British Survey was motivated in part by desire to apply the accrued wisdom of geology to the economic development of the nation. Once in office he certainly accomplished this goal, but in a distinctive and highly characteristic way which has been repeatedly misunderstood by historians who have accused him of neglecting economic geology at the Survey in favour of field research.[28] Some evidence would seem to support such an interpretation, however. In 1869, for example, despite his sympathy for the commercial importance of the undertaking and his earlier completion of a survey of Britain's iron ore deposits, Murchison refused to comply with a joint request from the Iron and Steel Institute of Great Britain and the British Association to conduct a follow-up evaluation of the nation's haematite reserves on the grounds that, without additional financing, such a specialised operation would jeopardise the Survey's regular duties. These low phosphorous ores were required for the new Bessemer process of steel manufacture, and with the patent due to expire the following year, rising demand was expected to overreach Britain's limited proven deposits.[29]

Such strictures of Murchison's performance fail to take into account three important factors. First, his primary domestic responsibility was to complete the mapping of the home islands – itself a task fundamental to the rational exploitation of mineral resources, as the repeated requests of Murchison's landowning friends for advice regarding the wisdom of opening mines on their properties attest. Second, he readily undertook specialised economic investigations such as the coal survey of 1866–71 when provided with the requisite resources. Finally, Murchison's scientific and economic interests focused on the empire, as his almost total identification by the 1860s with practical, large-scale colonial geological questions and his repeated apologies for appointing key personnel to imperial posts illustrate. This overseas diversion of talent exemplified Britain's investment of resources in empire. But Murchison's refusal to allow the Jamaican Survey to be deflected from its systematic work, or the Syrian Committee to apply his geologists to the improvement of the Ottoman peasantry, demonstrated his unwillingness to condone the disruptive use of his department for overseas projects he considered impractical, unscientific, or serving the ends of special interests rather than the general public.

As events in the West Indies, Straits Settlements, and Natal proved, Murchison would not even recommend geologists to conduct colonial surveys unless the salaries were sufficient to compensate men with the necessary expertise for the hardships of colonial fieldwork. He was, however, eminently successful at promoting the discovery of mineral resources of direct economic benefit to the colonies and the mother country while simultaneously initiating geological research in many new environments. The evolution of Murchison's theory on gold occurrence, its use as a guideline for mineral reconnaissance for some twenty years, and the influence on the global search for coal of his Palaeozoic systems amply demonstrate that theoretical research had direct economic application, and that the practical experience of the mining industry influenced the views of geologists.

Murchison, as he admitted in frustration himself, went further than many of his contemporaries in his concern with practical geology. But in the light of the similar imperial work accomplished by De la Beche, as well as the interest sustained by other geologists such as Buckland, Greenough, Conybeare, and Lyell in developing colonial mineral resources, it appears that failure to consider the empire in the accounting has been a major reason for the emphasis on the non-utilitarian bias of British geology. Studies of the professionalisation of British geology have for the most part conspicuously omitted consideration of the importance of colonial employment opportunities in the formation of career structures.[30] Many geologists dissatisfied by or marginal to the career options in Britain found themselves impelled into the periphery by centrifugal economic and social forces. Thus for young men seeking openings or adventure, for veterans suffering from inadequate pay and stifled advancement, for Scots – even for Germans in Prince Albert's day – the empire constituted an employment frontier.[31] Imperial service provided the same sort of extended employment network for scientists as it did for other Victorian middle-class professional groups.

Well before the founding of the Geological Society of London, Joseph Banks and Charles Greville had secured the appointment of mineralogists in India, New South Wales, and on various Admiralty exploring expeditions. The duties of these men included the practical assessment of economic minerals as well as the collection of exotic specimens for the cabinets of their metropolitan patrons. The East India, South Australia, New Zealand, and Australian Agricultural Companies all employed geologists at a relatively early stage to evaluate their holdings. If stratigraphy was exported from the metropolis as a finished product for *ex post facto* application, the reciprocal trade in colonial data, largely generated by practically-oriented research, confirmed, extended, and refined the validity of the European column. Colonial data also played a crucial role in shaping theoretical debates on such issues as glaciation, orogenesis, rifting, mineralisation, coal formation, palaeoclimatology, and evolution. Darwin himself still believed in the 1860s that 'the geology of distant countries would help the progress of science more than anything else'.[32] In geology as in politics, new worlds were called into being to redress the balance of the old.

Several of the first colonial geologists such as Williams and Logan had had extensive experience in the British mining industry before they took up their posts, and all of the subsequently appointed Survey veterans and School of Mines graduates were instructed in mining methods and exposed to actual practice before going abroad. With private enterprise effectively developing the mineral resources of the home islands, there seemed little need for geologists to provide the mining industry with more than accurate maps and occasional advice. But in the colonies – especially India, the imperial laboratory of economic development by state intervention – they were more aggressive about encouraging the government to use their science to inaugurate mining and promote systematic exploitation.

Nor was this simple opportunism: the evidence demonstrates that the geologists genuinely wished to develop the economic and strategic strength of their country's empire. The chemist Sir Humphry Davy had believed that England's prosperity resulted directly from the aid which the physical sciences provided her manufacturing industries, and not from the possession of overseas colonies,[33] but the geologists, as disciples of a pre-eminently territorial science, never made this distinction. Geology and political economy, both intimately concerned with the idea of progress, converged especially in their interest in exploiting mineral resources to increase the nation's wealth and advance human civilisation.[34] Contemporary geographical thought was similarly dominated by the identification and exploitation of resources for reasons at once practical and teleological.[35] The empire, with its empty continents, primitive peoples, and unknown resources, presented an ideal testing ground for these utilitarian theories of improvement. Geology and the ethos of empire benefited mutually from an identification of interest which was first and foremost a practical contract. Murchison's reputation as 'the gold-finder' and his success at convincing colonial authorities of the economic usefulness of geology illustrates that the appeal of natural science in the imperial context was largely its self-advertised ability to discover latent productivity in colonial environments.

The King of Siluria, as the man who could read the rocks, was invaluable to the imperial government, and his career mirrors on a personal level the mutually advantageous bargain struck between science and the forces of expansion in nineteenth-century Britain. Murchison's organisation of an emergency search for naval coal in the Crimean military theatre, some sixty years before the mass mobilisation of science in the cause of national defence which occurred during the First World War, likewise illustrates his innovative importance in developing a co-operative axis between science and the state. The fact that private individuals and government officials, including Clarendon, repeatedly confused his geological and geographical titles indicates that his status owed more to his unique personal position than to the office of Director-General. Incidents regarding Western Australia, Victoria, Nova Scotia, and British Columbia demonstrate that Murchison's opinion as to the mineral potential of particular colonies was believed to have a significant bearing on their ability to attract immigrants and investors, whether he spoke as Director-General or President of the RGS.

The dramatic slump in overseas activity at the Survey after Murchison's death also illustrates that Humboldt's dream of a truly global geology had been largely realised by Murchison precisely because he was the first geologist to combine access to a worldwide intelligence-gathering network with the imagination and drive to make maximum use of it. Though he has never received adequate credit for this particular achievement despite his ongoing efforts to influence the historical record, organising geology as a branch of human knowledge throughout much of the world was one of Murchison's most significant accomplishments. As the great German geologist Leopold von Buch remarked in admiration: 'God made the world, Sir Roderick arranged it.'[36] Humboldt himself told Murchison two years after he became Director-General: 'Celui qui de tous les Géologues vivants a embrassé le plus vaste sphère des connaissances précises sur la structure de notre planète, c'est Sir Roderick Murchison. C'est l'opinion que je proclame!'[37]

By contrast, Murchison's successor Andrew Ramsay showed little interest in overseas affairs, though he himself undertook a survey of Gibraltar in 1876. Ramsay conducted few analyses for the imperial departments, simply passing most specimens and reports on to the Geological Society. The disenchantment was due in part to the failure of geological surveys to make historic mineral strikes, successful though they were in discovering profitable lodes. In Australia, New Zealand, Canada, and South Africa, nearly all the dramatic copper, gold, and diamond discoveries were made by prospectors. Despite the claims of geologists that science could systematise discovery, luck still seemed to play a predominant part in mineral exploration. When a permanent survey was proposed for South Africa in 1872, prominent men in the Cape Parliament attacked the idea, arguing that 'poor colonies could not afford geological surveys; that rich mineral regions did without them; and that even famous geologists had held peculiar opinions, afterwards proved wrong.'[38] Even today there is much truth in the old adage that 'gold is where you find it', the availability of sophisticated new conceptual tools to help predict the location of mineral deposits notwithstanding.

In the long run, what E. M. Forster called 'the implacable offensive of science' largely succeeded in overcoming luck in the search for resources and empiricism in efforts to use them more efficiently.[39] The RGS, as much as the colonial geological surveys and botanic gardens, represented an attempt on the part of science to pioneer and regulate economic development. These institutions sought to replace outright ignorance, vague travellers' tales, and unsubstantiated claims with hard facts, eradicating the haphazardness which had always ruled the discovery and exploitation of new resources and markets. To governing authorities, landowners, and commercial interests who stood to gain from increasing and diversifying economic activity, this was a welcome innovation. But as incidents in Australia and South Africa demonstrated, the new scientific order menaced speculators and charlatans who battened on the public's gullibility about resources in little-known lands. In 1889 even the British South Africa Company would be floated and sold on the strength of the assumed mineral wealth of Rhodesia before accurate geological surveying proved the claims to be vastly inflated.[40] Nor did the British Survey succeed in stamping out the serpent of fraud in the home islands themselves without a struggle. As late as 1858 Murchison was forced to defend the landed proprietors of still-unmapped north-east Yorkshire from the blandishments of a swindler styling himself as a mining engineer who claimed to be able to guarantee coal discoveries in strata of entirely inappropriate age.[41]

Murchison resisted every new geological theory advanced after the publication of *Siluria* – his mind by the 1860s has been compared to 'a Silurian matrix, impervious and resistant'[42] – and displayed striking conservatism in his attitude towards an expanded national commitment to scientific education. Yet his vision and promotional ability were ahead of his time, and with his passing, governmental interest in geology markedly diminished. Fifteen years after his death, official thinking caught up with Murchison. During the 1880s, several Australian surveys were founded on a permanent basis and the South African colonies followed suit in the 1890s as geology in the white dominions settled into a local role more prosaic than its dramatic exploratory and imperial phase. Dependent territories newly added to the empire also began receiving surveys as part of the standard administrative superstructure. After the turn of the century the Imperial Institute, having assumed control of mineral reconnaissance from the Survey in 1887, accelerated analysis of colonial mineral products and directed temporary surveys in several dependencies.[43]

Science, imperialism, and gentlemanly capitalism

While Britain generally abstained from an aggressive policy of commercial promotion abroad, the activities of Pentland and Lloyd in Bolivia, Schomburgk in Siam, and Parkes and Swinhoe in China demonstrate that the nation's diplomats – especially those whose scientific training made them anxious to apply their findings and seek results by experimentation – often attempted to pursue a more forward line of action than the Foreign Office. D. C. M. Platt has observed that Clarendon's extension of commercial information collecting duties from the consular service to the diplomatic corps in 1857 was primarily intended to increase the knowledge of legations and gauge their performance, rather than to acquire more accurate statistical data. This interpretation also requires modification in light of the Foreign Secretary's repeated requests to Murchison to upgrade the scientific training of the two services and his long-standing reliance on geological reports to inform his policy decisions.[44]

The Foreign Office tradition of sharing information with the metropolitan scientific societies, and the continuation of this practice through the Geological Survey and the Royal Botanic Gardens, similarly suggests that, once the scientific dimension is taken into account, the government must be admitted to have played a more active role than has been generally acknowledged in the promotion of overseas trade, both in seeking new opportunities abroad and in advertising them at home. Within the British empire, colonial and Indian administrations also showed an increasing, though sporadic, willingness to employ science in the interests of economic development. The maintenance in London of private museums by the East India and New Zealand Companies, as well as the steady flow of specimens from these and other quasi-sovereign corporations into the public museums of the metropolis, likewise demonstrates a general commitment to the scientific exploration and publicity of the periphery's resources.

British natural scientists collaborated with the government in the extension of the 'empire of Free Trade', using their disciplines as instruments of espionage, reform, and development in colonies, client states, and unorganised regions. Science, like trade and religion, frequently led rather than followed the flag. For all of these aspects of British expansion, the empire provided a secure base of operations from which radiated a wider network of involvment and influence. The pattern of Murchison's own activities demonstrates that the boundary between formal and informal empire was permeable and had little meaning as a barrier to efforts to organise the world on the British pattern and for British ends. He was never systematic in the methods he advocated for extending British influence as long as the fact was accomplished. This ambivalence reflected the pattern of *ad hoc* reaction to specific events which characterised British policy and public opinion towards peripheral regions throughout the nineteenth century. A truer picture of Britain's relationship with the non-European periphery is, as Gallagher and Robinson suggested thirty-five years ago, one of a repertoire of devices ranging from the creation of influence to outright annexation, all employed to facilitate the integration of an expanding zone of producing regions and markets into the British-dominated world economy.[45]

Britain enjoyed the lion's share of international trade and a near monopoly of international investment during much of the nineteenth century. Committed to *laissez-faire*, the economic ideology which complemented that position, Britain's most persistent economic problem remained the discovery of protection-free new markets and sources of raw materials to absorb her capital, services, and manufactured goods and to supply her industries.[46] Geological and geographical explorations in and beyond the empire may be seen as a corollary of this horizontally expansive attitude to growth. President Sir Rutherford Alcock told the RGS in 1877 that Britain must seek new markets in Africa free of hostile tariffs, and the following year, while cautioning that new colonies were out of the question, he reiterated the necessity of organising and exploiting the continent.[47] Britain's pursuit of a strategy of expansion emphasising the commercial, financial, and service sectors, rather than the industrially oriented and vertically integrated style of economic development undertaken by Germany, was conditioned in part by her possession of an empire at the time industrialisation began, in part by her subsequent adoption of a Free Trade policy. Given the prevailing pattern of availability of the factors of production, with land, labour, and natural resources abundant abroad and the incentives for increasing industrial intensification at home relatively weak, drawing unexploited overseas territories into Britain's economic and cultural orbit represented a rational option for policy makers, entrepreneurs, and scientists alike.

For Murchison as for many other members of the elite, and indeed for the state which embodied their political and economic aspirations, the line of comparative advantage lay in expansion and empire. Yet in comparison with France and Holland, which possessed smaller overseas empires, and Prussia, which in this era had none, Britain's official commitment to exploratory science remained small. The continuing strength of the amateur natural history tradition long associated with hegemonic culture, combined with the dominance of the economic doctrine of *laissez-faire*, dictated that nineteenth-century British science would be largely sponsored and managed on a voluntary basis.[48] Whether or not the commitment to overseas adventure permanently skewed Britain's economic development is a larger question beyond the scope of this study. P. J. Cain and A. G. Hopkins, however, have recently suggested that conventional interpretations of the expansion and decline of Britain's empire as a function of the fluctuating strength of her industrial sector ignore the continuity and success of a 'gentlemanly capitalism' based on commercial and financial services as the principal motor of Britain's economic power and imperial ambitions from the mid-nineteenth century until well into the twentieth.[49]

The natural sciences in Britain – especially taxonomic field sciences such as geology, botany, and zoology – likewise profited from the nation's diversifying presence around the world during the same period. Parallels clearly exist between the paths chosen for disciplinary development by these British scientists and for national development by their counterparts in commerce, finance, services, and politics.[50] Indeed, this new professional class shared the cultural values of the landed elite which it supplanted as the dominant economic and social interest in Britain. Murchison, as a gentlemanly scientist, was a professional in function if not in status. His career neatly bridges the eras of amateur and professional science in Britain, just as it coincides with the transition between the landed and service phases of gentlemanly capitalism which have been identified with successive stages of British imperialism. In this respect Murchison's role as the principal proponent of mid-nineteenth-century British imperial science links Gramsci's theory of hegemony, Shapin's emphasis on multifunctionality in the sociology of scientific knowledge, and Cain and Hopkins' research regarding the relationship between the history of British capitalism and imperialism.

The pattern of Murchison's behaviour exemplifies the same approach to enlarging his personal ambit of power, as well as those of his institutional and disciplinary instruments, as do the activities of gentlemanly capitalists and the related policies of imperial Britain. Murchison's preference for an extensive search for alluvial gold rather than intensive exploitation of venous gold further demonstrates that, for British scientists as well as British capitalists, the export of well-tried methods and services into unexploited areas of opportunity offered the easiest option for growth. Instead of promoting the development of the new, increasingly 'scientific' facets of geology which he found so bewildering or encouraging similar emergent aspects of geography, Murchison masterminded a programme of overseas exploration based on the simple principles of survey and mapping which he himself had employed so productively. At the time of Murchison's death, geography, like geology in the era in which he had learned the science, required no formal training. One factor in the rise in popularity of geography may have been that the growing emphasis on technical skills and laboratory analysis in geology caused many potential upper-class scientific recruits to follow Murchison into intellectually less demanding activities such as original exploration or the collection of natural history specimens.[51]

Viewed in the light of the hypothesis that Britain's economy was dominated by her internationally oriented commercial, financial, and service sectors rather than manufacturing, Murchison's apparently negative influence upon the development of the School of Mines may be reinterpreted as an attempt to carry out a policy he believed best suited to creating new opportunities in the formal and informal empires which Free Trade had rendered essential.[52] His repeated explanations that imperial demands for personnel were retarding the progress of the British Survey, and the employment in the colonies of roughly half of the graduates of the School of Mines during its first two decades, support this assessment.[53]

Murchison's obsession with maps, the product of a three-dimensional vision of the world combining the concerns of geography and geology, related directly to Britain's expansive drive. As we have seen, Murchison's geological thought was tinged with concepts borrowed from the lexicon of the military and the history of empires. In fact, he viewed the progressive march of all the natural sciences in the same light. He told the British Association in 1838 that the purpose of its annual reports was to 'look forward to conquests to come . . . to survey the border territory, and reconnoitre the *debatable* land.'[54] Contemporary British scientists also used cartographic metaphors as a cognitive link between the process of scientific discovery and imperial expansion. Herschel conceived of the role of the British Association as being 'to point out unexplored paths – to inspect the map of science and chalk out districts for individual or combined diligence to explore, subdue, and fertilise.'[55] Sedgwick, in announcing his Cambrian System, spoke of a geological 'terra incognita' having become 'a known country, which though just discovered, already exhibits every symptom of regularity and improvement'.[56] Buckland described Britain's geologists as 'extending the progressive operations of a general inclosure act over the great common field of geology',[57] and the *Calcutta Review* described geologists as 'extending the limits of . . . dominion' of their science and accomplishing 'the steady reduction to "law and order" of the provinces thus acquired'.[58]

This vision of the goal of science as the classification of knowledge regarding the physical world in order to facilitate human control was ultimately rooted in the Enlightenment. In the nineteenth century, as the industrial revolution, commercial growth, and demographic pressure provided Europe with the means and the will to expand, scientific discovery acquired greater economic and political significance. The question as to whether the information which became crucial to the maintenance of the European sphere of control was previously known to other cultures became irrelevant: discoveries were adjudged in European terms. Priority disputes between scientists regarding economically important discoveries in the periphery could therefore acquire special meaning, as Murchison's determination to secure credit for his Australian and African predictions demonstrates.[59]

Throughout his career Murchison strove to ensure that the finest obtainable maps of every portion of the world were made available to the British public. Not only were accurate topographical maps required for the construction of their geological counterparts, but because they codified vast amounts of scientific, strategic, and commercial intelligence in an easily assimilable graphic representation, maps symbolised power and progress. Their possession allowed the great powers to co-ordinate their administrative, developmental, and defensive activities, monitor those of their rivals, and organise the exploitation of peripheral resources. As army engineers and naval hydrographers had long realised, maps were also essential to the deployment of the military technology of imperialism. Even sovereign nations were drawn ineluctably into the ambit of European cartography. Countries anxious to assert their status were forced to employ European scientists and engineers to survey their territories; likewise, foreign capitalists could only be convinced to invest in development schemes researched by such experts. While some of these men, such as Jameson in Ecuador, settled abroad in academic or technical posts whose tenure facilitated a diffusion of their specialised knowledge, many more, like Lennox in Turkey or Drew in Kashmir, exploited the periphery for its lucrative temporary appointments and virgin research opportunities in order to enhance careers focused on the metropolis. Their repatriation of credit, experience, and data helped perpetuate cultural dependence on European expertise.

Murchison's accomplishment of so much exploration under the aegis of the imperial government and his vast expansion of RGS membership provide rough indices of the degree to which the mania for maps had permeated the Victorian upper and middle classes, and the strength of the connection between this cartographic fascination and the quest for wider opportunities abroad. Both as scientific devices embodying theory and as persuasive images which may inspire new explorations or ways of viewing the physical world, maps serve a heuristic function.[60] Maps have always been to some extent economic tools, instruments for shifting trade patterns according to rational calculation, but as the activities of the Hakluyt Society acknowledged, geography had also had distinctly imperial connotations since at least Elizabethan times. The founders of the RGS had explicitly recognised the usefulness of their science to the British empire, and Murchison noted in his first term as the Society's President that 'travelling and colonizing are still as much the ruling passions of Englishmen as they were in the days of Raleigh and Drake'.[61] In 1857 he remarked in regard to the popularity of geography that Englishmen, 'whose colonies extend to the Antipodes', had 'more grounds than any other nation, for making themselves well acquainted with the surface of the earth, its productions, and inhabitants'.[62] Modern historians have also noted the interplay between geography and imperialism at the RGS.[63]

The lure of the blank spots on the map thus became a kind of spatial intoxication which influenced Europe's acquisitive attitude towards the periphery. In Australia, China, Central Asia, Africa, South America, and western North America, vast resources were postulated to exist because of the sheer size of unexplored territories. Murchison was prey to this seduction himself, especially after his Russian excursions and Australian prediction taught him to think at the macro-level of continental units and he began actually participating in European efforts to organise such regions. In the heady days of the great gold-rushes, however, such optimistic assumptions were not as naive as events would later make them appear, and overselling a piece of real estate frequently proved the best strategy for winning support for an exploring expedition.

Palmerston and Clarendon, the two statesmen with whom Murchison formed the closest working relationships, were both imbued with geopolitical breadth of vision equalling Murchison's scientific scope. Robert Gavin has shown that Palmerston's world view derived from the universal principles of political economy which he used to analyse foreign societies.[64] Murchison's Humboldtian style of scientific investigation provided a parallel set of assumptions for explaining alien environments around the globe. Both Palmerston and Clarendon were unusually interested in the use of maps, statistics, scientific evaluations, and commercial intelligence of all kinds to maintain British paramountcy and promote an expansion of foreign trade through the discovery and development of new resources which could be exchanged for manufactured goods.[65]

As members of the landed classes, these men also shared an interest in estate improvement. Murchison and Clarendon acted as co-trustees to rescue the estate of one of Murchison's Scottish cousins from mismanagement, and Palmerston had initiated extensive technical developments on his Irish properties as well as investing heavily in Cornish and Welsh mining ventures.[66] Murchison, Clarendon, and De la Beche had also been involved in the founding of London's Royal College of Chemistry in 1845. This institution, which merged with the School of Mines in 1853, was intended to apply science to the improvement of the competitive efficiency of agriculture, manufacturing, and mining throughout the empire.[67] Several incidents discussed here illustrate how easily the concern with organising the exploitation of resources was transferred abroad to create a vision of the empire and peripheral world as an undeveloped estate. In an earlier era, the utilitarian approach to science as an instrument of imperial development exhibited by Joseph Banks had also stemmed from the improving impulse of the landed gentry, and the Royal Institution had represented a similar congruence between landed and colonial interests.[68]

Murchison's financial history illustrated this transition in attitude, for after the sale of his estate his investments closely mirrored the changing geographical pattern of British foreign investment.[69] He concentrated on Spanish and French government funds and Peruvian mines in the 1820s, American state and municipal securities from the 1830s to the 1860s, British railways in the 1840s and 1850s, and Indian railways in the early 1850s.[70] These speculations were frequently imprudent: in 1862 Murchison admitted to Livingstone that he had once lost £10,000 in a single railway, and the following year he estimated that he had forfeited a total of three times this sum through bad investments.[71] Predictably, his investments went largely to countries within Britain's formal or informal empires. Murchison never announced his own position regarding Free Trade; but his apparent approval of Peelite and Whig support for this doctrine, as well as his consistent encouragement of a foreign and colonial policy tailored to it, suggest that, having quitted the ranks of the landholders, he adopted the cause of the bondholders. The corollary of Free Trade was international peace, and despite his military background the mature Murchison opposed wars as threats to the worldwide system of financial credit which underlay his own and Britain's prosperity. At the same time, wars disrupted the cultural exchange which guaranteed his scientific reputation.[72] Financial motives, often at odds with his steadfast patriotic stance, influenced Murchison's response to several international crises, including the Crimean War, the American Civil War, the Oregon and San Juan Boundary Disputes, and the Indian Mutiny. His reaction to the Mutiny, in particular, revealed a hypocritical disparity between his public and private confidence in empire. As other events proved, however, he was more careful than De la Beche to avoid any conflict of interest between private investments and public duty.[73]

Palmerston and Murchison were equally aware of and adept at manipulating the growing power of middle-class opinion in public affairs. Using the social aspirations of bourgeois supporters to consolidate his political position as a man who represented the nation at a level transcending party identities, Palmerston channelled Britain's expansive forces into world politics in order to justify his assertive policies.[74] Murchison behaved very similarly in hitching the car of science to the motive power of expansionism, exploiting external events and the shifting alliances among domestic interest groups to create the support necessary for prosecuting his projects. Murchison began his career as a deep-dyed Tory like Palmerston but, actuated by personal grievance rather than political conviction, he shifted his private allegiance to the Whigs during the second quarter of the nineteenth century. By the early 1860s, reasoning that wisdom lay in changing with the times, he had converted completely to Palmerston's special brand of ultra-patriotic Whiggery.[75]

Publicly, Murchison remained detached from party politics in order to manoeuvre as a servant of the nation at large, promoting an aggressive overseas expansion of science, commerce, finance, and population backed by diplomatic and military policy in order to consolidate his prestigious position as a link between these interest groups, the cultural status of science, and Britain's ranking as the premier world power. By manipulating the emotive issue of the slave trade, Palmerston achieved some of his greatest successes in foreign policy. Murchison, too, accomplished much African exploration by exploiting this topic. But his attitude to the Syrian Committee, the Universities Mission, and the opening of Japan demonstrated that his sympathy with the evangelical approach to solving overseas problems was limited to its utility in creating support for exploring or colonising schemes such as Livingstone's, or producing useful scientific results, as in China.

Murchison's relative aggressiveness regarding various regions, like Palmerston's, was based on careful economic and strategic assessment. Only in areas of potential political rivalry with great powers other than Russia – western North America, the Isthmus of Panama, the islands facing northern and eastern Australia, Abyssinia and the upper Nile basin – or intensifying commercial competition – China, Siam, Borneo, the Niger and Zambezi valleys – did he attempt to rally public opinion and nudge the government towards a more forward policy than it deemed expedient. As a public servant in his own right and the head of another institution dependent on good relations with the government, Murchison found it was not worth while to deviate from official policy unless he believed the issues at stake demanded it. His attitude towards Japan and Korea illustrates this calculated restraint.

Since the assumptions and methods of Palmerston, Clarendon, and Murchison were so similar, it is little wonder that the latter found the long years of Whig rule during the middle decades of the century the most congenial political atmosphere in which to pursue his goals. Murchison's friendship with Clarendon was critical in this regard, but his close association with other cabinet ministers such as Russell, Lewis, Argyll, and Herbert also paid off in support for exploring expeditions. In 1863 the unrivalled ascendancy of Palmerston and Murchison in their respective spheres was so nearly parallel that an observation regarding the Prime Minister which appeared in the *Annual Register* – 'the veteran statesman, with his great tact and knowledge of the world, his large experience and skilful management of affairs of men, was of all whom the times afforded, the person best adapted for the situation which he filled'[76] – might have been applied with equal validity to Murchison. Just before Murchison's death, a colleague summing up his career similarly remarked that his name was 'as thoroughly and respectfully identified throughout the Continent with English geography as once was that of Palmerston with English policy'.[77]

Geography

Murchison thus found a natural outlet for his insatiable ambition and promotional talents in supervising the reconnaissance of overseas natural resources. As he progressively withdrew from active fieldwork into the organisation of overseas research, he was essentially exporting his scientific expertise into Britain's official and unofficial empires in the same way that British commerce, finance, engineering, and evangelicalism moved abroad in search of new opportunities during this period. The encouragement of geographical exploration also represented a logical progression from the pursuit of geological knowledge: both were terrestrial sciences concerned with the occurrence of natural resources. The RGS, which even more fully than the Geological Survey embodied the connection between Murchison's aspirations and Britain's drive for world power, reaped the largest share of institutional prestige for its part in directing the national exploratory effort. Murchison was never as comfortable working through the Royal Society as through the more popular RGS and British Association. His influence with government ministers, coupled with the rise of these rival entities and official organisations such as the Geological Survey and Kew Gardens, caused the Royal Society to be gradually supplanted in its traditional role as the premier advising agency on matters regarding overseas scientific exploration.[78]

The RGS had hoped from the first to profit from serving Britain's interests abroad, but Murchison was primarily responsible for guiding it further in that direction. His sophisticated manipulation of publicity techniques transformed the ailing Society into a theatre of national suspense, a company of talented adventurers purveying high drama in exotic settings with himself as manager/director. Just as Murchison himself had thrilled to the challenge of 'geologising' in the wilds of Russia, Britons bored by their tame landscape found excitement in vicariously following the explorers on pilgrimages to distant and savage lands. The exploits of these geographical heroes were symbolic as well as real, for they demonstrated the fitness of the nation for dominion over palm, pine, and native races. By the late 1850s the RGS more perfectly represented

British expansionism in all its facets than any other institution in the nation. Its emergence as an extra-parliamentary venue for debate on empire and peripheral exploitation was partly inevitable, partly planned. Before the Royal Colonial Institute was organised on the model of the RGS in 1868, Britain lacked a specialised outlet for imperial sentiment.[79] But as interest in overseas expansion quickened as a result of widening opportunities for trade and investment, the great gold-rushes, and Palmerston's interventionist policies, activists tended to gravitate to the RGS and use it as a forum for public discussion. Murchison's assertion of authority coincided with this trend and was a deliberate move to catalyse and exploit it.

The promotion of expansion and empire never overcame geographical exploration as the main role of the RGS, but Murchison and other advocates of a partnership between science and national expansion encouraged the use of the Society in this regard because the institution and its leaders stood to profit by it. The President realised, however, from repeated *contretemps* with Joseph Hooker, Strangford's trenchant criticism of the Abyssinian debate, and occasional complaints of the evening assemblies being 'more like beargardens than meetings of gentlemen',[80] that there were limits to how far the Society could be popularised and politicised without backlash from scientists committed to a more sharply bounded definition of research and discovery. The RGS, because of the eclecticism of the inchoate discipline of geography, functioned as a neutral meting ground for men of many scientific and lay persuasions. Its popular approach, while distasteful to some purists, attracted many others because of its success in generating funding for expeditions and the welcome change it represented from the unrelieved seriousness of the specialist societies. As Murchison commented in 1852, congratulating the Fellows of the RGS on 'the dignity and usefulness' of their journals, 'although they contain some papers to which rigid critics may object as not coming strictly within our domain, it must be recollected that in a Society constituted like our own, the tastes and pursuits of many individuals who look at geography after their own manner, must be consulted.'[81]

Murchison's policy of grafting the interests of science to those of the state, and simultaneously developing the activity of exploratory research as a form of public entertainment and social display to generate support for this hybrid, created a conflict of interest which eventually transformed the RGS into a radically different entity. Because of the dilution – even, some argued, prostitution – involved, 'Big Science' was not acceptable to all scientists, but to Murchison the possibilities made calling it into being worth the risks. Geography's equivocal status as a synthetic science made it natural that its representative institution should form a major stage for this debate, for demarcating geography from other areas of cognitive interest formed part of the process of establishing the cultural boundaries of science itself. The RGS thus formed an ideal venue for Murchison's style of scientific statesmanship, permitting him to act not only as a mediator of the physical reality being discovered overseas, but as the negotiator of the goals and authority of geography.

As with geology, Murchison played a crucial role in establishing the contractual arrangements between the discipline and the government. This was reflected at the level of expeditionary appointments as well as that of metropolitan institutions. By the middle decades of the century, the practical sciences of geology and botany had achieved pre-eminence on official exploring expeditions, while the previously popular but more esoteric subject of terrestrial magnetism gradually dropped off in representation.[82] These shifts in access to new data correspondingly affected the relative patterns of progress and prestige among the field sciences. By the late 1860s, too, Murchison had not only succeeded with Markham's Abyssinian appointment in placing a geographical specialist on a military expedition, but was dignifying RGS explorers – Livingstone, for example, and Hayward – with the exalted title of 'Geographical Envoy'.[83] Expeditionary representation for other sciences remained secondary in this period, although they demonstrated willingness to exploit tactical opportunities created by Murchison's generalship.

Murchison's wrangles with the Sea Lords over the organisation and command of the Niger, Zambezi, and Franklin search expeditions, as well as the Japanese steamboat reconnaissance, demonstrate his desire to move the focus of British expeditionary activity beyond the empire, from its traditional location at the Admiralty into the Foreign Office. This struggle mirrored the era's shift of emphasis from maritime to terrestrial exploration, just as Murchison's rise to power at the RGS symbolised a concomitant change from the dominance of the Society by naval men like Sir John Barrow and Admiral W. H. Smyth.[84] Murchison knew that if authority was once transferred to the Foreign Office, the President of the RGS would replace the Admiralty Hydrographer as the government's director of expeditions. Here again, Murchison's bid for increased personal and institutional control also constituted an attempt to achieve greater authority for science in general over decisions regarding the deployment of resources to maintain Britain's dominant position. The institutional rivalry cut across personal relationships: John Washington's allegiance lay with the navy, despite his friendship with Murchison and close identification with the RGS. The senior service managed to blunt Murchison's power-play, and the final arrangements for these expeditions represented political compromises.

The RGS was more successful in winning indirect control over non-riverine explorations undertaken by the Foreign Office, such as those of Burton and Speke, and those conducted by the Colonial Office, such as A. C. Gregory's and Palliser's expeditions. In these cases much frustration was still felt at the government's retention of ultimate authority and unwillingness to act as a mere paymaster. Expeditions mounted by individual colonial governments remained outside the purview of the RGS. The Indian government, too, remained largely a law unto itself regarding exploration, although it proved more amenable to influence concerning expeditions at the least volatile edges of its sphere, such as Arabia and east Africa. The sponsoring of Hayward as a private traveller in direct contravention of the Calcutta government's Central Asian policy, however, suggests that even this sacrosanct preserve was not immune from challenge by the power-hungry Society.

During Murchison's era the relationship of the RGS to the imperial government was in some ways analogous to that held later by chartered corporations like the Royal Niger Company or the British South Africa Company. While it never undertook the administration of any territory – although this was suggested on more than one occasion[85] – it both supported private colonising ventures such as James Brooke's and functioned as a quasi-official organisation which governments opposed in principle to proliferating overseas commitments might use as an instrument of policy. The Society allowed the government to reconnoitre and project a British presence into strategically or commercially significant territories through an agency whose ambiguous status permitted denials of official responsibility.

If the Geological Society provided the organisational model for the RGS, the British Association was the pattern for its function as a social institution.[86] Murchison's experience at the Association had taught him the value of generating public support for science. As General Secretary during the 1830s, for instance, he had helped lever the government into accelerating the pace of the Ordnance Survey of Great Britain and funding the Ross expedition to Antarctica for the purpose of researching terrestrial magnetism.[87] The enhanced status bestowed on him by his soirées and deliberately courted alliances with Humboldt, Tsar Nicholas, and other luminaries had likewise shown him that social manoeuvring was as important as research in building a scientific reputation. Murchison's desire to secure his social position led him to work constantly to enhance the status of science and its practitioners and, when scientific controversy threatened his prestige, to seek support in the wider circles of elite society where his allegiance really lay. Murchison's well-honed social skills, his presidencies of various societies, his work at the Athenaeum Club and the British Museum, his international geological correlations, his activities as an ambassador of goodwill to foreign scientists, and his experience with various public subscriptions, all contributed to the development of his organisational abilities. When he 'retired' from field geology into his second career at the RGS and the Geological Survey, he had therefore perfected the arts of patronage and wire-pulling.

In a role similar to that of the British Association, the RGS provided a medium for its principal members to dramatise themselves by articulating the cultural identity of the middle classes and welding their assertiveness and economic power to the traditional expertise of the elite in foreign affairs. As Murchison's constant tinkering to adjust the balance between competing factions demonstrated, the function of the RGS reflected its social composition. The concepts of hegemony, multifunctionality, and gentlemanly capitalism found simultaneous expression once more in the linked utilitarian and cultural goals of the RGS. By satisfying the real need for accurate information about the peripheral world, encouraging and celebrating the national adventure of overseas expansion, and attempting to channel the benefits of this phenomenon to the interest groups represented in its expanding membership, the Society helped express the shared intellectual values, economic goals, and social purposes of the collaborating middle and upper classes throughout the empire in a useful and prestige conferring cognitive enterprise.

A Fellow of the Society succinctly expressed the workings of this contract to Murchison:

geography lays open to the Government and to the Capitalist the hidden resources of the remote parts of this great Empire and teaches the one how to govern it at the least cost, and the other how to apply profitably the surplus capital and labour of the Country to objects, which thru' the RGS may be made known sooner than thru' any other means.[88]

By simultaneously discovering similar opportunities outside the empire, the RGS thus created knowledge which was interpreted and dispensed for general use and the more particular aggrandizement of the two interest groups with decisive power in the Society's affairs — scientists and servants of the imperial state. Ellesmere's observation that exploring expeditions 'should be started with the fiat of the experienced and sustained with the money of the ignorant',[89] as well as Murchison's remark regarding finances for the North Australian Expedition that 'the British public ought to pay & *will* pay all such adequate sums',[90] encapsulated the attitude of the Society's governing core.

The admission of women to RGS meetings during Murchison's second term as President in 1852–53 — another device perfected by the British Association — also illustrates his skill at tapping new sources of support for the RGS. This initiative was primarily intended to raise additional entrance fees, but it was also calculated to create a new dimension of social interest and imbue Britain's 'fair sex' with geographical knowledge 'to be by them communicated to the sons of England'.[91] Murchison was genuinely anxious, however, that women be admitted as full members of London's scientific societies. Though he urged this upon the RGS Council in 1854 and awarded gold medals to Lady Franklin in 1859 and Mary Somerville in 1868, women were only grudgingly admitted to the Society in 1904 and given full fellowship privileges in 1913.[92] From the time of his reorganisation of the Geography Section of the British Association in 1850, Murchison also worked assiduously to strengthen the links between the RGS and the Association. Not only did the meeting of Section E serve as a gathering for the geographers between the Society's sessions at which celebrity explorers could be advantageously displayed to provincial audiences, but as the examples of the Zambezi and Niger expeditions amply demonstrate, the Association at large provided the mechanism for mobilising support for exploring projects from a wider spectrum of interest groups and sciences than were represented in the RGS itself.

The RGS functioned too as an agency by which marginal men like Matheson or Brooke, who had achieved extraordinary gains in fortune by exploiting opportunities in regions yet unorganised by European initiative, were integrated into the British establishment. For buccaneering entrepreneurs as well as civil and military officials operating in the empire and the periphery, the Society served as a representative metropolitan institution in a role analogous to that played by the Royal United Services Institution in purely military circles. Once the RGS was established as a validating agency to bring the exploits of explorers before public notice, exploration came to offer a means for professional advancement through alternative national service. This function was exhibited as accelerated promotions in established diplomatic or military careers, the foundation of new careers in various branches of official service, enhanced opportunities in private sectors concerned with peripheral resource, transport, or communication development, or in patterns of employment alternating between private, official, and quasi-official activities within Britain's sphere of influence. The RGS was the touchstone of all such endeavours, and winning the backing of its perennial President was an essential rite of passage for individuals anxious to profit from the leverage it had accrued. The Society was dedicated to the delineation of the *tabula rasa* upon which the 'men on the spot' engraved their careers, and through its meetings, publications, library, and map room, the varied group whose professional ambitions focused overseas found a venue for articulating its concerns, keeping abreast of developments, and influencing politicians and makers of public opinion.

The activities of Robert Mann of Natal, among others, demonstrate that forward-looking colonists used the metropolitan scientific societies to publicise resource opportunities and thus encourage the immigration and investment required for economic development. The choice of scientists such as William Logan of Canada to organise colonial displays at international exhibitions, like the appointment of Richard Daintree as Queensland's Agent-General, illustrates that by the middle decades of the nineteenth century scientists had achieved recognition as appropriate metropolitan representatives of colonial economic interests. Metropolitan science could also be misused in the interests of private profit, as the publication by the RGS of the New Zealand Company's doctored report on the Chatham Islands shows, and the manoeuvres of Edward Cullen and Trelawny Saunders demonstrate that self-serving individuals saw the Society as a venue for increasing their own visibility in hopes of obtaining lucrative overseas posts. In incidents regarding British Columbia, China, and Ecuador, the RGS, though ostensibly an apolitical scientific society, became an important venue for generating support for the schemes of aggressive financiers anxious to exploit opportunities opened up by Free Trade imperialism.

The attempt to maintain the settlement at Port Essington, the Palliser Expedition, and the North-west Australia Expedition reveal how special interests in the colonies, by casting their projects as vehicles for the simultaneous achievement of imperial and scientific goals, could mobilise Murchison to support expansionist aims thwarted by local administrations. The President and his Council, in turn, tailored such projects to the goals of the various interest groups represented in the RGS before presenting them to government. Murchison simultaneously manipulated his social connections to bypass recalcitrant officials and win the sympathy of politicians and bureaucrats who as RGS Fellows formed a coherent bloc amenable to his persuasion on issues affecting the Society's aims. Reciprocally, the Society provided a political service by advertising the government's willingness to accede to those demands which could be presented as of imperial concern.

Even when results glaringly failed to match predictions, the cause of geographical progress was served. While some good came of most expeditions, those with wholly negative economic results could be justified as having prevented further wasted efforts. Similarly, the funding of exploration was invariably cast as an investment in the future, a highly visible means by which colonial and imperial administrations might portray themselves as progressive. Explorers, RGS officials, supporting institutions, consenting administrators, professionally interested military and naval officers, involved merchants and financiers, mapmakers, and book publishers – all those groups, in short, which together formed the geographical interest – thus drew either direct financial profit or increased social prestige from the promotion of such expeditions. Similarly, as the activities of Chesney, Lynch, Earl, Brooke, Cullen, Synge, Baikie, Livingstone, Speke, and Baker demonstrate, members of this group tended to develop vested interests in particular regions. These led them to promote successive schemes for exploitation, most of which would have involved deepening Britain's official commitment, in order to vindicate their own pioneering efforts or ideas. The logic of geographical discovery, like the logic of scientific research, dictated that results be practically applied.

As the archetypal RGS President, Murchison exemplifies the 'men of knowledge' who were most successful at mobilising and exploiting the new national scientific institutions of early nineteenth-century Britain by harmonising the diverse interests of their constituents.[93] He was strategically placed as a member of the gentry to mediate between the aristocracy and the middle classes, well connected abroad, and adept at manipulating ideas and verbal symbols to build consensus. His famous Anniversary Addresses exemplified this integrating function, offering comprehensive accounts of geographical progress, obituaries of prominent Fellows that presented the ruling class as a unified and natural leadership, and desiderata regarding explorations still to be accomplished. Edited or co-authored by regional specialists like Galton, Rawlinson, Crawfurd, and Parish, colonial politicians such as Nicholson, and occasionally even statesmen of Clarendon's calibre, these addresses constituted

British imperial 'position papers' – statements of where the nation's exploratory, commercial, and philanthropic efforts were focusing, as well as strategic
warnings to rival powers such as France, Russia, or the United States. They
organised the welter of fresh geographical information just as succeeding
editions of *Siluria* correlated new Palaeozoic discoveries and data regarding the
occurrence of gold with the existing corpus of geological knowledge.
Murchison thus schooled the Fellows of the Society – and, through the news-
paper reports and internationally distributed extra printings of his addresses,
the larger geographical audience forming a penumbra beyond – in how to
regard British expansion.

Although a comprehensive prosopographical study of the RGS remains to
be done, preliminary research has demonstrated that the aristocratic and
military elements remained strong throughout the nineteenth century, setting
it markedly apart from the specialist societies as a social network. However,
David Stoddart's characterisation of the style of the Geographical Society as
dilettante, and its composition as that of a travellers' and explorers' club
supported by gentlemen and made intellectually respectable by scientists,[94]
ignores the unifying theme of national expansion, the diverse professional
allegiances of middle-class Fellows, and the significant political, economic,
and social services performed by the Society. R. C. Bridges strikes much closer
to the truth in his emphasis on the prominence in the Society's affairs during
the middle decades of the nineteenth century of members of an imperial
'service class' consisting largely of administrators and military officers.[95]
Despite its unusual 'social topography', the RGS, like the specialist societies,
exhibited a series of graduated zones of 'ascribed competence' among its
members, based on their expertise in and commitment to the discipline of
geography.[96] As Clements Markham, the most influential officer of the RGS
and the Hakluyt Society for the remainder of the century after Murchison's
death has testified, Murchison maintained a careful balance of representation
in the Council between regional experts, kindred sciences, the military
services, and fashionable society.[97]

Gladstone had realised as early as 1864 that the success of the RGS would
soon prove its undoing when he commented at a meeting: 'Gentlemen, you
have done so much that you are like Alexander, you have no more worlds to
conquer.'[98] Indeed, Murchison's geological career had suffered a similar fate,
and it was largely the lack of new stratigraphical opportunities in Europe that
had turned him to the promotion of overseas exploration.[99] He was fortunate
enough to preside over the RGS during the brief period when Europe's ability
and urge to expand roughly balanced the size of the overseas wilderness that
remained unexplored and unexploited. Significantly, Murchison's long dominance of geography, like Palmerston's of politics, coincided with the high
noon of Victorian prosperity.[100] During this period foreign trade and
investment played an important role in stimulating the economy, and both
men, labouring each in his own way to develop new markets for British goods

and capital, contributed to this phenomenon. The first impetus to this great era of growth in the world economy had in fact been the output of the gold-rushes which Murchison attempted to preside over as a scientific authority, so that his exploratory efforts surged across the seas on the same golden current that propelled the expansion of British commerce and credit. [101]

No one of Murchison's stature and skill was available after his death to synchronise the social role and actual function of the RGS. The unity of purpose at the Society therefore gradually deteriorated under internal and external pressures generated by the changing prospects for exploration and the unhampered exercise of British hegemony. Markham noted in retrospect:

the Society owes nearly everything to Sir Roderick Murchison; its prestige, its prosperity, its internal organization, its Charter, its house. His position in society and the world of science afforded him special opportunities for advancing our interests . . . His sound judgement and tact established concord in the Society for many years . . . The death of Sir Roderick was a catastrophe. At first there was a complete blank. It would be very long before anyone could acquire the same authority. [102]

In fact, no one ever did, though Markham himself came nearest to achieving this feat. Sir Henry Rawlinson, Murchison's chosen successor as President but a man who lacked his flair for public display, expressed the Society's debt to the leader who had conducted its fortunes 'from infancy to maturity' and the hopelessness of anyone's attempting to emulate his administrative qualities. [103] Vice-President Sir Henry Bartle Frere seemed also to be literally setting the seal on the end of an era when, in awarding a gold medal to Murchison in the last year of his life, he acknowledged that 'the history of Sir Roderick Murchison's connection with the Royal Geographical Society is, in fact, the history of the Society itself.' [104]

Figurehead presidents with direct personal links to the empire continued to hold office at the RGS until the First World War. [105] The Society also briefly experimented with overt support of imperialistic activities in east Africa during the late 1870s [106] and won an accession of members on account of the popularity of imperialism, but the scientific element at the RGS led by Francis Galton and Richard Strachey succeeded during the last quarter of the nineteenth century in transforming the Society into a rigorously defined professional institution. This seemingly paradoxical development was one aspect of geography's rapid disciplinary evolution in this period. With the pioneering function of primary exploration largely accomplished, geographers were forced to make a science of their heterogeneous pursuit in order to provide organising principles and a methodology capable of suggesting and containing further research. The fashionable, political, and military factions, as well as the attendant ideologies of imperialism, racism, and environmental determinism – all of which had helped to express and justify geography's social utility during its professionalising phase – were thus gradually purged from the Society. [107]

In both geography and geology Murchison promoted fact accumulation as the necessary antecedent to the definition of laws regulating the complex interrelationships between the physical and biological realms. Geography continued to lack any comprehensive theoretical framework for synthesising these data before the gradual assimilation of Darwin's evolutionary ideas towards the end of the century. But the Humboldtian concept of geography to which Murchison adhered, stressing exploration and delineation of the physical landscape rather than the history of human environmental adaptation, provided a set of working parameters for much useful research. At the RGS as at the Geological Survey, ironically, the imperial emphasis which he created was undermined during the very apogee of imperialism by redefining tendencies which he had managed to contain or suppress.

We have seen that charges that Murchison damaged Britain's long-term scientific, educational, and industrial interests by neglecting economic geology at the Survey, by maintaining the School of Mines as an exclusively geological academy, and by diverting human resources abroad which might have been better employed at home, will not bear scrutiny. It has also been argued with the wisdom of hindsight that he delayed the development of geography through fixating the discipline's principal institution on original exploration and employing it to help focus public attention overseas.[108] Again, there is the same danger of succumbing to Whiggish 'presentism' in this analysis as in the *post hoc* dismissal of empire as an historical aberration. Given Britain's possession of an empire whose resources required investigation, unclaimed areas of unmapped wilderness around the world, an assertive public, and his own ambitions, Murchison's course of action constituted the best option for balancing the disparate economic and social forces his institutions represented.

Exploration is deliberate reconnaissance. Its goals and evaluations are conditioned by the categories of existing knowledge and current objectives of the civilised centre which conducts it. The activities of RGS expeditions, like the Society's evolving function as an institution and the changing composition of its membership, demonstrate the complexity of exploration as a cultural activity. Through their reports, collections, sketches, and personal accounts, Victorian scientific explorers synthesized the information gathered at the periphery. These impressions helped shape the destiny of the newly explored places and their inhabitants, and at the same time ineluctably altered Britain's culture.[109]

In 1871 Frere remarked of Murchison:

he has become the common reference of geographers and scientific travellers of our own and of all other countries. It is no exaggeration to say that, during the past thirty years, no geographical expedition of any consequence has been undertaken in our own, or, I believe I might say, in any other country, without some previous reference to him for advice and suggestion.[110]

Joseph Hooker and Lord Strangford had likewise testified to Murchison's geographical omnipotence and indispensability. [111] By arbitrating the timing and destination of most mid-Victorian exploring expeditions, choosing the leaders and dictating the nature of the information they gathered, rewarding their efforts, frequently exercising editorial control over the publication of their results, and meting out praise and criticism for the content and style of their narratives, Murchison ensured that the canon of taste in travel literature largely reflected his own. His views, in turn, were conditioned by the assumptions and goals of the elite to which he belonged.

The RGS served as the contextual envelope in which such knowledge of overseas frontiers was generated and disseminated. Just as scientific knowledge was socially moulded by an inner clique at the British Association and the Geological and Zoological Societies, [112] Murchison and his colleagues at the RGS filtered an awareness of the resources, geography, natural history, and ethnology of the non-European world into the British consciousness for the first time on a large scale. In doing so they popularised exploration and influenced the attitudes Britons developed about their empire and the alien environments being revealed abroad. The motifs and expressive devices which their style of thought sanctioned came to represent the imperial hinterlands, so that in the minds of metropolitan rulers the Australian bush, the Canadian forests, the Himalayan plateaux, and the African savannahs became precisely what was written of them. [113] The geologists' attempts to discover a 'lost world' inhabited by otherwise extinct flora and fauna – first in Tasmania, then in New Zealand, and finally in Africa – was one significant feature of this process. They were motivated by the perceived resemblance of tropical and southern hemispheric environments to those of the former epochs disclosed by palaeontology, by anxiety to test Lyell's uniformitarian implication that ancient species might somewhere survive in unexplored parts of the modern world, and by the desire to annex the landscapes of the past to the domain of British science. [114]

The picture that gradually emerged from the activities of the explorers was one of undisciplined nature, uncivilised peoples, primitiveness, limitlessness, and bizarre extremes. Implicit in the challenge of these strange continents, however, lay opportunity, excitement, the illusion of infinite resources, exotic fields for personal and national expression. The disposition towards these new lands propagated by Murchison and his colleagues was aggressively interventionist. As Lewis Mumford observed: 'The same underlying animus prompted the explorer and the scientist, the pioneer and the inventor.' [115] While the scientists often expressed regret at the destruction of native peoples and natural environments, the majority viewed this process as inevitable and the European right to drastic modification and manipulation of peoples, species, and entire biotas as unquestionable. [116]

As Geikie's comparison of the researches of Murchison and Livingstone suggests, [117] science subjugates the chaos of the unexplained by ordering phenomena into categories. It dissects, analyses, reconstructs, and defines the apparently haphazard and irregular. Because it pursues a retreating frontier of ambiguity in order to annex it, science is self-perpetuating. The discontinuity where discovery occurs constitutes a shifting zone of penetration, resolution, and consolidation, where the imposed symmetries of science express man's need for security as well as his perceptions of nature's inherent order. In seeking understanding, then, science seeks to establish dominion over objects and processes. Inasmuch as it succeeds, knowledge *is* power. Exploration, like natural science, creates exploitable new knowledge in the laboratory of the wilderness: until the raw materials it finds are made known, for instance, they can serve no purpose as resources. New information therefore constitutes actual as well as cultural capital, and it represents a prize for competing individuals, interest groups, and nations seeking the honour of discovery and control of its beneficial consequences. Murchison's obsession with extending the bounds of Siluria reflects this, just as, at other levels, do the factional tensions within the RGS, the competition between British and continental geographical societies, and the economic, imperial, and cultural rivalry among the great powers themselves.

Once science displaced art in the early nineteenth century as the dominant cultural mode and cognitive method of the new industrial civilisation,[118] logic dictated its extension overseas, where the marshalling of empirical data helped discover new resources and rationalise the development and administration of alien territories. The introduction of European technology in colonies and peripheral countries required the implementation of new ways of organising human and physical resources as well as the importation of actual machinery to process and transport usable raw materials. The systematics and techniques of science classified the components of the periphery to facilitate their smooth integration into the economic, political, and cultural systems of Europe. European expansion, of which imperialism was one aspect, thus constituted an important vector for the diffusion of innovations. Reciprocally, the peripheral experience acted as a forcing house for further innovation because of its variety and intensity, feeding back to the metropolis to suggest modifications in technological design, industrial practice, scientific theory, and administrative modes.

In expediting resource mobilisation through the overseas deployment of scientific skills and the metropolitan processing of resulting data, Murchison vastly accelerated and intensified the work begun by De la Beche's Survey on the one hand, and Banks, Barrow, and the unreconstructed RGS on the other. In this respect, he may be considered an innovator in Joseph Schumpeter's sense of an extraordinary individual who qualitatively transforms objective circumstances by exploiting perceived possibilities.[119] From an economist's viewpoint, his activities represent attempts to develop productivity at both the intensive and extensive margins of empire, encouraging systematic resource exploration and exploitation within existing colonies while simultaneously searching for new opportunities further afield. In Gramscian terms, Murchison exemplifies the organic fusion of the personal with the cultural, economic, and political realms, all of which served to enlarge Britain's sphere of influence and reinforce the hegemony of his own class.

Science has never been value-neutral or wholly objective; especially in the context of empire, it implies control, as an instrument both of administration and of knowledge.[120] British natural science in the nineteenth century conquered new lands by reducing them to regularity and providing the requisite information to exploit them. Its importation of overseas data into the libraries and museums of the metropolis constituted in one sense a gigantic looting operation which helped maintain British ascendancy. The content of science, like the style of an empire, is shaped by its cultural context. British geology and geography, as well as other sciences not treated here, were significantly influenced by Britain's possession of a colonial empire. Imperial concepts, metaphors, data, and career opportunities informed the development of these disciplines, and their institutions in varying degrees expressed this ideological matrix. The Victorian empire, in turn, evolved within an increasingly scientific environment. Geology and geography provided its rulers with new tools of administration and development, instruments of expansion, and methods for conceptualising the world. Murchison's career illustrates both sides of this bargain, for he was its key negotiator.

The related disciplines of geology and geography exemplify the 'sub-imperialism' of scientists – their desire for new data, careers, classificatory conquests, and power in administrative affairs – meshing with the needs of the imperial government. The mediation provided by natural science gave Europeans intellectual as well as actual authority over colonial environments by classifying and ultimately containing their awesome dimensions. This new level of control, linked with the technology representing its practical application, also conferred prestige on the metropolitan power as a civilising force, helping to legitimate imperial rule *vis-à-vis* subject races, domestic masses, and rival great powers.[121] At several levels, natural science helped imperialism reshape non-European environments in the image of Europe, and by so doing became an important component in its grid of cultural, political, economic, and military domination.

ABBREVIATIONS USED IN THE NOTES
AND BIBLIOGRAPHY

APS, MuP	American Philosophical Society, Philadelphia, Murchison Papers
BGSA	British Geological Survey Archives, Keyworth, Nottinghamshire
BL(AM), LaP	British Library (Additional Manuscripts), London, Layard Papers
BL(AM), MuP	British Library (Additional Manuscripts), London, Murchison Papers
BLO, BAP	Bodleian Library, Oxford, British Association Papers
BLO, MSS.Clar.	Bodleian Library, Oxford, Clarendon Papers
CUL, MaP	Cambridge University Library, Mayo Papers
CUL, SeP	Cambridge University Library, Sedgwick Papers
EUL, GeP	Edinburgh University Library, Geikie Papers
EUL, MuP	Edinburgh University Library, Murchison Papers
GSL/G	Geological Society of London, Geikie Papers
GSL/M(A-Z)(no.)	Geological Society of London, Murchison Papers (incoming correspondence)
GSL/M/J(no.)	Geological Society of London, Murchison Papers (fair-copy journals)
GSL/M/N(no.)	Geological Society of London, Murchison Papers (field notebooks)
HLD, HeP	Hocken Library, Dunedin, Hector Papers
HRO, LyP	Hertfordshire Record Office, Hertford, Lytton Papers
IOLR, BCC	India Office Library and Records, London, Board of Control Collections
IOLR, IBD	India Office Library and Records, London, India and Bengal Despatches
JMA	John Murray Archive, London
JRGS	*Journal of the Royal Geographical Society*
MLS, ClP	Mitchell Library, Sydney, Clarke Papers.
MUL, LoP	McGill University Library, Montreal, Logan Papers

NAZ, LiP	National Archives of Zimbabwe, Harare, Livingstone Papers
NLS, MuP	National Library of Scotland, Edinburgh, Murchison Papers
NMW, DeP	National Museum of Wales, Cardiff, De la Beche Papers
PGS	*Proceedings of the Geological Society of London*
PRGS	*Proceedings of the Royal Geographical Society of London*
QJGS	*Quarterly Journal of the Geological Society of London*
RBA (date), TS	*Report of the (no.) Meeting of the British Association for the Advancement of Science; Held at (city or town) in (date), Transactions of Sections*
RBG, JHP	Royal Botanic Gardens, Kew, Joseph Hooker Papers
RBG, WHP	Royal Botanic Gardens, Kew, William Hooker Papers
RGS, BaP	Royal Geographical Society, London, Baikie Papers
RGS, CMB	Royal Geographical Society, London, Council Minute Books
RGS, CO	Royal Geographical Society, London, Colonial Office Correspondence
RGS, DL	Royal Geographical Society, London, David Livingstone Collection
RGS, FO	Royal Geographical Society, London, Foreign Office Correspondence
RGS, GC	Royal Geographical Society, London, General Correspondence
RGS, LB	Royal Geographical Society, London, Letter Books
RGS, MuP	Royal Geographical Society, London, Murchison Papers
RGS, NAE	Royal Geographical Society, London, North Australia Expedition Papers
RHO	Rhodes House, Oxford
RS, HeP	Royal Society, London, Herschel Papers
RS, MuP	Royal Society, London, Murchison Papers
TGS	*Transactions of the Geological Scoeity of London*
TLW	Alexander Turnbull Library, Wellington
UCL	University College, London
UMO, PhP	University Museum, Oxford, Phillips Papers
UW, BaP	University of the Witwatersrand, Johannesburg, Bain Papers

NOTES

Introduction

1 For a pioneering overview of the history of science in the context of European expansion, see Basalla, 1967; for science and British imperialism, MacLeod, 1982.
2 Fieldhouse, 1981: 3.
3 C. Lloyd, 1970: 115, 149–157.
4 Geikie, 1875. For shorter biographical notices see Eyles, 1971; Gilbert and Goudie, 1971; Rudwick, 1974b; see also Thackray, 1972.
5 Rudwick, 1985; Secord, 1986a.
6 Secord, 1982; Stafford, 1984, 1988a, 1988b.

1 The King of Siluria

1 R. I. Murchison, MS. journal entitled 'Events of the Author's Life from 1792 to 1845', EUL, MuP; Geikie, 1875, i: 1–13; K. Murchison, 1940.
2 Geikie, 1875, i: 14–48.
3 R. I. Murchison, MS. journal, 1792–1845, EUL, MuP; Geikie, 1875, i: 49–68.
4 Geikie, 1875, i: 68–87; for military observations, see, e.g. GSL/M/N173, 1817, entry for 31 July.
5 GSL/M/J5, 1818–25, ff. 1–15; Geikie, 1875, i: 88–90.
6 R. I. Murchison, field diary and cash book, 1820, BGSA, GSM 1/124; Geikie, 1875, i: 91; GSL/M/J5, 1818–25, ff.32–37.
7 Geikie, 1875, i: 92–95, 117.
8 Berman, 1977; GSL/M/N9–15, 1824–25.
9 Minutes of General Meetings, Royal Institution, G3, 1822–28, ff.100, 254.
10 GSL/M/J6, 1825–26, ff.7–8.
11 Geikie, 1875, i: 124–128.
12 Murchison, 1829a.
13 Murchison, 1829b; Geikie, 1875, i: 126–135.
14 Geikie, 1875, i: 137–155.
15 Secord, 1982: 421.
16 Gilbert and Goudie, 1971: 506–507; Page, 1976.
17 Geikie, 1875, ii: 40.
18 Secord, 1986a: 47–57.
19 Geikie, 1875, i: 172.
20 Secord, 1986a.

21 Rudwick, 1972; Secord, 1986a: 57–68.
22 Davies, 1969: 214, 245, 253; Hooykaas, 1970; Page, 1976; Rudwick, 1985: 390–391, 449.
23 Rudwick, 1963.
24 Cannon, 1978: 73–110; J. Browne, 1983: 42–62.
25 Rudwick, 1976.
26 Murchison, 1868b: cxciii.
27 Geikie, 1875, i: 194.
28 Mill, 1930: 1–34.
29 Murchison, 1859b: cxvi.
30 Mill, 1930; Stoddart, 1980.
31 Mill, 1930: 17.
32 Morrell and Thackray, 1981; MacLeod and Collins, 1981.
33 Murchison, 1839b; Thackray, 1978.
34 Phillips to Murchison, 22 Feb. 1835, GSL/M/P14/11.
35 Secord, 1982.
36 Rudwick, 1974b: 583.
37 Secord, 1986a.
38 Rudwick, 1985; Geikie, 1875, i: 244–288.
39 Secord, 1982; 1986a: 122–123; Rudwick, 1985: 350–351.
40 Geikie, 1875, i: 243.
41 GSL/M/J8–14, 1840–45; Murchison *et al.*, 1845; Geikie, 1875, i: 288–302, 315–357. See also Thackray, 1979.
42 Murchison and Verneuil, 1841; Rudwick, 1985: 352–368.
43 Murchison, 1840–41: 362, 373–375.
44 Geikie, 1875, i: 328–329.
45 Murchison *et al.*, 1845, i: 343.
46 ibid., 1845, i: 471–506.
47 Geikie, 1875, i: 330–331; Murchison, 1843b.
48 GSL/M/J12, 1841, f.517.
49 ibid., f.531.
50 Conybeare, 1846: 350.
51 Murchison *et al.*, 1845, i: 89–123.
52 Murchison, 1842.
53 Geikie, 1875, i: 356–357.
54 For example Crosse, 1892, ii: 49.
55 Geikie, 1875, i: 277, 359; Secord, 1986a: 123–124.
56 Murchison, 1843b.
57 Geikie, 1875, ii: 13–15.
58 Page, 1976: 160; Murchison to Peel, 27 June 1844, BL(AM) 40547, Peel Papers, ff.284–285.
59 Murchison to Herschel, n.d. [1842], RS, HeP, HS.12.440; Gladstone to Murchison, [copy] 24 Jan. 1844, BL(AM) 44527, Gladstone Papers, f.168.
60 See respectively Murchison to Featherstonhaugh, 24 Jan. 1844, CUL, SeP, Add.MS.7652.II.kk.36; Murchison to J. Forbes, 23 Dec. 1844, St Andrews University Library, J. D. Forbes Papers, 1844/88.
61 GSL/M/J15, 1843–44, pt.2, ff.1–2.
62 Murchison *et al.*, 1845; Thackray, 1979.

63 GSL/M/J16, 1844–45, pt.2, f.54.

64 Mill, 1930: 55.

65 Geikie, 1875, ii: 49–53.

66 GSL/M/J17, 1845, ff.74–77.

67 Murchison to Featherstonhaugh, 26 May [1846], CUL, SeP, Add.MS.7652.
    II.kk.26.

68 Wileman, 1971.

69 GSL/M/J18, 1845–47, pt.2, f.13.

70 Murchison, 1846; Geikie, 1875, ii: 63–65.

71 For diplomacy, see below, Chapters 4 and 5; for commerce, Murchison *et al.*,
    1845, i: 438.

72 Geikie, 1875, ii: 42.

73 Geikie, 1875, ii: 84–88; GSL/M/J20, 1848, ff.124–125.

74 Geikie, 1875, ii: 91.

75 GSL/M/J19, 1847, f.422.

76 GSL/M/J21, 1848, f.417.

77 Rudwick, 1985: 439–441; Secord, 1986a: 122–123.

78 GSL/M/J21, 1848, ff.464, 467; below, pp. 99–100.

79 Geikie, 1875, ii: 101–107; Secord, 1982: 439–441.

80 Ellesmere to Murchison, 11 Dec. 1853, BL(AM) 46126, MuP, ff.93–4.

81 For the peerage, see GSL/M/J22, 1849–52, f.25; for the baronetcy, Page, 1976:
    160.

82 M. Hall, 1984: 188–192; MacLeod, 1983.

83 Murchison, 1852a: lxiii–lxiv; see also Murchison to Aberdeen, 13 May 1853,
    BL(AM) 43250, Aberdeen Papers, ff.64–65.

84 *Hansard's Parliamentary Debates*, House of Lords, 3rd ser., cxxi: 1181–1192.

85 Stapleton to Murchison, 13 June and 13 July 1853, BL(AM) 46128, MuP,
    ff.129–130, 135–136.

86 Murchison, 1854; Thackray, 1981.

87 See Chapter 2.

88 Porter, 1973.

89 For example GSL/M/J26, 1863–65, ff.28–29, 35; Vignoles to Murchison, n.d.
    1848, BL(AM) 46128, MuP, ff.238–239.

90 Secord, 1982: 414.

91 Huxley to Murchison, 12 Apr. 1860, BL(AM) 46126, MuP, ff.450–451; reply,
    14 Apr. 1860, Imperial College, Huxley Papers, 23.154.

92 GSL/M/J14, 1841–43, ff.779–780.

93 Secord, 1982.

94 For the quotations, see respectively Geikie, 1875, i: 303; ii: 86, 151, 106.

95 Porter, 1978.

96 Mill, 1930.

97 S. Beaver, 1982.

98 Murchison, 1866c: 42.

99 Cf. Stoddart, 1980; D. N. Livingstone, 1984: 291.

100 RGS, 1854.

101 Marshall-Cornwall, 1976.

102 Middleton, 1986.

103 Murchison, 1857b: clxxxix.

104 For example Kenrick to Murchison, 3 Nov. 1861, BL(AM) 46127, MuP, ff.7–8.

105 McCartney, 1977.

106 E. Bailey, 1952: 47; see also Reeks, 1920; Flett, 1937; H. Wilson, 1985.

107 Secord, 1986b.

108 De la Beche and Playfair, 1848.

109 Stafford, 1984.

110 Geikie, 1875, ii: 184–191; Reeks, 1920, pt.1: 82–89; E. Bailey, 1952: 50–54.

111 McCoy to Sedgwick, 22 Feb. 1856, CUL, SeP, Add.MS.7652.II.g.27.

112 Porter, 1978: 823–824; Rudwick, 1985.

113 Murchison, 1839a: xxxix; *Morning Post* article entitled 'Museum of Practical Geology', 11 May 1851, BGSA, GSM 1/202.

114 *QJGS*, 11, 1855: xxiv.

115 Horner to Murchison, 19 Apr. 1855, GSL/M/H29/19.

116 Murchison to Sedgwick, 16 Oct. 1842, in Geikie, 1875, i: 376–382; McCoy to Sedgwick, 22 Feb. 1856, CUL, SeP, Add. MS.7652.II.g.27.

117 Secord, 1986a: 273, 296–298.

118 Rudwick, 1985: 194.

119 Secord, 1986b.

120 For Campbell's candidacy, see Reeks to Geikie, 3 Jan. 1870, EUL, GeP, Gen.525.

121 Murchison to Clarendon, 17 Apr. 1855, BLO, MSS.Clar.Dep.C.103, ff.523–524.

122 Salter to Sedgwick, 22 Apr. 1855, CUL, SeP, Add.MS.7652.II.L.35.

123 Parish to Shaw, 3 June 1855, RGS, Parish Collection.

124 Gowing, 1978; Murchison, Memo., 10 Dec. 1856, BGSA, GSM 2/2.

125 Murchison, 1867c: 433; 1861c: 151.

126 Smyth, Reeks, and Rudler, 1864: 124.

127 Murchison to the Committee of the Privy Council on Education, 4 Apr. 1858, BGSA, GSM 1/7, ff.231–235.

128 For example Aberdeen to Murchison, 18 Oct. 1858, BL(AM) 46125, MuP, ff.3–4; Wellington to Murchison, 24 May 1860, GSL/M/W29/3.

129 Stafford, 1984.

130 Hammond to Murchison, 3 Jan. 1857, BGSA, GSM 1/7, f.356; reply, 6 Jan. 1857, ff.35–36.

131 Flett, 1937: 74–76.

132 Layard to Murchison, 7 Feb. 1866, BGSA, GSM 1/8, ff.397–398; reply, 22 Feb. 1866, ff.384–389.

133 Dutton, 1977; see below, p. 84.

134 Murchison to Grant, 9 Mar. 1866, National Library of Scotland, Grant Papers, MS.1790, ff.21–22.

135 *The Times*, 2 Dec. 1865, p. 5, col. 3; Eyre to Murchison, 14 Aug. 1867, 5 Feb. 1869, BL(AM) 46126, MuP, ff.112–113, 120–121.

2 The Antipodes

1 MacKay, 1985; Vallance, 1975.

2 Murchison, 1832: 365–367.

3 Mangles to Murchison, n.d. 1832, BL(AM) 46127, MuP, ff.182–183.

4 Murchison to Clarke, 17 Dec. 1838, MLS, ClP, MSS. 139/43, ff.665–668.

5 King to Murchison, 24 Jan. 1834, GSL/M/K3/1; Branangan, 1985. John Murchison is commemorated by the name of the town of Murchison, Victoria.

6 McLeay to Murchison, 3 Apr. 1841, BL(AM) 46127, MuP, ff.160–161; Murchison, 1842: 645.

7 Stafford, 1988a; MacLeod, 1982.

8 See below, p. 68; Raleigh Club Minute Book, entry for 11 May 1829, RGS; GSL/M/J6, 1824–26, f.11.

9 Fitzpatrick, 1949: 192–205; Hoare, 1969.

10 Hoare, 1974: 136–137.

11 Murchison to Franklin, 12 Feb. 1842, Scott Polar Research Institute, Franklin Papers, MS. 248/222.

12 Murchison, 1842: 646.

13 Franklin to Strzelecki, 19 July 1842, in Havard, 1941: 73–74.

14 See Jukes, 1847; C. Browne, 1871: 127–268.

15 Franklin–Strzelecki correspondence, MLS, CY Reel 1759, MLA 1604; Jukes to Sedgwick, 28 Nov. 1842, CUL, SeP, Add.MS.7652.I.F.Portfolio II.90.

16 Franklin to Murchison, n.d. Nov. 1842, in Heney, 1961: 132; Murchison, 1852a: lxxxii; Strzelecki, 1845. See also Branagan, 1986.

17 Franklin to Murchison, 30 Jan. 1845, GSL/M/F17/1.

18 For the scientists' loyalty, see Franklin–Buckland correspondence, MLS, CY Reel 1759, MLA 75; Murchison, 1845a: xlvi; for the send-off, Murchison to Morgan, 16 Nov. 1851, Yale University, Beinecke Library, Morgan MSS.

19 Elie de Beaumont, 1831; Murchison, 1845b; Greene, 1982: 69–121.

20 Murchison, 1867c: 448–475.

21 GSL/M/J7, 1826–38, ff.112–113.

22 Murchison, 1844: xcix–c.

23 Murchison, 1846.

24 Smith to Murchison, 27 Feb. 1852, EUL, MuP, Gen.523/6, also in Silver, 1986: 135–136; Murchison, 1850c: 63.

25 Murchison to Grey, 5 Nov. 1848, in 'Further Papers', 1852–53: 521, also in Silver, 1986: 122–123.

26 Murchison, 1852a: lxxxiii.

27 Hawes to De la Beche, 13 July 1849, BGSA, GSM 1/5, f.181; reply, 4 Aug. 1849, ff.188–191; see also Mozley, 1965.

28 De la Beche to Hawes, 12 Aug. 1850, BGSA, GSM 1/6, ff.17–18; Archbold, 1981.

29 Hawes to De la Beche, 5 July 1851, BGSA, GSM 1/6, ff.98–99; reply, 16 July 1851, ff.104–106.

30 W. Morrell, 1940: 200–259, 282–308; Blainey, 1969: 1–22.

31 Strzelecki to Murchison, 17 Jan. 1854, EUL, MuP, Gen.523/6.

32 Jervis, 1944: 369–370. For fuller discussion, see Stafford, 1988a; cf. Grainer, 1982: 163–204; Newland, 1983; Silver, 1986.

33 For example Mundy, 1852, iii: 310–311, 428–429.

34 Davison, 1861: ii.

35 Cf. Clarke, 1852, 1860; and Murchison, 1852b. For Murchison's defence of his prediction, see correspondence in EUL, MuP, Gen.523/6; MLS, ClP, MSS. 139/43; Murchison to Newcastle, 8 July 1853, in 'Further Papers', 1852–53: 522, also in Silver, 1986: 145–147.

36 Clarke to Murchison, 7 July 1851, EUL, MuP, Gen.523/6; Clarke, 1860; Jervis, 1944: 414, 423.
37 Clarke to Murchison, 30 May 1855, EUL, MuP, Gen.523/4/7.
38 Hargraves to Murchison, 17 Jan. 1855, EUL, MuP, Gen.523/4/22.
39 Secord, 1986a; Rudwick, 1985.
40 Grainger, 1982: 215–221; Newland, 1983: 100–105.
41 Holland to Murchison, n.d. [1854], EUL, MuP, Gen.523/6.
42 Murchison, 1850a: 397; Stafford, 1988c.
43 Herschel to Murchison, n.d. Jan. 1851, 10 Jan. 1852, EUL, MuP, Gen.523/6; for Peel, see above, p. 15.
44 Murchison, 1844: lxix.
45 Murchison 1867c: 448–475.
46 Murchison to Phillips, n.d., EUL, MuP, Gen.1999/1/18.
47 Murchison, 1850c: 61.
48 Murchison, 1853: cxxvii.
49 Fitzroy to Shaw, 7 Nov. 1853, RGS, Fitzroy Collection.
50 Humboldt to Murchison, 16 Aug. 1853, EUL, MuP, Gen.523/4/47.
51 Wyld, 1853; Brayley to Murchison, 3 Feb. 1852, EUL, MuP, Gen.523/6.
52 Jukes, *et al.*, 1852; Jukes to Murchison, 1 July 1852, EUL, MuP, Gen.523/6; Stafford, 1988c.
53 Jukes, 1850; Bayliss, 1978.
54 Murchison to Whewell, 19 Feb. 1852, quoted in Geikie, 1875, ii: 131.
55 Murchison to Pakington, 2 Mar. 1852; see also Murchison to Grey, [draft] 27 Nov. 1851 – both in EUL, MuP, Gen.523/6, and printed, with errors, in Silver, 1986: 134–135, 140–142.
56 For Australian mountebank John Calvert, see Calvert, 1853a; Murchison, 1854: 436.
57 Murchison to Sopwith, 24 Mar. 1852, APS, MuP, B:M93p.
58 Odernheimer to Murchison, 22 June 1853, GSL/M/O1/1; GSL/M/J22, 1849–52, f.138; King to Murchison, 24 Nov. 1853, EUL, MuP, Gen.523/6.
59 Odernheimer, 1855; Murchison, 1867c: 463.
60 Herschel to Murchison, 2 Jan. 1853, EUL, MuP, Gen.523/4/43.
61 For example Terry to Murchison, 21 May 1853, RGS, GC.
62 G. Stephen to Sedgwick, 30 Mar., 4 Apr. 1853, CUL, SeP, Add.MS.765-2.II.X.61,.62.
63 Nicol to Murchison, 22 Nov. 1852, GSL/M/N8/6a+b.
64 W. Morrell, 1940; Blainey, 1969.
65 Murchison, 1867c: 454–457. For waves of translation, see Chorley, Dunn, and Beckinsale, 1964: 226–231, 287, 355; Davies, 1969: 243–254.
66 Murchison, 1842: 645; 1865b: cxli.
67 Murchison, 1844: xcvii–ciii.
68 Earl, 1845; Murchison, 1854: 450–451.
69 Howard, 1931–32; Graham, 1967: 402–443; C. Lloyd, 1970: 160–164; A. Shaw, 1966: 315–317.
70 Murchison, 1845a: lviii, lx.
71 Murchison, 1852a: lxxxii–lxxxvi; 1853: cxxiv–cxxx.
72 Haug to Murchison, 23 Apr. 1853, RGS, NAE, 2/7.

73 Shaw to Newcastle, 24 Aug. 1853, RGS, LB, v. 1850–59, ff. 91–95; Newcastle to Ellesmere, 23 Dec. 1853, RGS, GC.

74 Calvert, 1853b; Saunders, 1853.

75 Murchison to Saunders, 6 Oct. 1853, Dixon Library, Gregory Papers, MS. Q421, ff. 86–88.

76 Murchison to Shaw, 4 Oct. 1853, RGS, MuP.

77 Murchison to Shaw, 6 Oct. 1853, RGS, MuP; Murchison to Haug, 6 Oct. 1853, LB, v. 1850–59, ff. 95–96.

78 Murchison to Shaw, 12 Jan. 1854, RGS, MuP; Holland to Murchison, [two letters] n.d. 1854, EUL, MuP, Gen. 523/6.

79 Saunders to RGS, 2 Jan. 1854, RGS, NAE, 5/7; Howard, 1931–32: 139–140.

80 J. Stokes, 1856.

81 Shaw to Merivale, 14 Jan. 1854, Dixon Library, Gregory Papers, MS. Q421, ff. 17–18.

82 Haug to Murchison, 24 Jan. 1854, GSL/M/H44/1.

83 Archbold, 1981.

84 Murchison to Shaw, 14 Aug. 1854, RGS, MuP.

85 Wilson to Shaw, 23 Nov. 1853, RGS, NAE, 5/1; Murchison, 1867c: 462–463.

86 Wilson to Murchison, 10 Aug., 16 Sept., wiith enclosure, 1854, RGS, NAE, 5/1; 14 Dec. 1854, BL(AM) 46128, MuP, ff. 317–318; Wilson to Sturt, 15 Dec. 1854, RHO, Sturt Papers, MSS. Austral. S. 5, f. 198; Murchison to Sturt, 22 Dec. 1854, ff. 143–144; reply, 24 Feb. [1855], GSL/M/S85/2.

87 Washington to Sturt, 8 Oct. 1854, RHO, Sturt Papers, MSS. Austral. S. 5, no. 44, ff. 122–123.

88 Cumpston, 1972: 17–18.

89 Council Minutes, v. 1841–57, entry for 27 June 1855, BLO, BAP.

90 Denison to Murchison, 21 May 1855, in Denison, 1870, i: 309–310.

91 Denison to Murchison, 16 Aug. 1856, in Denison, 1870, i: 362–363; Denison to Sturt, 22 May 1855, RHO, Sturt Papers, MSS. Austral. S. 5, no. 44, ff. 137.

92 A. and F. Gregory, 1884: 99–194; Cumpston, 1972; Wallis, 1976: 51–88.

93 A. Gregory, 1855–56.

94 J. Stokes, 1856.

95 Wilson to Murchison, 11 July 1856, RGS, Journ. MSS. Aust.; A. Gregory, 1858.

96 Murchison, 1857b: clxxvi; see below, p. 72.

97 Murchison to Clarke, 30 Dec. 1856, MLS, CIP, MSS. 139/43, ff. 705–706.

98 Wilson to Murchison, 23 Dec. 1856, RGS, Journ. MSS. Aust.; see below p. 90.

99 Murchison to Shaw, 19 Apr. 1858, RGS, Journ. MSS. Aust., referee's comments. The name of these mountains has not endured, but another range in the region still bears Murchison's name.

100 Fitton, 1856–57. The oldest sedimentary rocks of this region are actually much earlier deposits.

101 J. Wilson, 1858.

102 Murchison, 1857b: clxxix.

103 Murchison, 1858b: cxcvi.

104 Cumpston, 1972: 60–66.

105 Austin, 1856

106 F. Gregory, 1858–59.

107 DuCane to Shaw, 23 June 1858, with enclosure, and n.d. [received 14 June 1859], RGS, GC.

108 Embling to Murchison, 8 Jan. 1858, with Strzelecki's comments, RGS, GC.

109 F. Gregory to Shaw, 24 Jan. 1860, MLS, F. Gregory Papers, MSS.A306, ff.19–22; W. Burgess, L. Burgess, and Sanford to Murchison, 30 May 1860, ff.23–24; W. Burgess to Murchison, 31 May 1860, RGS, GC.

110 Murchison to Shaw, 7 June, 23 July 1860, RGS, MuP; W. Burgess to Murchison, 18 July 1860, GSL/M/B69/1.

111 Ashburton to Shaw, [two letters] 8 Aug. 1860, RGS, Ashburton Collection.

112 Shaw to Fortescue, 24 Aug. 1860, RGS, LB, v.1859–79, ff.23–27.

113 Shaw to Treasury, 24 Jan., 20 Mar. 1861, RGS, LB, v.1859–79.

114 Murchison to Fortescue, 8 June 1861, RGS, LB, v.1859–79; Murchison, 1861b: clxxv.

115 RBA 1861, TS, 1862: 197.

116 F. Gregory, 1861.

117 Murchison to Shaw, 12 May 1862, RGS, Journ.MSS.Aust., referee's comments; F. Gregory, 1862.

118 JRGS, v.33, 1863: cx–cxi.

119 See below, p. 54 and chapter 7.

120 W. Morrell, 1940: 294–296; Blainey, 1969: 161–168.

121 Murchison, 1859b: ccxix–ccxx; Denison, 1870, i: 455.

122 MacDonnell to Murchison, 26 Oct. 1860, BL(AM) 46127, MuP, ff.142–143; 24 Dec. 1863, GSL/M/M26/1; MacDonnell, 1864: 17, 59; Murchison, 1864b: cxlix–cl.

123 Bowen to Murchison, 30 Apr. 1860, BL(AM) 46125, MuP, ff.197–198; Murchison, 1861b: clxxvi; Bowen, 1889, i: 212–213; see below, p. 54.

124 Sturt to Shaw, 24 May 1858, 23 Aug. 1860, RGS, Sturt Collection; Sturt to Murchison, 9 Feb. 1861, BL(AM) 46128, MuP, ff.161–162.

125 Murchison, 1842: 645

126 Murchison to Mitchell, 19 Jan., 29 Dec., and n.d. 1853, MLS, Mitchell Papers, A294, ff.133–135, 351–354; A295¹, ff.152–154; Murchison, 1853: cxxx.

127 Murchison, 1859b: ccxvi.

128 Nicol to Murchison, 9 July 1854, GSL/M/N5/6; Secord, 1986a: 271–272.

129 Murchison to Sedgwick, 1 Aug. 1854, CUL, SeP, Add.MS.7652.D.notebook V.68; McCoy to Sedgwick, 25 June 1857, 7652.II.J.50.

130 Murchison, 1867a; Vallance, 1981.

131 Darragh, 1987.

132 Elliot to De la Beche, 6 Sept. 1853, BGSA, GSM 1/6, ff.377–378; Reeks to Elliot, 20 Sept. 1853, ff.278–279.

133 Barkly to Bulwer-Lytton, 27 Jan. 1859, MLS, A2346, p. 6385; Murchison to Herschel, 3 Nov. 1863, RS, HeP, HS.12.430.

134 'Second Progress Report', 1856–57; 'Geological Survey', 1859–60. See also Stafford, 1988a. The chief gold deposits of Victoria occur in Ordovician rocks – the stratigraphic classification established after Murchison's death as a compromise between the Cambrian and Silurian Systems.

135 Barkly to Stanley, 12 July 1858, in 'Geological Survey', 1859–60.

136 For example Murchison, 1867c: 464–467.

137 Murchison, 1862b.

138 Murchison, 1863a; Darragh, 1987.

139 Thomson to Murchison, 20 Apr. 1866, BL(AM) 46128, MuP, ff.195–196.

140 Barry to Murchison, 4 Jan. 1870, BL(AM) 46125, MuP, ff.117–120; reply, 15 Mar. 1870, BGSA, GSM 1/9, ff. 179–180.

141 Denison to De la Beche, 1 Mar. 1849, BGSA, GSM 1/14, f.34; De la Beche to Hawes, 27 July 1850, GSM 1/6, ff.11–13.

142 Clarke, 1852; Clarke to Murchison, 18 Feb. 1852, EUL, MuP, Gen.523/6; Blainey, 1969: 36–37.

143 Jackson to Murchison, 2 July 1852, EUL, MuP, Gen.523/6.

144 Murchison to Merivale, 7 Jan. 1856, 14 Dec. 1858, BGSA, GSM 1/7, ff.266–268, 577–578; reply, 6 Nov. 1856, ff.354–355.

145 Murchison, Memorandum, 10 Dec. 1856, BGSA, GSM 2/2; Murchison, 1859b: ccxviii.

146 Smyth, Reeks, and Rudler, 1864: 127.

147 Young to Murchison, 11 Dec. 1859, GSL/M/Y1/1; Gould to Clarke, 16 Sept. 1860, MLS, CIP, MSS.139/38, ff.301–308; 9 Dec. 1863, quoted in Moyal, 1976: 149–150; Gould to Murchison, 20 June 1861, BGSA, GSM 1/15, f.160.

148 Gould, 1862; Murchison, 1862b, 1868a.

149 Johns, 1976: 57.

150 Murchison, 1867c: 461.

151 *JRGS*, v.31, 1861: civ–cx; Stuart, 1861.

152 Murchison, 1863b: cxxiv.

153 *JRGS*, v.32, 1862: c–cii; Murchison, 1863b: clxiii–clxvii.

154 Murchison to Shaw, 19 March 1856, RGS, Journ.MSS.Aust., referee's comments; Austin, 1856.

155 Murchison to Markham, 29 Jan. 1864, RGS, Journ. MSS.Aust., referee's comments; Hargraves, 1863–64. Gold is widespread in Western Australia: the Murchison Goldfield helped set the colony on a firm financial footing in the 1890s.

156 Johns, 1976: 46–48.

157 Murchison to Markham, 16 Jan. 1868, RGS, Journ.MSS.Aust., referee's comments; Clarke, 1867–68; Daintree, 1872.

158 MacMaster to Murchison, 22 Aug. 1861, BL(AM) 46127, MuP, ff.166–167.

159 Crawfurd to Murchison, 17 June 1863, BL(AM) 46125, MuP, ff.487–488.

160 Murchison, 1871b: cxcii.

161 Merivale, 1863; Geikie, 1875, ii: 294.

162 For the Athenaeum, see Murchison to Bulwer-Lytton, 2 May 1859, HRO, LyP; for the reunion, *The Times*, 27 Jan. 1858, p. 12, col. 2; Davison, 1861: 270–274; Murchison, 1867c: 461–462.

163 King to Murchison, 28 Aug. 1852, EUL, MuP, Gen.523/6; see above, pp. 41–44.

164 Stokes to Murchison, n.d. [1858], BL(AM) 46128, MuP, ff.144–145.

165 See above, p. 45; *PRGS*, v.3, 1858–59: 91; Murchison to Bulwer-Lytton, 29 Dec. 1858, HRO, LyP.

166 Murchison, 1859b: ccxx–ccxxi; 1864b: cl; Bowen, 1889, i: 214–215.

167 Murchison, 1864b: cli; Reese, 1968: 13–14, 20.

168 See above, p. 44.

169 MacDonald, 1857.

170 Ruland to Murchison, 31 Jan. 1860, BL(AM) 46127, MuP, ff.483–484; Murchison to J. Hooker, 1 Feb. 1860, RBG, JHP, Eng. letts., MOO-MYL, 1847/1900, v.96, f.410.
171 Seemann, 1862; see below, p. 148.
172 Murchison, 1863b: clxxii.
173 Murchison, 1867c: 286.
174 Oldroyd, 1967; Rudwick, 1972: 208–209.
175 Dieffenbach, 1843, 1846.
176 G. Bell, 1976: 59, 64.
177 ibid.: 27–29, 45.
178 Dieffenbach, 1841; G. Bell, 1976: 71–74, 85–86.
179 Murchison, 1843a: 142–149; the quotation is from 149.
180 Miller to Murchison, 5 Mar. 1845, GSL/M/M15/2.
181 Mantell, 1848.
182 Smyth, Reeks, and Rudler, 1864: 127.
183 Clarke to Murchison, 31 May 1851, EUL, MuP, Gen.523/6; C. Forbes, 1855.
184 J. Stokes, 1851.
185 Brunner, 1850
186 W. Morrell, 1940: 260–281.
187 Clarke to Murchison, 18 Feb. 1852, EUL, MuP, Gen.523/6.
188 Heaphy, 1854.
189 Marshall-Cornwall, 1976: 22–23.
190 Wyld, 1853: 29.
191 Murchison, 1857b: cl–cli; Scherzer to Murchison, 16 Dec. 1860, BL(AM) 46128, MuP, ff.34–35.
192 Denison to Murchison, 9 Nov. 1858, in Denison, 1870, i: 454; Murchison, 1859b: ccxvii.
193 Hochstetter, 1864.
194 H. Haast, 1948.
195 Somerset to Murchison, 25 Apr., 2 June 1860, BL(AM) 46128, MuP, ff.89–91.
196 *PRGS*, v.4, 1859–60: 222; Murchison, 1861b: clvii–clviii.
197 Thomson, 1858.
198 See below, pp. 71–79, 120; Hector to Murchison, 17 Jan., 22 Oct. 1861, GSL/M/H13/1,2; Otago Office to Murchison, 29 Oct. 1861, GSL/M/06/1.
199 Burnett, 1936: 51–54.
200 Lindsay, 1863a, 1863b.
201 Featherston to Murchison, 6 Dec. 1861, with enclosures of *New Zealand Government Gazette* (Wellington), 2 Nov. and 6 Dec. 1861, BGSA, GSM 1/8, ff.201–202, 1/15, ff.174–175.
202 Triphook to Murchison, 2 May 1855; reply, 14 May 1855, BGSA, GSM 1/7, f.187.
203 Murchison to Featherston, 25 Feb. 1862, BGSA, GSM 1/8, ff.204–205; reply, 12 Aug. 1865, ff.347–348; Murchison to W. Mantell, 25 Feb. 1862, TLW, Mantell MS. Papers, 83, f.352.
204 Geikie, 1875, ii: 255.
205 Murchison, 1862a.
206 Murchison to Haast, 12 Jan. 1862, TLW, Haast Papers, MS.37, folder 122.
207 H. Haast, 1948.

208 Murchison to Haast, 16 Nov. 1863, TLW, Haast Papers, MS.37, folder 122.

209 J. Browne, 1983: 130–131; Crosby, 1986: 165, 167, 217–268.

210 Murchison to Haast, 16 Nov. 1863, TLW, Haast Papers, MS.37, folder 122.

211 J. Haast, 1864.

212 J. Hooker to Haast, 18 Feb. 1864, in H. Haast, 1948: 331.

213 Murchison–Haast correspondence, 1864–67, quoted in H. Haast, 1948: 506, 1035–1045; J. Haast, 1865; Murchison, 1864b: cli–cliv, clx–clxxx; 1865d; Oldroyd, 1973.

214 Murchison to Haast, 18 Aug. 1864, TLW, Haast papers, MS.37, folder 122, also partially quoted in H. Haast, 1948: 1038.

215 Hooker to Haast, 3 May 1865; Murchison to Haast, 26 May 1867 – both quoted in H. Haast, 1948: 505–506.

216 For example see J. Haast, 1870; Murchison, 1871b: cxciv–cxcv; for the medal, H. Haast, 1948: 853.

217 Hector to Clarke, 10 Nov. 1863, MLS, ClP, MSS.139/39, ff.185/190; Burton, 1965.

218 Hector to Murchison, [draft] 17 Oct. 1863, National Museum of New Zealand, Hector Papers.

219 Murchison, 1863b: clxx–clxxi. Otago's alluvial gold does in fact derive from veins in early Palaeozoic rocks broken down by orogeny and erosion – Derry, 1980: 85.

220 Hector, 1864; McKerrow, 1864; J. Haast, 1864; see also *PRGS*, v.9, 1864–65: 32–34.

221 Hector, 1865a.

222 Manten, 1968.

223 Hector, 1865b; Oldroyd, 1973.

224 The Canadian Survey, founded in 1842, was not expanded to include all the territories in British North America until 1868–69.

225 Hector, 1879, 1886.

226 J. Hooker to Murchison, 29 Dec. 1865, GSL/M/H58/1.

227 Hutton, 1869.

228 Burton, 1965: 26–28; C. Fleming, 1987.

229 Murchison, 1867a.

230 Oldroyd, 1972.

231 C. Lloyd, 1970: 160–164.

3 The Americas

1 Bayliss, 1978: 202; C. Browne, 1971: 37–119; Jukes–Sedgwick correspondence, 1839–41, CUL, SeP, Add.MSS.7652.I.B,D.

2 Jukes, 1842; Murchison, 1843a: 126–127.

3 C. Browne, 1871: 131–132; Bayliss, 1978: 203, 207.

4 Henwood, 1841; Murchison, 1842: 671.

5 Murchison, 1867c: 426.

6 H. Bell, 1936, i: 252.

7 Section C. minute book, entry for 1840, BLO, BAP.

8 Logan, 1842; Lyell, 1843.

9 Murchison to Logan, 25 Feb. 1843, MUL, LoP, 1207/11/10, 108.

10 Harrington, 1883; Zaslow, 1975.

11 Bagot to Stanley, 18 Feb. 1842, BGSA, GSM 1/2, f.26.

12 Geikie, 1875, i: 368.

13 Murchison to Logan, 4 July 1842, MUL, LoP, 1207/11/10, 86.

14 Logan, MS. history of Canadian Geological Survey, Aug. 1850, McGill University, McLennen Library, Dept of Rare Books and Special Collections.

15 Zaslow, 1975: 3.

16 De la Beche to Logan, 2 Dec. 1844, MUL, LoP.

17 ibid., 4 May 1846.

18 Vecchi, 1978: 46–59.

19 Murchison, 1867c: 9–14, 424–426; Secord, 1986a: 285–290.

20 Murchison, 1858b: clxxvii–clxxix.

21 Murchison, 1852b.

22 Murchison to Logan, n.d. Feb. 1856, MUL, LoP.

23 Ramsay to Logan, 2 June 1859, MUL, LoP.

24 Logan to Murchison, 29 Nov. 1867, GSL/M/L12/2.

25 Logan, 1863; Murchison, 1863a.

26 For example Murchison to Logan, 3 May 1846, MUL, LoP, introducing E. Verneuil.

27 Dawson to Murchison, 18 May 1870, GSL/M/D7/2.

28 Murchison, 1867c: 469; Evans to Murchison, n.d. 1861, GSL/M/E24/2.

29 Ross to Murchison, 28 Sept. 1867, GSL/M/R14/1; cf. Murchison, 1867a, 1868a.

30 Reeks to Logan, 10 Nov. 1855, MUL, LoP, 1207/11/10, 36.

31 Murchison, 1868b: clxvi.

32 See above, p. 7.

33 Murchison, 1867c: 439–442.

34 Back to Murchison, 10 July 1867, GSL/M/B40/1.

35 Ellesmere to Murchison, 7 Jan. 1852, GSL/M/E19/3.

36 Murchison to Shaw, 18 Jan. 1856, RGS, MuP.

37 Murchison, 1853: lxxxii; Robinson to Shaw, 13 June 1853, RGS, GC.

38 P. Sutherland, 1853; see below, p. 158.

39 Murchison to Romaine, 14 May 1860, BGSA, GSM 1/8, ff.98–99; reply, 12 June 1860, ff.99–100; Murchison, 1861b: cxxxvii–cxxxviii.

40 See, respectively, Murchison to Cooke, 6 Nov. 1863, EUL, DC/4/101–103, f.12; Murchison, 1865b: clxxxii.

41 Murchison, 1866b: cxc.

42 GSL/M/J7, 1826–38, ff.244–246.

43 Murchison, 1845a: lxviii–lxx.

44 C. Murchison to Mrs Featherstonhaugh, 8 July 1846, CUL, SeP, Add.MS.765-2.II.kk.56.

45 Murchison, 1850a: 420.

46 W. Morrell, 1940: 120; Gough, 1971: 132–134.

47 Merivale to De la Beche, with enclosures, 22 Apr. 1853, BGSA, GSM 1/6, ff.234–237; reply, 28 Apr. 1853, ff.238–239.

48 Arrowsmith to Shaw, 13 June 1853, RGS, Journ. MSS.N.America, referee's comments on MS. submitted by Capt. Rooney; *JRGS*, 24, 1854: 248–249.

49 Brown, 1868–69; Smyth, Reeks, and Rudler, 1864: 135.

50 Hammond to Shaw, 19 Oct. 1859, RGS, FO.

51 *PRGS*, 12, 1867–68: 188; Murchison, 1850c.

52 Murchison to Featherstonhaugh, 10 and 21 Aug. 1861, CUL, SeP, Add. MS.7652.II.kk.85–86.

53 Franklin to Murchison, 29 July 1861, BL(AM) 46126, MuP, ff.273–277.

54 Palmerston to Murchison, 3 Oct. 1861, BL(AM) 46127, MuP, ff.342–343.

55 See Spry, 1963, 1968: especially xv–cxxxviii.

56 Murchison to Labouchere, 6 Jan. 1857, RGS, LB, v.1850–59, ff.245–249; also printed in Spry, 1968: 495–500.

57 Spry, 1968: xxiii–xxv, xxxiv, 505–512; Memorandum, 16 Mar. 1857, regarding Ball to Murchison, 6 Mar. 1857, RGS, MuP; Ball to Wrottesley, with enclosures, 6 Mar. 1857, RS, MM.4.38.

58 Murchison, 1857b: clxxx–clxxxii.

59 Synge, 1852.

60 Murchison, 1852a: cxv–cxvii.

61 Synge to Shaw, 24 Aug. 1852, RGS, GC.

62 ibid., 17 Nov. 1853, 13 Feb. 1855.

63 Nicolay to Shaw, 27 Feb. 1857, RGS, GC; Spry, 1968: xxii.

64 Kernaghan to Murchison, 4 Apr. 1857, HLD, HeP, MS.443, f.682; Spry, 1968: xxii.

65 Bannister to Murchison, 28 Apr. 1857, HLD, HeP, MS.443, f.522.

66 W. Grant, 1857; Grant to Shaw, 16 Dec. 1857, RGS, GC.

67 K. Sutherland, 1858.

68 Palliser to Murchison, 7 May 1857, BL(AM) 46127, MuP, ff.338–339.

69 Murchison to Hector, 7 May 1857, HLD, HeP, MS.443, f.655; [13 May 1857], BGSA, GSM 1/7, ff.401–405, also printed in 'The Journals', 1863: 656–657.

70 Isbister, 1855; Murchison to J. Hooker, 17 July 1855, RBG, JHP, Eng. letts. MOO-MYN, 1847/1900, v. 96, f.395.

71 Beaufort to De la Beche, 9 Apr. 1847, BGSA, GSM 1/13, f.12; Gough, 1971: 98–106, 184; for Labuan, see below, pp. 147–148.

72 Murchison, 1867c: 440.

73 Nicolay to Shaw, 15 Dec. 1857, RGS, GC.

74 Hammond to Murchison, 18 Jan. 1858, RGS, FO.

75 Panmure to Murchison, with enclosure, 8 Feb. 1858, RGS, GC.

76 Murchison to Hammond, 2 Mar. 1858, RGS, LB, v.1850–59, ff.278–280.

77 Murchison to Sharpey, 26 Jan. 1858, RS, MC.5.330.

78 Murchison to Bauermann, 10 Mar. 1858, BGSA, GSM 1/7, ff.524–528.

79 Murchison, 1858b: clxxvi–clxxvii; W. Morrell, 1940: 119–127.

80 Hector to Murchison, 6 Jan. 1858, RGS, GC; Murchison, 1858b: clxxvi–clxxviii.

81 Hector to Murchison, 9 June 1858, RGS, GC.

82 Hammond to Murchison, 19 July 1858, RGS, FO.

83 Carnarvon to Palliser, 31 July 1858, HLD, HeP, cited in Burnett, 1936: 39.

84 Palliser to Labouchere, 13 Mar. 1858, in Spry, 1968: 513–520.

85 Murchison to Bulwer-Lytton, 10 July 1858, HRO, LyP.

86 Murchison to Shaw, 29 Jan. 1859, RGS, MuP.

87 Murchison to Bulwer-Lytton, 31 Jan. 1859 [draft], RGS, MuP.

88 Palliser and Hector, 1858–59; for Bury, see Reese, 1968: 14–20, 34–35.

89 Murchison to Hector, 16 Feb. 1859, HLD, HeP, MS.443c, f.633.

90 Hector to Murchison, n.d. [Jan. 1859], HLD, HeP, MS.443c, f.636; Merivale to Murchison, 24 Mar. 1859, BGSA, GSM 1/8, f.13; reply, 2 Apr. 1859, 1/7, f.600.

91 *JRGS*, 29: xcvii–c.
92 Murchison, 1859b: cxcv–cciii; Bauermann, 1860.
93 W. Morrell, 1940: 128–134.
94 Mayne, Palmer, and Begbie, 1861.
95 Hector to Murchison, 18 May 1859, HLD, HeP, MS.443c, f.631.
96 Palliser to Bulwer-Lytton, 20 May 1859, in 'Further Papers', 1860: 430–433, and Spry, 1968: 521–526.
97 Palliser, Hector, and Sullivan, 1860.
98 Murchison, 1861c.
99 'Further Papers', 1860, also partially printed in Spry, 1968: 521–581.
100 Hector, 1861; 'The Journals', 1863, also in Spry, 1968: 1–493.
101 Murchison, 1861b: clxix–clxxii.
102 Murchison to Murray, 19 July 1861, BGSA, GSM 1/8, ff.206–208.
103 For example W. Grant, 1861.
104 Smyth, Reeks, and Rudler, 1864: 135.
105 W. Kelly, 1861–62.
106 'Copies of Correspondence', 1863; 'Copy or Extracts', 1864.
107 Synge, 1862–63; W. Morrell, 1940: 129.
108 Murchison, 1864b: cxxx.
109 Murchison to Markham, 15 Apr. 1864, RGS, Journ. MSS.N.America, referee's comments.
110 Palmer, 1864.
111 Murchison, 1868b: clxxi–clxxii.
112 Thornton–Clarendon correspondence, Jan.–June 1870, BLO, MSS.Clar.Dep. C.481, ff.3–90, C.476(3), ff.125–165.
113 Stephen to Jackson, 19 Jan. 1841, RGS, CO.
114 Murchison, 1844: xciv–xcvii.
115 Schomburgk, 1845a.
116 Murchison, 1845a: xlix; Schomburgk, 1845b, 1846.
117 Parish to Shaw, 31 Jan. 1854, RGS, Parish Collection.
118 Murchison to Russell, 25 June 1855, BGSA, GSM 1/7, ff.215–217.
119 Murchison to Elliot, with enclosure, 26 Nov. 1855, BGSA, GSM 1/7, ff.237–238, 253–254; Merivale to Murchison, 15 Dec. 1855, ff.261–262.
120 Merivale to Murchison, 11 Jan. 1856, BGSA, GSM 1/7, ff.274–275.
121 Murchison to Merivale, 14 Feb. 1856, BGSA, GSM 1/7, ff.275–277; Murchison to Labouchere, 5 Apr. 1856, ff.294–295.
122 Murchison, 1856.
123 Murchison to Labouchere, 4 Apr. 1857, BGSA, GSM 1/7, ff.387–388; Ball to Murchison, 16 May 1857, f.407.
124 Murchison to Keate, 2 Dec. 1857, BGSA, GSM 1/7, ff.502–504.
125 Murchison, 1859b: cliv–clv; Wall and Sawkins, 1860; Wall, 1860.
126 Murchison to Elliot, 16 Oct. 1857, BGSA, GSM 1/7, f.444.
127 Murchison to Merivale, 2 Oct. 1858, BGSA, GSM 1/7, ff.554–555.
128 Barrett to Phillips, 25 Jan. 1859, UMO, PhP, 1859/6.
129 Murchison to Carnarvon, 4 May 1859, BGSA, GSM 1/7, ff.601–602.
130 Murchison to De la Beche, 20 Mar. 1841, NMW, DeP; McCartney, 1977: 22–25.
131 Howard de Walden to Murchison, 20 Nov., 20 Dec. 1853, BL(AM) 46126, MuP, ff.423–430.

132 Rogers to Murchison, with enclosures, 30 June 1860, BGSA, GSM 1/15, ff.114, 118–120.

133 Murchison to Rogers, 6 July 1860, BGSA, GSM 1/8, ff.103–104.

134 For the quotation, see Murchison, 1859c; see also Murchison, 1859b: ccx–ccxi; Murchison to Newcastle, 3 Dec. 1859, BGSA, GSM 1/8, ff.51–52.

135 Murchison, 1861a.

136 Rogers to Murchison, with enclosures, 27 Sept. 1865, BGSA, GSM 1/8, ff.351–352, 1/16, ff.39, 46; reply, 1 Oct. 1865, 1/8, ff.354–355; Forster to Murchison, 13 Jan. 1866, f.381.

137 Sawkins, 1869.

138 Murchison to Elliot, 19 Oct. 1867, BGSA, GSM 1/8, f.535.

139 'Boundaries', 1857.

140 Merivale to Murchison, 9 Feb. 1858, BGSA, GSM 1/7, f.520.

141 Holmes and Campbell, 1857–58.

142 Sawkins to Murchison, 4 Aug. 1868, BGSA, GSM 1/9, ff.58–59.

143 Foster, 1869.

144 Sawkins and Brown, 1875.

145 Ward to De la Beche, with enclosures, 18 Apr. 1848, BGSA, GSM 1/13, f.67; Addington to De la Beche, 12 Apr. 1849, f.136.

146 Murchison, 1844: xci–xciv.

147 Murchison, 1845a: lxix–lxx.

148 'Copies of the Reports', 1851.

149 MS. 'Extracts from the Journals of Mrs. W. H. Smyth', entry for 5 Nov. 1850, RGS, Smyth Collection; Palmerston to Shaw, 21 Mar. 1851, FO; Fitzroy, 1850.

150 Cullen, 1851.

151 Murchison, 1850c; Wyld, 1853: 12.

152 Murchison to Shaw, 27 Mar. 1852, RGS, MUP; Gisborne and Forde, 1853.

153 Warburton to Murchison, n.d. 1851, BL(AM) 46128, MuP, ff.257–258.

154 Cullen–Shaw correspondence, 1850–52, RGS, GC.

155 Haldane to Shaw, 14 June 1852, RGS, GC.

156 Murchison, 1853: cxix–cxx.

157 Fitzroy, 1853.

158 Prevost, 1854.

159 Fitzroy to Shaw, 10 May 1856, RGS, Fitzroy Collection.

160 F. Kelly, 1856.

161 Murchison, 1857b: clxxv–clxxvi.

162 Whittington to Shaw, 1 Dec. 1857, RGS, GC.

163 Murchison, 1858b: clxxix–clxxx.

164 Murchison, 1859b: ccviii–ccix.

165 Pim, 1861–62.

166 Oliphant, 1865.

167 Collinson, 1867–68; Murchison, 1867a; Pim, 1868.

168 Baker, 1868.

169 Murchison, 1868b: clxxii–clxxiii.

170 Crichton to Murchison, 7 Apr. 1826, GSL/M/C19/1.

171 Murchison, 1850c.

172 For the fossils, see Murchison, 1867c: 424; for the mission, Pentland, 1827.

173 Pentland to Phillips, 26 June 1836, UMO, PhP, 1836/25.

174 Parish to Bates, 17 Apr. 1876, RGS, Parish Collection.
175 Parish to Shaw, 18 Apr. 1858, RGS, Parish Collection.
176 Parish to Washington, 22 Feb. 1839, with enclosure of Murchison's referee's comments, RGS, Parish Collection.
177 Cf. Murchison, 1839b: 583; 1867c: 425.
178 Herschel to Murchison, 13 Feb. 1842, EUL, MuP, Gen.523/4/28.
179 RGS, CMB, v. 1841–53, entry for 25 Mar. 1850, f.246; Henderson and Paynter, 1850.
180 Henwood to Murchison, 26 Oct. 1850, EUL, MuP, Gen.523/6.
181 Baikie to Parish, 22 Nov. 1852, RGS, BaP; *JRGS*, 22, 1852: lvi; Murchison, 1853: cxxi–cxxii.
182 Murchison, 1857b: clxxxvi–clxxxviii.
183 For example Murchison, 1866b: clxxxv–clxxxviii.
184 For example Jackson, 1858.
185 Grey to Murchison, n.d. 1853, RGS, Journ.MSS.S.America; Murchison, 1853: cxxii–cxxiii.
186 J. Lloyd, 1853, 1854; Murchison, 1867c: 475; Parish to Shaw, 16 May 1855, RGS, Parish Collection.
187 D. Forbes, 1861a.
188 For the quotation, see Murchison, 1867c: 425; see also ibid., 467–474; D. Forbes, 1866.
189 D. Forbes, 1861b.
190 D. Forbes to Murchison, 13 Dec. 1860, GSL/M/F10/1.
191 Gerstenberg to Shaw, 15 June, 18 Aug., 13 Nov. 1860; Levinsohn to Shaw, 5 Apr. 1860 – all in RGS, GC; Jameson, 1861.
192 Neale, 1866.
193 *PRGS*, 3, 1858–59: 93–98.
194 *PRGS*, 4, 1859–60: 33.
195 Williams, 1962; Brockway, 1979: 103–139.
196 C. Markham, 1861; Murchison, 1866b: clxxvi–clxxviii; Markham to Murchison, 24 Jan. 1862, BL(AM) 46127, MuP, ff.184–185.
197 *PRGS*, 6, 1861–62: 75–79.
198 For example *PRGS*, 6, 1861–62: 248–249; 10, 1865–66: 107.
199 Fitzroy to Murchison, 26 Feb. 1862, BL(AM) 46126, MuP, ff.164–165.
200 Wheelwright, 1861, 1867.
201 Murchison, 1871b: cxcvii.
202 Murchison to Layard, 22 Feb. 1866, BGSA, GSM 1/8, ff.384–389.
203 Hammond to Murchison, 6 Nov. 1866, BGSA, GSM 1/8, f.427; reply, 14 Nov. 1866, ff.429–431.
204 Murchison, 1867c: 303–307.
205 E. Thornton, 1867.
206 Murchison to Bates, 3 July 1871, APS, MuP, BM93p; Reeks to Arēas, 17 July 1871, BGSA, GSM 1/9, ff.263–264; *PRGS*, 15, 1870–71: 365–367.

4 The Middle East

1 GSL/M/J7, 1826–38, ff.215–216.
2 Murchison, 1868b: cxxxiv–cxxxvii.

3 See Strickland and Hamilton, 1842; Hamilton, 1842.
4 Campbell, 1866–67.
5 'Report', 1834.
6 Chesney and Ainsworth, 1837; Chesney, 1850.
7 Headrick, 1981.
8 Lynch, 1839.
9 Lynch to Washington, 28 July 1839, RGS, GC.
10 F. Bailey, 1942; Platt, 1968: 185–225. Layard, 1903, i: 71–72, 328–331.
11 Chesney to Washington, 6 Feb. 1838, RGS, Chesney Collection.
12 Chesney to Washington, 13 Apr. 1838, RGS, Chesney Collection.
13 Chesney to Ponsonby, 10 Sept. 1839 [copy], RGS, Chesney Collection.
14 Palmerston to Washington, 7 Aug. 1840, RGS, FO.
15 Hamilton to Palmerston, 8 July 1848, RGS, LB, v. 1844–50, f.268; Addington to Hamilton, 15 Aug. 1848, FO.
16 GSL/M/J8, 1840–45, f.5.
17 Eddisbury to De la Beche, 20 Dec. 1848, BGSA, GSM 1/13, f.109; Loftus to De la Beche, with enclosure, 5 Jan. 1849, ff.113–114; 'Correspondence Respecting', 1865.
18 Loftus to Smyth, 4 May 1850, 4 Sept. 1851, CUL, SeP, Add.MS.7652.VQ.31, 32; W. Beaver, 1976.
19 Loftus, 1855, 1856.
20 Murchison, 1845a: lvi–lvii.
21 Murchison, 1844: ciii–civ.
22 Ellis to Murchison, n.d. 1851, GSL/M/E7/2.
23 Hicks, 1853.
24 Chikhachëv to Murchison, 19 Mar. 1847, GSL/M/T4/1.
25 Chikhachëv, 1849; Murchison, 1853: ciii–civ; Geikie, 1875, ii: 148–150.
26 Chikhachëv to Murchison, 5 Oct. 1853, GSL/M/T4/6.
27 Murchison, 1867c: 368, 398.
28 Murchison to C. Murchison, 20 Sept. 1857, EUL, MuP, Gen.523/1/15.
29 Holland to Murchison, 31 Dec. [1853], EUL, MuP, Gen.523/6.
30 Holland to Murchison, 12 Jan. and n.d. 1854, EUL, MuP, Gen.523/6.
31 Brunnow to Murchison, 8 Feb. 1854, BL(AM) 46125, MuP, ff.279–280.
32 Murchison to Clarendon, 4 Feb. 1855, BLO, MSS.Clar.Dep.C.103, ff.516–519.
33 Murchison, Memorandum, 24 June 1861, BGSA, GSM 1/8, ff.170–174.
34 Burgoyne to Murchison, 4 June 1855, BL(AM) 46125, MuP, ff.330–331; reply, 5 June 1855, Royal Engineers Museum, Burgoyne Papers; Murchison, 1863b: cxiv.
35 Murchison to Clarendon, 17 Apr. 1854, 4 Feb. 1855, BLO, MSS.Clar.Dep. C.103, ff.231–232, 516–519; Herbert to Murchison, 16 Mar. 1854, BL(AM) 46126, MuP, ff.396–397.
36 Murchison to Shaw, 14 Jan. 1854, RGS, MuP; Jochmus, 1854.
37 Lloyd to Murchison, n.d. 1854, BL(AM) 46127, MuP, ff.88–89.
38 For example Murchison to Clarendon, 10 May, 2 July 1854, BLO, MSS.Clar. Dep.C.103, ff.233–235, 301–308.
39 Crosse, 1892, ii: 49–50.
40 Denison to Murchison, 21 May 1855, in Denison, 1870, i: 309–310.
41 Murchison to Clarke, 15 Aug. 1855, MLS, ClP, MSS.139/43, ff.695–698.
42 GSL/M/J17, 1845, f.100.

43 Murchison to Murray, 13 Nov. 1854, JMA.

44 Murchison to Clarendon, 11 Apr. 1853, BLO, MSS.Clar.Dep.C.103, ff.229–230; Page, 1976: 160–161.

45 Warrington to Sedgwick, 13 Apr. 1846, CUL, SeP, Add.MS.7652.I.E.Box III.155.

46 *QJGS*, 6, 1850: 367; 7, 1851: 65.

47 Hay to Murchison, 4 Oct. 1853, GSL/M/H47/1.

48 Ellesmere to Clarendon, 7 Apr. 1854, BLO, MSS.Clar.Dep.C.23, f.4.

49 Stratford de Redcliffe to Clarendon, 11 Jan. 1855, BLO, MSS.Clar.Dep.C.40, ff.22–25; see also Lane-Poole, 1888, ii: 352.

50 Murchison to Clarendon, 26 Apr. 1855, BLO, MSS.Clar.Dep.C.103, ff.539–540; Sandison, 1855.

51 Murchison to Wodehouse, 8 May 1855, BGSA, GSM 1/7, ff.199–201.

52 Hammond to Murchison, with enclosure, 18 June 1855, BGSA, GSM 1/7, ff.211–213; Murchison to Clarendon, 11 May 1855, BLO, MSS.Clar.Dep.C.42, ff.378–379.

53 Murchison to Hammond, 11 June 1855, BGSA, GSM 1/7, ff.202–204.

54 Murchison to Poole, 21 June 1855, BGSA, GSM 1/7, ff.213–215.

55 Poole to Murchison, 21 July 1855, BGSA, GSM 1/7, ff.233–235; reply 2 Aug. 1855, ff.235–236.

56 Murchison to Clarendon, 7 Oct. 1855, BLO, MSS.Clar.Dep.C.103, ff.549–551.

57 Palmerston to Clarendon, with enclosure, 9 Aug. 1855, BLO, MSS.Clar.Dep. C.31, f.396, C.42, ff.429–436.

58 Poole, 1856.

59 Clarendon to Murchison, 12 Nov. 1856, GSL/M/C52/1.

60 Murchison, 1856.

61 'Memorandum', 1857.

62 Murchison to Hatherton, 21 Jan. 1856, Staffordshire Record Office, Hatherton Papers, D260/M/F5/27/29/10.

63 Murchison to Clarendon, 8 May 1856, BLO, MSS.Clar.Dep.C.57, ff.59–62.

64 *PRGS*, 1, 1856–57: 483.

65 Rawlinson, 1856–57.

66 Backhouse to Washington, 17 Nov. 1837, RGS, FO.

67 For example *QJGS*, 15, 1859: 605–606; 17, 1861: xi.

68 Murchison, 1857b: clv; Monteith, 1857.

69 GSL/M/J24, 1857–59, ff.54–55.

70 Murchison, 1859b: clxix–clxx.

71 Oakley to Murchison, 15 Mar. 1859, BL(AM) 46127, MuP, ff.281–282.

72 For example Murchison, 1865b: clxxiv–clxxv.

73 Goldsmid, 1863; Murchison, 1863b: cli–clvi.

74 Chesney, 1853.

75 Lane-Poole, 1885: 422–460.

76 Murchison, 1852a: cxvii.

77 Burgoyne to Murchison, 13 Jan. 1853, RGS, MuP; Argyll to Murchison, 25 May 1853, BL(AM) 46125, MuP, ff.43–44; Murchison, 1853: civ–cvi.

78 Murchison, 1858b: clxxxi–clxxxiii.

79 Russell to Murchison, 23 July 1861, Public Record Office, Russell Papers, 30/32/116/228.

80 Murchison to Huxley, 30 July 1861, Imperial College, Huxley Papers, 23.157; reply, 31 July 1861, 23.159.

81 Lennox to Murchison, 27 May 1865, BL(AM) 39115, LaP, ff.343–345.

82 Murchison to Layard, 13 June 1865, BL(AM) 39115, LaP, ff.339–342.

83 Musurus to Murchison, 6 July 1865, BL(AM) 46127, MuP, f.257.

84 Lennox to Murchison, 4 Oct. 1867, BL(AM) 46127, MuP, ff.62–63.

85 British Association Section E Minutes, v.1865–73, f.2, RGS.

86 W. Beaver, 1976.

87 Murchison, 1858b: clxiv–clxvi.

88 Geikie to Murchison, 21 Jan. 1866, GSL/M/G2/6.

89 Murchison, 1852a: cxiv–cxv; 1853: cvi–cvii.

90 Palgrave, 1864; Murchison, 1864b: cxlvi.

91 Strangford, 1869, ii: 110; Pelly, 1865.

92 Palgrave, 1869.

93 Bauermann, 1869; Murchison, 1869a.

5 *The Indian empire and Central Asia*

1 MacLeod, 1975; Dionne and MacLeod, 1979.

2 Layard, 1903, i: 328–330.

3 Mill, 1930: 44.

4 Hotchkis to Hastings, 20 May 1779, IOLR, Home Misc./358(1), ff.51–61; Berman, 1977: 88–92; Weindling, 1979: 251–254.

5 [Anon.], 1816.

6 Hardwick to Harrington, 10 Aug. 1818, BL(AM) 9894, f.29; Blanpied, 1973.

7 Fermor, 1951.

8 Larwood, 1958.

9 Conybeare, 1833: 395–396, 413.

10 Moore, 1982.

11 Murchison, 1868b: cxlix.

12 Carr to Murchison, 9 Nov. 1828, GSL/M/C3/2.

13 Murchison, 1833: 457; Ward, 1834.

14 Christie to Murchison, 1 Mar. 1831, GSL/M/C9/1.

15 Burnes to Murchison, 2 Mar. 1835, GSL/M/B38/1; White to Murchison, 18 Oct. 1834, GSL/M/W5/1.

16 IOLR, IBD, E/4/761, ff.415–416; Murchison, 1843a: 135–136.

17 Moore, 1982: 403–404.

18 Fox, 1947; Kumar, 1985; IOLR, IBD, E/4/760, ff.348–352.

19 McClelland to Lyell, 10 Feb. 1841, enclosure in IOLR, BCC, 99802A, draft no. 957, 1845; see also 'Copy of Dr. M'Clelland's Report', 1863.

20 Lyell to McClelland, 10 May 1843; Murchison to McClelland, 22 Sept. 1842, both enclosures in IOLR, BCC, 99802A, draft no. 957, 1845.

21 Murchison, 1843a: 134–135; India Marine to Home Marine, 14 Oct. 1843, IOLR, BCC, 99802A, draft no. 957, 1845.

22 Buckman to Murchison, 12 Jan. 1844, GSL/M/B35/1; Sykes to Murchison, 12 Mar. 1845, GSL/M/S38/1.

23 Murchison *et al.*, 1845, i: 6.

24 Williams to De la Beche, 8 June 1845, NMW, DeP.

25 Fox, 1947; Fermor, 1951; Kumar, 1985.

26 IOLR, IBD, E/4/762, ff.991–994; De la Beche to Melvill, 9 Sept. 1841, BGSA, GSM l/1, ff.224–225.

27 Williams–De la Beche correspondence, 1846–47, MNW, DeP; Rao, 1975.

28 IOLR, IBD, E/4/788, ff.62–66.

29 Bernstein, 1960.

30 IOLR, IBD, E/4/767, ff.f1081–1082; /770, ff.640–641; /772, f.1243; /773, f.147; /775, f.119; /791, f.790.

31 Vicary to Murchison, 8 Sept. 1847, 30 Mar. 1848, GSL/M/V6/1–2.

32 Vicary, 1847.

33 Owen to Murchison, 12 Mar. 1849, GSL/M/O7/14.

34 Postans, 1844; Murchison, 1845a: xlix.

35 Vicary, 1851.

36 Murchison, 1859b: cxxii.

37 Lyell to Murchison, 30 Nov. 1843, n.d. [1845], GSL/M/L17/26, 28; J. Fleming to Murchison, 5 June 1845, GSL/M/F8/2.

38 IOLR, Home Misc.761(4), ff.727–764.

39 J. Fleming–Murchison correspondence, 1853, GSL/M/F8/3–8, partially published in Moore, 1982: 405–408.

40 A. Fleming, 1853a, 1853b.

41 Murchison, 1852a: cx.

42 Hogg to De la Beche, 29 Mar. 1853, BGSA, GSM 1/7, f.44; reply, 19 Apr. 1853, ff.47–48; Melvill to De la Beche, 24 Dec. 1853, f.105; reply, 8 Mar. 1854, f.104; see also E. Stokes, 1959.

43 Murchison, Geological Survey notebook, entry for 7 Aug. 1855, GSL/M/N167, 1855–56, ff.24–25.

44 Geikie, 1875, ii: 310.

45 For example Murchison, 1858a: clxxxix; Murchison to Merivale, 12 Aug. 1867, BGSA, GSM 1/8, ff. 509–510.

46 Conybeare to Murchison, n.d. Jan. 1856, GSL/M/C17/9.

47 Murchison, 1863a; Denison to Murchison, 26 Oct. 1862, in Denison, 1870, ii: 216–218.

48 Geikie, 1875, ii: 280–281.

49 Murchison to Oldham, 25 Sept. 1856, BGSA, GSM 1/7, ff.333–335.

50 Murchison, 1858a.

51 C. Browne, 1871: 566.

52 Murchison to Stanley, n.d., BGSA, GSM 1/7, f.568; Murchison, 1868a.

53 Murchison to Stanley of Alderley, 25 Jan. 1856, BGSA, GSM 1/7, ff.301–307.

54 Williams to Murchison, 1 Sept. 1856, GSL/M/W41/1.

55 Melvill to Murchison, 28 Oct. 1858, BGSA, GSM 1/7, f.565; reply, 2 Nov. 1858, ff.566–567. Murchison suggested that Oldham's staff perform this work.

56 Dufferin to Murchison, 16 June 1865, BGSA, GSM 1/8, ff.344–345; reply, 27 Nov. 1865, f.363.

57 Rogers to Murchison, 4 May 1864, BGSA, GSM 1/8, ff.297–298; reply, 23 June 1864, f.303.

58 Murchison to Hammond, 26 Dec. 1867, BGSA, GSM 1/8, ff.12–14.

59 Adderley to Murchison, with enclosure, 13 June 1868, BGSA, GSM 1/9, ff.52–53, 1/17, f.46; reply, 20 June 1868, 1/9, ff.54–55.

60 GSL/M/J15, 1843–44, f.238.
61 R. Murchison–C. Murchison correspondence, Aug.–Sept. 1857, EUL, MuP, Gen.423/1/6–15.
62 Murchison to Denison, 28 Aug. 1862, in Geikie, 1875, ii: 285.
63 Ellenborough to Murchison, 28 Feb. 1858, BL(AM) 46126, MuP, ff.97–98.
64 See above, p. 45.
65 GSL/M/J24, 1857–59, ff.49–52.
66 Denison to Murchison, 8 Nov. 1863, 8 Mar. 1864, in Denison, 1870, ii: 287–288, 333–334; see above, p. 31.
67 *JRGS*, 22, 1852: lviii–lxi.
68 Murchison, 1858b: clxxxvii.
69 Herschel to Murchison, 13 Feb. 1842, EUL, MuP, Gen.523/4/8.
70 Batten to Sedgwick, 20 Jan. 1841, CUL, SeP, Add.MS.7652.I.F.Portfolio II.80.
71 R. Strachey, 1851a, 1851b; H. Strachey, 1853; Council Minutes, v.1841–57, entry for 31 Jan. 1852, BLO, BAP.
72 Moore, 1982: 418; Murchison, 1867c: 18.
73 J. Hooker to Murchison, n.d. [May–July 1852], EUL, MuP, Gen.523/3.
74 Murchison, 1852a: cvi–cx.
75 See above, note 42.
76 Murchison, 1867a; Murchison to J. Hooker, 7 Apr. 1866, RBG, JHP, Eng. letts. MOO-MYL, 1847/1900, v.96, f.420.
77 MacLeod, 1975.
78 Murchison to J. Hooker, 19 Jan. 1854, RBG, JHP, Eng. letts., MOO-MYL, 1847/1900, v.96, f.392.
79 Murchison to J. Hooker, 13 and 20 July 1859, RBG, JHP, Eng. letts., MOO-MYL, 1847/1900, v.96, ff.404, 409; reply, 19 July 1859, ff.406–408; Murchison to Lyell, 23 Aug. 1861, APS, Darwin–Lyell correspondence B:025.L; Huxley, 1918, ii: 146–147; Murchison, 1858b: clxxxiii–clxxxvii.
80 For example Murchison, 1859b: cxxxvii–cxxxviii.
81 Godwin-Austen, 1864a, 1864b; Murchison, 1864b: clx–clxxx.
82 Godwin-Austen, 1867; Murchison, 1867b: cxlix.
83 Murchison, 1867c: 18.
84 Vigne to Murchison 28 June 1849, BL(AM) 46128, MuP, ff.236–237.
85 Canning to Murchison, 29 Aug. 1859, in *PRGS*, 4, 1859–60: 30.
86 Windham to Murchison, 19 Nov. 1861, BL(AM) 46128, MuP, ff.321–322.
87 Murchison to Harkness, 23 Nov. 1861, CUL, SeP, Add.MS.7652.IV.B.32.
88 Murchison to Geikie, 21 Nov. and two of 23 Nov. 1861, GSL/M/M52–54.
89 Murchison to Wood, 16 Jan. 1862, BGSA, GSM 1/8, ff.199–200; Reeks, 1920, pt.2: 52–53.
90 Murchison, 1862a.
91 Alder, 1963: 106.
92 Murchison, 1868a.
93 Drew, 1875.
94 Murchison *et al.*, 1845, i: 310–313.
95 For background, see Gleason, 1950.
96 GSL/M/J9, 1840–45, ff.222–223; Murchison, 1842: 644; see above, pp. 14–16.
97 Murchison, 1853: lxxxvii–lxxxix; 1865b: clxiv; Derry, 1980: 66.
98 Murchison, 1867c: 368.

99 Wodehouse to Clarendon, 29 Aug. 1856, BLO, MSS.Clar.Dep.C.57.
100 Atkinson, 1859, 1861.
101 Murchison, 1858b: clvii–clviii; L. Atkinson to Murchison, 30 Dec. 1869, BL(AM) 46125, MuP, f.74.
102 Murchison, 1850c.
103 Murchison, 1857b: cliv.
104 Murchison to Hay, 1 June 1863, NLS, MuP, MS.14466, ff.9–10; Murchison, 1863b: clii–cliii.
105 Sheil to Murchison, 11 June 1864, BL(AM) 46128, MuP, ff.65–66; Vámbéry, 1863–64; Murchison, 1865b: clxiv–clxv.
106 Denison to Murchison, 23 July 1865, in Denison, 1870, ii: 400–402; Vámbéry to Murchison, 13 Dec. 1864, BL(AM) 46128, MuP, ff.228–229.
107 Ellenborough to Murchison, 24 Feb. 1865, BL(AM) 46126, MuP, f.98.
108 For Saunders, see Murchison to Northcote, 15 Oct. 1867, BLO, MS. Autogr.b.3, f.92.
109 Murchison to Murray, 27 Jan. 1866, JMA.
110 For example Hippius, 1865.
111 See respectively Murchison, 1865b: clii; 1863b: cli.
112 Murchison to de Grey and Ripon, 23 Jan. 1867, BL(AM) 46126, MuP, ff.13–14.
113 *PRGS*, 13, 1868–69: 20.
114 Duthie, 1983; A. Thornton, 1954.
115 Rawlinson, 1866.
116 Rawlinson, 1865.
117 Rawlinson, 1866–67.
118 Murchison, 1867b: cxxxii–cxlix.
119 Rawlinson, 1868–69.
120 Strangford, 1869, ii: 332–335; see also 263.
121 Rawlinson, 1866–67.
122 Richards, 1869: 123.
123 Baker, 1868: 108.
124 *PRGS*, 9, 1864–65: 276.
125 *PRGS*, 11, 1866–67: 163–166.
126 *PRGS*, 12, 1867–68: 169–170.
127 Kiernan, 1953–55; A. Thornton, 1956; Alder, 1963.
128 *JRGS*, 40, 1870: cxxvi.
129 Richards, 1869: 123.
130 Murchison, 1868b: clxxvii–clxxviii.
131 Alder, 1963: 165; Smith, 1981: 686–687.
132 Mayo to Murchison, 12 Mar. 1869, CUL, MaP, Add.MS.7490/34/93.
133 Rawlinson, 1868–69; Montgomerie, 1869; Forsyth, 1868–69; *PRGS*, 13, 1868–69: 200–203.
134 Murchison, 1869b: clxxvii–clxxxi.
135 Frere, 1870.
136 Murchison to Clarendon, 16 Nov. 1869, BLO, MSS.Clar.Dep.C.510, folder 1; see also A. Thornton, 1953–55: 209.
137 Forsyth, 1870; Saunders, 1870.
138 Khanikov, 1870; Chikhachëv, 1870.

139 Grant Duff to Murchison, n.d. Oct. 1869, GSL/M/D59/1.
140 Mayo to Murchison, 29 July, 4 and 19 Aug. 1869, CUL, MaP, Add.MS.7490/ 36/176, 182, 203; Showers to Clarendon, 18 Aug. 1869, BLO, MSS.Clar. Dep.C.510, folder 1.
141 Hayward to Murchison, 11 Sept. 1869, in *PRGS*, 14, 1869–70: 40–41.
142 *PRGS*, 14, 1869–70: 2–7.
143 Hayward to Murchison, 21 Oct., 17 Nov. 1869, RGS, C.Asia.
144 Murchison to Osten Sacken, 16 Nov. 1869, RGS, LB, v.1859–79.
145 Forsyth to Murchison, 1 Dec. 1869, GSL/M/F27/1.
146 Kiernan, 1953–55.
147 Hayward, 1870.
148 Hayward–Rawlinson correspondence, RGS, C.Asia.
149 Mayo to Murchison, 30 Nov. 1869, BL(AM) 46127, MuP, ff.195–196.
150 Mayo to Murchison, 27 Sept. 1869, CUL, MaP, Add.MS.7490/36/260.
151 Mayo to Murchison, 30 Nov. 1869, BL(AM) 46127, MuP, ff.195–196.
152 Forsyth to Murchison, 18 Nov. 1869, RGS, C.Asia.
153 Shaw to Forsyth, 22 July 1869, enclosure in ibid.
154 Montgomerie, 1869–70.
155 Shaw to Murchison, 22 Feb. 1870, GSL/M/S55/1; Murchison to Murray, 23 Feb. 1870, JMA.
156 Mayo to Murchison, 30 Nov. 1869, BL(AM) 46127, MuP, ff.195–196; 8 Feb. 1870, CUL, MaP, Add.MS.7490/38/54.
157 Shaw, 1869–70.
158 Hayward to Showers, 17 Feb., 27 Apr., 8 and 11 May 1870; Hayward to Murchison, 14 Mar. 1870 – all in RGS, C.Asia; Hayward, 1871.
159 Osten Sacken, 1870; Morgan, 1869–70.
160 *JRGS*, 40, 1870: cxxiv–cxxviii.
161 Murchison, 1870b: clx–clxvii.
162 Mayo to Murchison, 17 May 1870, CUL, MaP, Add.MS.7490/39/129; Alder, 1963: 107–109.
163 Hayward to Murchison, 21 May 1870, in *JRGS*, 41, 1871: 11.
164 Mayo to Murchison, 18 July 1870, CUL, MaP, Add.MS.7490/40/206.
165 Mayo to Murchison, 9 Sept. 1870, CUL, MaP, Add.MS.7490/40/267; Murchison, 1871c.
166 Forsyth, 1871; Rawlinson, 1871.
167 Showers to Murchison, 7 Oct. 1870, RGS, C.Asia; *PRGS*, 15, 1870–71: 4–10.
168 Compare his comments in Rawlinson, 1865; Hayward, 1871.
169 Drew to Murchison, 21 Dec. 1870, in *PRGS*, 15, 1870–71: 117–120.
170 Mayo to Murchison, 30 Nov. 1870, CUL, MaP, Add.MS.7490/41/341.
171 Shaw to Murchison, 8 Dec. 1870, 20 Feb., 18 Mar. 1871, in *PRGS*, 15, 1870–71: 175–180; Murchison, 1871b: clxxxi–clxxxiv.
172 Montgomerie, 1871.
173 Murchison, 1871b: ccix.
174 Duthie, 1983.
175 Mayo to Wade, 6 Aug. 1871, CUL, MaP, Add.MS.7490/44/180.
176 Blanford, 1878.
177 Kiernan, 1953–55.
178 Mill, 1930: 131–132.

6 The Far East

1 Berman, 1977: 81–82.
2 Collins, 1918: 1–35; Greenberg, 1951.
3 Graham, 1978: 117.
4 Murchison, 1842: 644.
5 Murchison, 1843a: 136.
6 *PGS*, 3, 1842: 615–616; 4, 1843: 50; *QJGS*, 18, 1862: ix–x.
7 Murchison, 1844: lxix.
8 Murchison, 1845a: liv, xci.
9 Hamilton to Aberdeen, with enclosure, 8 Dec. 1842, BL(AM) 43240, Aberdeen Papers, ff.76–77, 80–81.
10 Graham, 1978: 10, 102, 232.
11 Jackson to Staunton, 23 May 1844, RGS, LB, v.1844–50; CBM, v.1841–53, entry for 25 Jan. 1844, f.68.
12 Murchison, 1844: cvi.
13 Murchison to Martin, 25 June 1846, EUL, MuP, DC.4/101–3, f.6; the quotation is in Graham, 1978: 419.
14 Matheson to Murchison, 27 June 1845, GSL/M/M40/1.
15 Geikie, 1875, i: 197.
16 Matheson to RGS, 13 May 1853, RGS, GC.
17 GSL/M/J25, 1860–62, ff.65–82.
18 Greenberg, 1951: 183.
19 *QJGS*, 9, 1853: vii; Chorley, Dunn, and Beckinsale, 1964: 162.
20 Shaw to Greenough, 15 Jan. 1850, UCL, Greenough Collection, 4; Gutzlaff, 1848.
21 Murchison, 1850b.
22 De la Beche to Bowring, 8 Jan. 1849, UCL, MS. Ogden 62/38, Bowring Collection; Macleod to De la Beche, 18 Mar. 1854, BGSA, GSM 1/13, f.165.
23 Playfair to Murchison, 6 July 1855, BGSA, GSM 1/413; *QJGS*, 16, 1860: ix.
24 Murchison to Bowring, 15 May 1856, UCL, MS. Ogden 62/52, Bowring Collection.
25 For example Bowring to Murchison, 26 July 1858, RGS, GC.
26 De la Beche to Hobart, 31 Dec. 1850, BGSA, GSM 1/6, ff.33–34.
27 Platt, 1968: 267–268.
28 Parkes to Phillips, 11 Aug. 1855, RGS, GC; Parkes, 1856.
29 Murchison, 1852a: cx–cxi.
30 Murchison, 1853: cviii–cix.
31 Cobbold, 1856.
32 Murchison, 1857b: clx; Davis, 1856–57, 1858.
33 Lockhart, 1858; *QJGS*, 9, 1853: 353; Murchison, 1859a.
34 Quoted in Graham, 1978: 298.
35 Murchison, 1858b: clxxxix.
36 Murchison, 1867c: 18, 400.
37 Murchison, 1858b: cxci.
38 Murchison to Herschel, 26 Mar. 1859, RS, HeP, HS.12.420.
39 Parish to Shaw, 2 Dec. 1858, RGS, Parish Collection.
40 Osborn, 1858–59; Markham to Shaw, 18 Feb. 1859, RGS, Markham Collection.

41 Osborn to Shaw, 22 Jan. 1859, RGS, GC.
42 Collinson to Shaw, 13 Aug. 1858, RGS, GC.
43 W. Morrell, 1940: 59–64.
44 Murchison, 1864b: cxlii–cxlv; see above, pp. 71–72.
45 Oliphant, 1860.
46 Davis to Murchison, 19 Mar. 1859, BL(AM) 46126, MuP, ff.5–6; Davis, 1858–59.
47 Elgin to Murchison, 21 May 1859, GSL/M/E18/2.
48 Blackney, 1860.
49 Murchison, 1859b: clxx–clxxv.
50 Crawfurd to Murchison, 5 Dec. 1859, RGS, Crawfurd Collection; Michie, 1859–60; for Lindsay, see Graham, 1978: 69–71.
51 Crawfurd to Shaw, 29 Feb. 1860, RGS, GC.
52 Murchison to Ellice, 2 Nov. 1860, NLS, MS. 15039, ff.120–121.
53 Murchison, 1861b: cxlii–cxliii; the quotation is from Murchison, 1867b: cxxii.
54 Murchison, 1861b: clxvi–clxvii.
55 Duckworth, 1862; H. Robinson, 1862.
56 Sarel, 1862; Barton, 1862.
57 Ashburton, 1862: cxliv–cxlvii.
58 Parkes to Shaw, 2 Dec. 1862, RGS, GC; Parkes to Murchison, 29 Oct. 1863, GSL/M/P28/1.
59 C. Grant, 1863.
60 Grant to Murchison, 25 Dec. 1862, BL(AM) 46126, MuP, ff.317–318; see also W. Morrell, 1940: 60.
61 See above, p. 79; for the financing of the China trade, see Greenberg, 1951.
62 Murchison, 1863b: clv; 1865b: cliv.
63 For example Murchison, 1864c; 1865b: cxlix.
64 Murchison to Bates, 10 Dec. 1867, APS, MuP, BM93p; Bickmore, 1868.
65 Cotton, 1867.
66 Sprye to Murchison, 24 July 1860, RGS, GC.
67 ibid., 26 Oct. 1860.
68 Sprye, 1860–61.
69 Sprye to Murchison, 12 June 1862, BL(AM) 46128, MuP, ff.109–110.
70 Sprye to Shaw, 24 May 1862, RGS, GC.
71 Schomburgk, 1861.
72 Murchison, 1861b: clxvi.
73 Moyle and Hillier, 1857.
74 Schomburgk to Bowring, 7 July 1858, UCL, MS. Ogden, Bowring Collection.
75 Wade to Murchison, 24 Aug. 1867, BL(AM) 46128, MuP, ff.244–248.
76 Murchison, 1869b: clxxxii–clxxxiii; T. Cooper, 1869–70, 1870–71.
77 RGS, CMB, v.1867–71, entry for 25 Nov. 1867; Waugh *et al.*, 1869; *PRGS*, 13, 1868–69: 3–8.
78 Murchison, 1871b: clxxxv; Sladon, 1871.
79 Swinhoe, 1864; *QJGS*, 20, 1864: xiii.
80 P. Hall, 1987: 43.
81 De la Beche and Playfair, 1848: 550–560; Gordon, 1849.
82 Collingwood, 1868.
83 Swinhoe, 1870.

84 Williamson, 1869.

85 J. Markham, 1870.

86 Kingsmill, 1869.

87 Murchison, 1871b: clxxxvi–clxxxix.

88 Platt, 1968: 265–305.

89 Collins, 1918: 106–107.

90 ibid.; Reeks, 1920, pt 2.

91 Beasley, 1951: 49–54.

92 ibid.: 143–144; Graham, 1978: 290.

93 Archbold, 1981.

94 Phinn to Murchison, 25 Feb. 1856, BGSA, GSM 1/7, ff.330–331; reply, 29 Feb. 1856, ff.284–285.

95 Murchison, 1852a: cxi.

96 GSL/M/J23, 1853–56, ff.94–95.

97 Murchison to Hatherton, 21 Jan. 1856, Staffordshire Record Office, Hatherton Papers, D260/M/F/5/27/29/10.

98 Murchison, 1867c: 440; see above, p. 69.

99 Murchison to Clarendon, 11 Feb. 1857, with enclosure of Hatherton to Murchison, 4 Feb. 1857, BLO, MSS.Clar.Dep.C.80, ff.236–242.

100 Beasley, 1951: 187–190.

101 Murchison, 1857b: xcviii, clxxxiv.

102 Cullen to Shaw, 10 Nov., 11 Dec. 1858, RGS, GC. See above, pp. 85–86.

103 Murchison, 1859b: clxxv–clxxvii.

104 Hodgson, 1860–61.

105 Oliphant, 1863.

106 Murchison, 1861b: clxvii.

107 Alcock, 1861.

108 Alcock, 1862.

109 Murchison, 1863b: clvii–clviii; 1864c.

110 Bullock, 1864.

111 Murchison, 1861b: clxxii.

112 Murchison to Egerton, 13 Nov. 1867, BGSA, GSM 1/8, f.543; Mitford, 1868.

113 For example Collingwood, 1868.

114 Young, 1864–65; Beasley, 1951: 187.

115 Murchison, 1865b: clxxii.

116 Murchison to Denison, 11 June 1865, in Geikie, 1875, ii: 303–304.

117 Williamson, 1869.

118 Murchison, 1844: xcviii; 1845a: lviii–lxi.

119 Memorandum entitled 'Borneo–Desiderata', 1835, RGS, S.E. Asia; Brooke, 1838; Brooke to Washington, 5 Nov. 1839, RGS, Brooke Collection.

120 Lay, 1840.

121 Buckland, 1841: 494–495.

122 Lingwood, 1985.

123 Graham, 1967: 394–399.

124 H. Williams to De la Beche, 9 Apr. 1845, BGSA, GSM 1/12, f.73; De la Beche and Playfair, 1848: 550.

125 Tarling, 1982; see above, p. 74.

126 Woolley to De la Beche, with enclosure, 4 Oct. 1850, BGSA, GSM 1/14, ff.73, 84; Motley, 1853.

127 Tarling, 1982.

128 Brooke, 1852.

129 Brooke to Shaw, 21 May, 5 Aug. 1858, RGS, Brooke Collection; Murchison, 1869b: cxlv.

130 Murchison, 1852a: cxi–cxii; Crawfurd, 1853.

131 Earl, 1845; see above, p. 41.

132 GSL/M/J24, 1857–59, ff.9–11; Murchison, 1858b: cxcii–cxcv.

133 Murchison, 1860; Parkinson to Murchison, 5 Jan. 1859, RGS, GC.

134 Crawfurd to Shaw, 25 Jan. 1859, RGS, GC.

135 Murchison to Clarendon, 27 July 1853, RGS, LB, v.1850–59, ff.89–90; *PRGS*, 1, 1855–56: 97.

136 Wallace, 1905, i: 327–332, 376–377.

137 Wallace, 1855–56; Crespigny, 1855–56.

138 Crespigny, 1857–58; Murchison, 1859b: clxxvii–clxxix.

139 St John, 1862; Murchison to Shaw, 15 Nov. 1860, RGS, MuP.

140 Washington to Murchison, 22 Mar. 1860, BGSA, GSM 1/8, ff.75–76.

141 Royle, 1846; Murchison to Washington, 2 Apr. 1860, BGSA, GSM 1/8, ff.76–77.

142 See, e.g., Piddington to Torrens, 26 Feb. 1841, BGSA, GSM 1/1, ff.197–211.

143 Murchison, 1859a.

144 Murchison to Groot, 28 July 1863, BGSA, GSM 1/8, ff.263–264; Groot, 1863; *QJGS*, 20, 1864: xiii.

145 Murchison, 1865b: cl.

146 Murchison, 1865b: clxxii–clxxiii.

147 Murchison, 1870b: clxix–clxx; *PRGS*, 14, 1869–70: 204.

148 Murchison, 1871b: cxcvi.

7 *Africa*

1 Murchison, 1839b, i: 8, ii: 583, 653.

2 Geikie, 1875, i: 260–61.

3 Murchison, 1842: 645.

4 GSL/M/J14, 1842–43, f.783; Stanger, 1843.

5 C. Lloyd, 1970: 112–122.

6 Gavin, 1959; Robinson and Gallagher, 1968: 27–52.

7 Murchison, 1844: cvi–cxx.

8 Murchison, 1845a: lxi–lxvii.

9 For example Murchison, 1850a: 405; see also Wyld, 1853: 44–45.

10 Burman, 1968; Rogers, 1937.

11 D. Livingstone, 1851; Galton, 1852.

12 Vogel, 1855; E. Forbes, 1852.

13 Stafford, 1988b.

14 Bain, 1845.

15 Murchison to Richards, 24 Oct. 1844, UW, BaP.

16 Bain to Richards, 21 June 1847; Bain to Owen, 25 Sept. 1848 – both in Rogers, 1937: 29–31.

17 Herschel to Murchison, 21 Mar. 1852; Murchison to Pakington, 4 Apr. 1852 – both in UW, BaP.

18 De la Beche to Desart, 30 Apr., 10 Nov. 1852, UW, BaP.

19 Murchison, 1852a: cxvii–cxxv; for the theory of persistent biological provinces, see J. Browne, 1982: 93–102.

20 *JRGS*, 22, 1852: liii; Jones to Bain, 24 Nov. 1852, UW, BaP.

21 Bain, 1852; Sharpe and Salter, 1856.

22 Murchison, 1867c: 17; Bain, 1857.

23 Rogers, 1937: 41; Merivale to De la Beche, with enclosures, 18 Nov. 1854, BGSA, GSM 1/7, ff.156–157; reply, 2 Dec. 1854, f.158; Smyth to De la Beche, 3 Jan. 1854, GSM 1/444.

24 Merivale to De la Beche, with enclosure, 31 July 1854, BGSA, GSM 1/6, ff.386–388.

25 'Correspondence on the subject', 1854.

26 Murchison, 1859b: cxl–cxli.

27 A. Bain to J. Bain, 6 and 26 June 1864, University of Cape Town Library, A. G. Bain Papers, BC543 13(r), (s).

28 Rogers, 1937: 50–54; Smyth, Reeks, and Rudler, 1864: 136–137; Murchison, 1858a.

29 Rogers, 1937: 52.

30 For example Rubidge, 1855.

31 Murchison, 1860.

32 Wyld, 1853: 45.

33 Peel to De la Beche, 22 Jan. 1853, BGSA, GSM 1/7, f.9.

34 Hattersley, 1968; Biographical notes, Natal Archives Depot, Pietermaritzburg, Kit–Bird collection, A79, v.11: 74–81.

35 Murchison to Shaw, 3 Apr. 1855, Journ.MSS.Africa, referee's comments; P. Sutherland, 1855a.

36 Forbes to Murchison, 15 Feb. 1856, BL(AM) 46126, MuP, ff.166–167.

37 Murchison to the Admiralty, with enclosure, 3 July 1855, GSM 1/7, ff.220–222.

38 P. Sutherland, 1855b.

39 Murchison to Shaw, 7 May 1856, RGS, MuP; Sutherland to Barrow, 24 Jan. 1856, BL(AM) 35306, Barrow Papers, ff.214–215.

40 Sutherland to Colonial Secretary of Natal, 10 May 1856, in *Natal Government Gazette*, 20 May 1861.

41 Murchison, 1857b: cxxv.

42 Murchison to Rogers, 1 May 1866, BGSA, GSM 1/8, ff.404–405; Lugg, 1949: 2.

43 'Despatch and Report', 1867. These coals are now considered predominantly Permian.

44 Murchison, 1867a.

45 For example Murchison to Fryar, 12 Dec. 1867, BGSA, GSM 1/9, ff.7–8.

46 *Natal Government Gazette*, 2 Oct. 1867: 425–426.

47 See above, Ch. 3, note 124.

48 Elliot to Murchison, with enclosure, 27 Feb. 1868, BGSA, GSM 1/9, f.25, 1/17, f.16; reply, 6 Mar. 1868, 1/9, ff.25–27.

49 Murchison to Lennox, 2 Apr. 1868, BGSA, GSM 1/9, f.39.

50 'Correspondence on the subject', 1868.

51 Murchison, 1868b: clxxxii; Mann, 1869a; Rogers to Murchison, with enclosures, 21 Oct. 1868, BGSA, GSM 1/17, ff.65, 69.
52 Murchison, 1868b: clxxxii–clxxxvii. The Zimbabwean craton is Precambrian.
53 Wilkinson, 1868–69; Murchison to the editor, *The Times*, 22 Aug. 1868; Mann, 1869b.
54 P. Sutherland, 1869, 1868.
55 Keate to Buckingham, with enclosures, 14 Dec. 1868, [copy], RGS.
56 Murchison to Markham, 12 Mar. 1869, RGS, Journ.MSS.Africa, referee's comments.
57 Murchison, 1869b: clxxvi–clxxviii.
58 Erskine, 1869.
59 Wallis, 1976: 160–220.
60 Baines, 1868; Wallis, 1976: 119–152.
61 Baines, 1871.
62 Rogers, 1937: 78–84.
63 W. Morrell, 1940: 313–348.
64 P. Sutherland, 1870.
65 Flint, 1974: 13, 18, 24.
66 Murchison, 1852a: cxxi.
67 Murchison to Derby, 10 July 1852, RGS, LB, v.1850–59, f.71–72; McLeod, 1853.
68 Murchison to Shaw, n.d. Dec. 1852, RGS, MuP.
69 ibid., 18 Oct. 1850.
70 Murchison, 1853: cix–cxii.
71 See above, p. 89.
72 Murchison later helped McLeod win consular posts in Mozambique, the Niger delta, and the Seychelles.
73 Murchison to Clarendon, 11 Dec. 1853, BLO, MSS.Clar.Dep.C.103, ff.156–157, 161–163.
74 ibid., 10 May 1854, ff.233–235.
75 ibid., 29 May 1854, ff.260–263; reply, 29 May 1854, BL(AM) 46125, MuP, ff.386–387.
76 Baikie, 1855.
77 Murchison to Clarendon, 12 Mar. 1855, BLO, MSS.Clar.Dep.C.103, ff.520–522.
78 ibid., 17 Apr. 1855, ff.525–528.
79 ibid., 17 Apr. 1855, ff.523–524.
80 Baikie to Murchison, 9 Jan. 1856, BL(AM) 46125, MuP, ff.82–83.
81 *PRGS*, 1, 1855–56: 16.
82 Murchison to Clarendon, two of 29 Jan. 1856, RGS, LB, v.1850–59, ff.175–176; BLO, MSS.Clar.Dep.C.103, ff.685–686.
83 *JRGS*, 26, 1856: clxvi–clxx; Beechey, 1856: ccx.
84 Murchison to Clarendon, with enclosures, 10 June 1856, BLO, MSS.Clar.Dep. C.103, ff.679–684.
85 ibid., with enclosure, 31 July 1856, ff.687–690.
86 Murchison, 1866b: cxcvii.
87 Gavin, 1959: 33–41, 214.
88 Baikie to Shaw, 8 Aug. 1856, RGS, BaP.

89 Baikie, 1857.

90 Manuscript of Baikie, 1857, dated 6 Aug. 1857, RGS, BaP.

91 D. Livingstone, 1857a.

92 Murchison to Shaw, 17 Oct. 1856, RGS, MuP.

93 Murchison to Clarendon, 17 Oct. 1856, quoted but wrongly ascribed, in Nzemeke, 1982: 64.

94 BAAS petition to Clarendon, [draft], 29 Oct. 1856, RGS, MuP.

95 Lewis to Clarendon, 3 Nov. 1856, BLO, MSS.Clar.Dep.C.48, ff.284–285.

96 Murchison to Clarendon, 19 Nov. 1856, BLO, MSS.Clar.Dep.C.103, ff.698–701.

97 ibid., 2 Dec. 1856, ff.702–703; n.d., 1857, C.80, ff.271–272; reply, 29 Nov. 1856, RGS, MuP.

98 Murchison to Clarendon, with enclosure, 12 Feb. 1857, BLO, MSS.Clar.Dep. C.80, ff.232–235.

99 ibid., 20 Apr. 1857, ff.253–256.

100 Murchison to Baikie, 21 Apr. 1857, BGSA, GSM 1/7, ff.394–398.

101 Vogel to Murchison, 10 July 1864, GSL/M/V17/1.

102 Murchison, 1857b: clxv–clxvi.

103 Murchison to J. Hooker, 30 Dec. 1857, RBG, JHP, Eng. letts., MOO-MYL, 1844/1900, v.96, f.400.

104 Baikie and May, 1857–58; Murchison, 1858b: cci–cciv.

105 May, 1860; Murchison, 1859b: clxxxviii–cxc. These formations are actually Precambrian.

106 Baikie to Murchison, 18 Oct. 1859, RGS, MuP.

107 Galton, referee's comments on printed prospectus of 'The African Agricultural & Commercial Co.', dated 1 Jan. 1847, enclosure in Mcqueen to Ashburton, 5 June 1860, RGS, GC.

108 May, 1860; Murchison, 1861b: cxxvi–cxxviii, cxli.

109 Baikie, 1862–63; Nzemeke, 1982: 47–135.

110 Murchison, 1864b: clxxxiii.

111 Quoted in Nzemeke, 1982: 126.

112 Murchison, 1865b: cxxii–cxxiv.

113 Murchison, 1866b: cxx.

114 Baikie, 1866–67.

115 Murchison, 1868b: clxxiv.

116 Nzemeke, 1982: 315–316, 331–332;[anon.], 1907–08, 1907–09.

117 Beechey to Murchison, 15 Apr. 1856, GSL/M/B48/1.

118 Murchison to Shaw, 31 May 1856, RGS, MuP.

119 Murchison, 1858b: ccvii–ccviii; 1859b: clxxix–clxxxvi.

120 Huxley, 1918, i: 406; J. Hooker to Murchison, 12 July 1862, RGS, Hooker Collection; reply, 12 Aug. 1862, RBG, JHP, Eng. letts., MOO-MYL, 1847/1900, v.96, f.416.

121 Grant to Murchison, 1 Feb. 1860, GSL/M/G30/1.

122 Petherick, proposal for White Nile expedition, 1860, RGS, Nile Journ. MSS.

123 Murchison, 1860; Petherick, 1859–60.

124 Bridges, 1963a: 185–187.

125 Murchison to Shaw, 22 Nov. 1862, RGS, MuP.

126 Murchison to Layard, 2 May 1863, BL(AM) 39106, LaP, ff.1–2.

127 *PRGS*, 7, 1862–63: 212–224.

128 Murchison, 1863b: clxxii–clxxxv. Gold is mined today near Lake Victoria and in southern Ethiopia – Derry, 1980: 61.

129 Murchison–Grant correspondence, 1864–70, NLS, Grant Collection, MS. 1790, 17931; Casada, 1974.

130 Speke to Murchison, 7 Mar. 1864, RGS, Speke Collection.

131 Council memorandum to Palmerston, 26 Mar. 1864, RGS, LB, v.1859–79, ff.80–83; Murchison, 1864b: clxxix–cxc; 1865c.

132 Murchison, 1863b: cxvi.

133 Bridges, 1973: 229.

134 Speke to Layard, 31 May 1864, BL(AM) 39110, LaP, ff.152–153.

135 Wylde to Layard, 11 June 1864, BL(AM) 38990, LaP, ff.249–250.

136 Murchison to Layard, 1 July 1864, BL(AM) 39111, LaP, ff.1–2.

137 Murchison, 1865b: cix–cxii.

138 Seaver, 1957: 239.

139 Livingstone to Buckland, 10 Oct. 1843, BL(AM) 42581, Owen Papers, ff.53–58.

140 D. Livingstone, 1855–57, 1857a. For the section, see Livingstone, 1857b; Stafford, 1988b.

141 Gladstone to Murchison, 4 Nov. 1854, BL(AM) 44530, Gladstone Papers, f.105.

142 Murchison to Livingstone, 16 Oct. 1855, RGS, MuP; Blaikie, 1910: 174.

143 Livingstone to Murchison, 5 Aug. 1856, in Wallis, 1956, i: xviii–xxi.

144 Murchison to Clarendon, 19 and 24 Oct. 1856, BLO, MSS.Clar.Dep.C.103, ff.691–697.

145 Livingstone to Murchison, 12 Dec. 1856, NAZ, LiP, LI 1/1/1, ff.572–579; *PRGS*, 1, 1856–57: 233–249.

146 Murchison to Clarendon, 5 Jan. 1857, BLO, MSS.Clar.Dep.C.80, ff.209–211.

147 ibid., with enclosure, 27 Jan. 1857, ff.222–230.

148 ibid., 11 Jan. 1857, ff.212–215.

149 ibid., two of 17 Mar. 1857, respectively ff.243–246, 247.

150 Cf. Livingstone to Clarendon, 19 Mar. 1857, BLO, MSS.Clar.Dep.C.80, ff.203–206, with Livingstone to Sedgwick, 6 Feb. 1858, [copy], NLS, Livingstone Collection, MS. 10779(20), ff.42–45.

151 Palmerston to Clarendon, 29 Dec. 1857, BLO, MSS.Clar.Dep.C.69, ff.702–705.

152 Murchison to Clarendon, 13 Apr. 1857, BLO, MSS.Clar.Dep.C.80, ff.248–250.

153 Murchison, 1857b: clxviii–clxxii.

154 Murchison to Murray, 31 July 1857, JMA; D. Livingstone, 1857b: 500.

155 Murchison to Clarendon, 24 Oct. 1857, BLO, MSS.Clar.Dep.C.80, ff.259–260; Wood to Clarendon, 4 Nov. 1857, C.70, ff.563–564.

156 Murchison to Clarendon, 19 Nov. 1857, BLO, MSS.Clar.Dep.C.80, ff.261–262; Laird to Murchison, 20 Nov. 1857, BL(AM) 46127, MuP, ff.44–45.

157 Lewis to Clarendon, 8 Dec. 1856, BLO, MSS.Clar.Dep.C.48, ff.290–291.

158 Murchison to Clarendon, 13 Dec. 1857, BLO, MSS.Clar.Dep.C.80, ff.263–264.

159 ibid., with enclosure, 16 Dec. 1857, ff.275–282.

160 ibid., 13, 14 Jan. 1858, C.103, ff.783–787.

161 Baker–Murchison correspondence, Jan.–May 1858, RGS, Baker Collection.

162 Murchison to Clarendon, 17 Dec. 1857, BLO, MSS.Clar.Dep.C.80, ff.265–266; Laird to Murchison, 19 Dec. 1857, BL(AM) 46127, MuP, ff.46–47.

163 Sharpey to Shelburne, with enclosures, n.d. [Jan. 1858], RS, MC.17.326, MM.14.16,17.

164 Murchison to Clarendon, 30 Dec. 1857, BLO, MSS.Clar.Dep.C.80, ff.273–274.

165 Murchison to J. Hooker, 30 Dec. 1857, RBG, JHP, Eng. letts., MOO-MYL, 1847/1900, v.96, f.400; Sabine to Murchison, 6 Feb. 1858, BL(AM) 46128, MuP, ff.20–21.

166 Murchison to Clarendon, 13 Jan. 1858, BLO, MSS. Clar.Dep.C.103, ff.783–785.

167 Clarendon to Murchison, 7 Feb. 1858, BL(AM) 46125, MuP, ff.390–391.

168 *PRGS*, 2, 1857–58: 116–142.

169 Murchison to Clarendon, 14 Feb. 1858, BLO, MSS.Clar.Dep.C.103, ff.793–794.

170 Tabler, 1963, i: ix–xii.

171 D. Livingstone and C. Livingstone, 1865: 9.

172 Murchison to Thornton, [1 Mar. 1858], BGSA, GSM 1/7, ff.520–523; also in Wallis, 1956, ii: 429–430.

173 Murchison, 1858b: cc.

174 R. Thornton, journal entries for 5 and 6 Apr. 1858, RHO, Thornton Papers, MSS.Afr.S.29; R. Thornton, 1859.

175 Livingstone to Malmesbury, 17 Dec. 1858, RGS, DL, 3/4/1; Murchison to the editor, *The Times*, 14 Jan. 1859.

176 Livingstone to Murchison, 5 Feb. 1859, in Seaver, 1957: 341–342.

177 Livingstone to Murchison, 9 May 1859, NAZ, LiP, LI 1/1/1, ff.1014–1017.

178 D. Livingstone, 1859–60a, 1859–60b; *PRGS*, 3, 1858–59: 99–106.

179 Livingstone to Murchison, 8 Aug., 6 Nov. 1859, NAZ, LiP, LI 1/1/1, ff.1046–1049, 1078–1081.

180 Jeal, 1974: 278–281; R. Thornton to G. Thornton, with enclosures, 16 Apr. 22 May, 22 Nov. 1861, in Tabler, 1963, ii: 220–229.

181 R. Thornton, 1862, 1864, 1865; Murchison, 1864b: cxxiii–cxxv; G. Thornton–Murchison correspondence, 1863–65, in Tabler, 1963, ii: 305–335.

182 Bedingfeld, 1859–60.

183 Murchison, 1863b: clxxxix–cxc.

184 For example Livingstone to Murchison, 23 Sept. 1861, NAZ, LiP, LI 1/1/1, ff.1317–1332.

185 Murchison, 1864d.

186 Seaver, 1957: 452, 463; Murchison to Murray, 30 Jan. 1865, JMA; D. Livingstone and C. Livingstone, 1865.

187 Murchison to Markham, 15 Aug. 1864, RGS, DL 3/16/6.

188 Jeal, 1974: 344–355; C. Livingstone to Murchison, 11 Oct. 18643, BL(AM) 46127, MuP, ff.86–87.

189 Murchison, 1865b: clxvi–clxvii, clxxvii–clxxviii.

190 Murchison to Kirk, 31 May 1866, APS, MuP, BM93p.
191 Murchison to Clarendon, 9 Dec. 1868, BLO, MSS.Clar.Dep.C.510, folder 1; *PRGS*, 13, 1868–69: 24–25.
192 M. Baines to T. Baines, 24 May 1867, NAZ, Baines Collection, BA 7, ff.697–700.
193 Bridges, 1963a: 283.
194 Geikie, 1875, ii: 346–347.
195 Murchison, 1870b: clxxviii.
196 Baker, 1863.
197 Murchison to Layard, 24 June 1865, BL(AM) 39115, LaP, ff.456–457; Murchison to the editor, *The Times*, 11 July 1865.
198 Baker, 1866, *PRGS*, 10, 1865–66: 2–6; see also Murchison, 1864b: clx–clxxx; Stafford, 1988b.
199 Murchison, 1866b: cxcii.
200 Baker, 1867; Gladstone to Murchison, 23 July 1866, copy furnished by Charles J. Sawyer, Bookseller, London.
201 Murchison to Layard, 15 Feb. 1866, BL(AM) 39118, LaP, ff.322–324.
202 Baker to Murchison, 8 March 1867, described and quoted in *Sotheby's sale catalogue*, 24 July 1978: 126.
203 Stanley to Murchison, 6 Aug. 1867, BL(AM) 46128, MuP, ff.123–124.
204 Baker, 1868.
205 Murchison, field diary, undated entry [1867], BGSA, GSM 1/124.
206 Benthall to Murchison, 24 Sept. 1867, BL(AM) 46125, MuP, ff.153–154.
207 Hamilton to Murchison, 5 Oct. 1867, BL(AM) 46126, MuP, ff.365–366.
208 Newton to Murchison, 3 Oct. 1867, BL(AM) 46127, MuP, ff.271–274.
209 Murchison to Hooker, 4 Oct. 1867, RBG, JHP, Eng. letts., MOO-MYL, 1847/1900, v.96, f.428; reply, 6 Oct. 1867, GSL/M/H58/2.
210 C. Markham, 1868a.
211 Baker to Murchison, 24 Nov. 1867, [copy], RGS, Baker Collection; Baker–Murchison correspondence, 1867–69, described and quoted in *Sotheby's sale catalogue*, 24 July 1978: 128; *PRGS*, 12, 1867–68: 25.
212 C. Markham, 1868b.
213 C. Markham, manuscript history of the RGS, n.d., RGS, Markham Papers, f.385.
214 See above, p. 123.
215 Strangford, 1869, ii: 208.
216 ibid., ii: 335–339.
217 Renard to Murchison, 11 Apr. 1868, BL(AM) 46127, MuP, ff.422–423.
218 *PRGS*, 12, 1867–68: 124.
219 Murchison, 1868b: clxxxix.
220 C. Markham, 1868c, 1869.
221 Blanc, 1869.
222 Blanford, 1869, 1870a, 1870b.
223 Clarendon to Baker, 23 Apr. 1869, [copy], BLO, MSS.Clar.Dep.C.476, folder 6, f.285.
224 Murchison, 1869b: clxxvi, cxciii–cxciv.
225 Murchison to Buxton, 28 Apr. 1870, Royal Commonwealth Society Library, London.

226 Murchison to White, 13 July 1869, Science Museum Library Archive, London, MS 1032.
227 *PRGS*, 15, 1869–70: 88–105.
228 Helly, 1969.
229 Stafford, 1988b.
230 Bridges, 1973.

8 *The architect of imperial science*

1 For general discussion of the sociology of scientific knowledge, see Shapin, 1982; for De la Beche's efforts to link geological surveying to social reforms, Secord, 1986b: 227–234.
2 Joll, 1977; Berman, 1975.
3 Geikie, 1875, i: 263.
4 Shapin, 1982: 187.
5 See above, p. 1.
6 MacLeod, 1975; Dionne and MacLeod, 1979.
7 MacDonagh, 1961.
8 See, respectively, Brockway, 1979; Ritchie, 1967: 189–280; Cawood, 1979.
9 D. MacKay, 1985: 3–27; the quotation is at p. 123.
10 Secord, 1986b.
11 E. Bailey, 1952.
12 The quotations are from, respectively, GSL/M/J23, 1853–56, f.116; Secord, 1982: 419; [E. Forbes], 1854: 582.
13 For an introduction to the little-studied role of the Scots in British expansion, see Cage, 1985.
14 See, e.g., GSL/M/J24, 1857–59, f.170.
15 Murchison, 1866a.
16 Murchison, 1862a.
17 Rudwick, 1963.
18 Airy to Forbes, 3 Mar. 1859, University of St Andrews, J. D. Forbes Papers, Letterbook VIII, 1859/6.
19 C. Browne, 1871: 546–547.
20 Rudwick, 1985: 206, 252; Secord, 1986a: 120.
21 MacLeod, 1982.
22 Stafford, 1984; cf. H. Wilson, 1985: 140.
23 Blainey, 1970; Burton, 1965.
24 Kenwood and Lougheed, 1971: 23–24, 102.
25 ibid.: 152.
26 See above, p. 19; Rupke, 1983: 18.
27 Roberts, 1976; Guntau, 1978; Weindling, 1979, 1983; J. Morrell, 1983; Rudwick, 1985; Secord, 1986b: 233–234; Laudan, 1987: 47–56.
28 Flett, 1937: 58, 64; E. Bailey, 1952: 59; Porter, 1978: 825–829.
29 Jones to Murchison, 12 Nov. 1869, BGSA, GSM 1/9, ff.146–147; Murchison to Cole, with enclosure, 31 Dec. 1869, GSM 2/22.
30 O'Conner and Meadows, 1976; Porter, 1978; cf. Porter, 1982: 194, which acknowledges this factor for the period 1850–1900.
31 Stafford, 1984: 20–21.

32 Quoted in Jervis, 1944: 449.
33 Foote, 1954: 439–440.
34 Rupke, 1983: 255–266; Rudwick, 1974a.
35 D. N. Livingstone, 1984.
36 von Buch to Murchison, 15 Apr. 1847, GSL/M/B33/7.
37 GSL/M/J24, 1857, f.5.
38 Quoted in Rogers, 1937: 71.
39 Forster, 1941.
40 Flint, 1974.
41 Zetland to Murchison, 12 Jan. 1856, GSL/M/Z2/4; Murchison to Committee of the Privy Council on Education, 6 Apr. 1858, BGSA, GSM 2/7.
42 Chorley, Dunn, and Beckinsale, 1964: 181.
43 Worboys, 1980: 143–191; H. Wilson, 1985: 140–142.
44 Platt, 1968: 109; see above, p. 31.
45 Gallagher and Robinson, 1953.
46 Platt, 1968: 147, 259–267.
47 Alcock, 1877: cxcvi–cxcvii, cci–cii; *PRGS*, 22, 1877–78: 20.
48 J. Morrell, 1971; Allen, 1976.
49 Cain and Hopkins, 1986–87.
50 For botany, see Brockway, 1979; for ornithology, Farber, 1982: 27, 33–46, 147–150; for biogeography, J. Browne, 1983: 77–81, 130–131; for zoology, Desmond, 1985; for the service class in advocacy of imperialism, Bridges, 1973.
51 Porter, 1978: 820.
52 Gowing, 1978; above, pp. 26–27.
53 Stafford,, 1984: 20–21.
54 Murchison, 1839a: xliii.
55 Herschel to Harcourt, 5 Sept. 1831, quoted in Orange, 1972: 165.
56 Sedgwick, 1835.
57 Buckland, 1840: 229.
58 [Anon.], 1859: 124.
59 For the sociology of discovery, see Brannigan, 1981.
60 Latour, 1985.
61 Murchison, 1844: cxxvii.
62 Murchison, 1857b: cxcvii–cxcviii.
63 For example Bridges, 1973; D. N. Livingstone, 1984.
64 Gavin, 1959: 10–34.
65 Bagehot, 1907; H. Bell, 1936, i: 86, 325–326, 401; Platt, 1968.
66 Murchison to Clarendon, 13 July, 7 Oct. 1855, BLO, MSS.Clar.Dep.C.103, ff.547–551; H. Bell, 1936, i: 24, 46–47, 325–326.
67 Clarendon, 1846; Roberts, 1976.
68 D. MacKay, 1985: 16; Berman, 1977: 75–99.
69 Kenwood and Lougheed, 1971: 41–44.
70 For Spain, see above, p. 6; for France, Mackenzie to Murchison, 9 Nov. 1829, GSL/M/M27/1; for Peru, above, p. 88; for America, above, pp. 69–71; for domestic railways, Young to Murchison, 24 Sept. 1846, GSL/M/Y3/2; Holland to Murchison, n.d. [1854], EUL, MuP, Gen.523/6; for India, above, p. 117.
71 Wallis, 1956, ii: 377–379; GSL/M/J22, 1849–52, f.36.
72 See above, p. 17.

73 See above, pp.39, 66.
74 H. Bell, 1936, i: 256–257; Gavin, 1959: 6–7, 41–45.
75 GSL/M/J5, 1818–25, f.28; see also above, pp. 15–16.
76 Quoted in H. Bell, 1936, ii: 341.
77 Yule, 1872: 163.
78 For the Royal Society's role in exploration, see M. Hall, 1984: 199–215.
79 Reese, 1968: 15, 67.
80 Milne to Murchison, 11 Nov. 1862, RGS, GC; see also Bridges, 1973: 226.
81 Murchison, 1852a: cxxv.
82 Cawood, 1979.
83 For Hayward, see above, p. 128; for Livingstone, Murchison to Wilberforce, 1 Jan. 1866, BLO, MSS.Wilberforce, C.19, ff.36–39.
84 C. Lloyd, 1970: 142–143, 156–162.
85 See above, pp. 87, 167.
86 See above, p. 10; Morrell and Thackray, 1981: 2–34.
87 ibid.: 331–333, 355–370.
88 Lee to Murchison, 3 Dec. 1852, RGS, GC.
89 See above, p. 69.
90 Murchison to Sturt, 22 Dec. 1854, RHO, Sturt Papers, MSS. Austral.S.5, ff.143–144.
91 Murchison, 1857b: cxcviii.
92 Murchison to RGS Council, 26 Nov. 1854, RGS, MuP; Stoddart, 1980.
93 Morrell and Thackray, 1981: 17.
94 Stoddart, 1980.
95 Bridges, 1973; see also Lochhead, 1980: 255–276.
96 The phrases are Martin Rudwick's, used to describe the Geological Society – see Rudwick, 1982.
97 C. Markham, manuscript history of the RGS, n.d., RGS, Markham Papers, ff.83–90.
98 *PRGS*, 14, 1869–70: 214.
99 Cf. Geikie's remark above, p. 17, with Gladstone's.
100 For correlation of the changing level of exploratory activity and the fluctuations in the British economy, see D. Hall, 1976: 160–190.
101 For Palmerston's role, see Cain and Hopkins, 1986–87, I: 523, II: 11–12; for the gold-rushes, W. Morrell, 1940: 411–415.
102 C. Markham, manuscript history of the RGS, n.d., RGS, Markham Papers, ff.391–393.
103 *PRGS*, 15, 1870–71: 52–53.
104 ibid.: 241.
105 Mill, 1930; Emery, 1984.
106 Bridges, 1963b.
107 Stoddart, 1980; Lochhead, 1980; D. N. Livingstone, 1984; for the contrasting situation in France at this time, see D. V. McKay, 1943.
108 Stoddart, 1980; Lochhead, 1980: 268.
109 See Goetzmann, 1966: ix–xii.
110 *PRGS*, 15, 1870–71: 242.
111 See above, pp. 118, 123.
112 Morrell and Thackray, 1981; Rudwick, 1985; Secord, 1986a; Desmond, 1985.

113 On the normalising and legitimising role of exploration literature in the imperial context, see Pratt, 1985.
114 Stafford, forthcoming.
115 Mumford, 1944: 252.
116 Crosby, 1986.
117 See above, p. 182.
118 Foote, 1954; Cannon, 1978: 1–28; Morrell and Thackray, 1981: 12–16.
119 Elster, 1983: 112–130.
120 Ranger, 1976, discusses the deployment of anthropology in this context; for organised knowledge in general as a component of political supremacy, see Said, 1979.
121 For the relationship between pure and applied sciences in the imperial context, see Pyenson, 1984, 1985.

BIBLIOGRAPHY

Note. Memoirs read before the Royal Geographical Society, most of which appear in both its *Proceedings* and its *Journal*, are cited whenever possible according to the year of their full publication in the *Journal*, while their earlier discussions and abridged textual versions are included in the citation for the *Proceedings*. Papers presented to the British Association as well as to the RGS are cited accordingly, following their primary citations. Papers presented only to the British Association are dated according to the year of publication of the *Report* of the Association containing them: the publication date succeeds the year of the meeting at which papers were actually presented. For example, Murchison, 1861c is printed in the *Report of the British Association for the Advancement of Science; held at Oxford in June and July 1860* (London, 1861).

[Anon.] 1816. *Inquiries into Geology, for the Purpose of Obtaining Information . . . in Such Parts of the World as May Afford Additional Intelligence, Particularly the East Indies; Circulated by the Geological Society in Great Britain.* Madras.

[Anon.] 1859. 'Geology in India', *Calcutta Review*, 32, March: 122–161.

[Anon.] 1907–08. *Southern Nigeria. Reports on the Results of the Mineral Survey, 1904/5—1907/8.* London.

[Anon.] 1907–09. *Northern Nigeria. Reports on the Results of the Mineral Survey, 1904/5—1907/9.* London.

Alcock, Rutherford. 1861. 'Narrative of a journey in the interior of Japan, ascent of Fusiyama, and visit to the hot sulphur-baths of Atami, in 1860', *JRGS*, 31: 321–356; *PRGS*, 5, 1860–61: 132–135; also in *RBA 1861, TS*: 183–184.

1862. 'Narrative of a journey through the interior of Japan, from Nagasaki to Yeddo, in 1861', *JRGS*, 32: 280–293; *PRGS*, 6, 1861–62: 196–197, 200–206.

1877. 'Address to the Royal Geographical Society', *JRGS*, 47: cxxxiii–cciii.

Alder, G. J. 1963. *British India's Northern Frontier, 1865–95: A Study in Imperial Policy.* London.

Allen, David E. 1976. *The Naturalist in Britain: a Social History.* London.

Archbold, N. W. 1981. 'Western Australian geology: an historical review to the year 1870', *Journal of the Royal Society of Western Australia*, 63: 119–128.

Ashburton, Lord. 1862. 'Address to the Royal Geographical Society', *JRGS*, 32: civ–clxxii.

Atkinson, T. W. 1859. 'On the volcanoes of Central Asia', *RBA 1858, TS*: 75.

1861. 'On the caravan routes from the Russian frontier to Khiva, Bokhara, Kokhan, and Garkand, with suggestions for opening up a trade between Central Asia and India', *RBA 1860, TS*: 153.

Austin, R. 1856. 'Report of an expedition to explore the interior of Western Australia', *JRGS*, 26: 235–274; *PRGS*, 1, 1855–56: 30–31.

Bagehot, Walter. 1907. 'The Earl of Clarendon', *Biographical Studies* (new edition, ed. by R. H. Hutton), pp. 373–377. London.

Baikie, William Balfour. 1855. 'Brief summary of an exploring trip up the rivers Kwòra and Chàdda (or Benuè) in 1854', *JRGS*, 25: 108–121.

    1857. 'On recent discovery in Central Africa, and the reasons which exist for continued and renewed research', *RBA 1856, TS*: 105–107.

    1862–63. 'Report on the countries in the neighbourhood of the Niger', *PRGS*, 7: 66–68.

    1866–67. 'Notes of a journey from Biba, in Nupe, to Kano, in Hausa, performed in 1862', *PRGS*, 11: 49–50.

Baikie, William Balfour, and Daniel T. May. 1857–58. 'Extracts of reports from the Niger Expedition', *PRGS*, 2: 83–101.

Bailey, Edward. 1952. *The Geological Survey of Great Britain*. London.

Bailey, F. E. 1942. *British Policy and the Turkish Reform Movement*. Cambridge, Mass.

Bain, Andrew G. 1845. 'On the discovery of the fossil remains of Bidental and other reptiles in South Africa', *TGS*, 2nd. ser., 7(3): 53–59.

    1852. 'On the geology of Southern Africa', *TGS*, 2nd. ser., 7(4): 175–192.

    1857. 'Geology of South Africa', *Eastern Province Monthly Magazine* (transcripts from A. G. Bain Collection, University of Cape Town Library, BC543, 11(c), (d)).

Baines, Thomas. 1868. 'On Walvisch Bay and the ports of south-west Africa', *RBA 1867, TS*: 113.

    1871. 'Exploration of the gold region between the Limpopo and Zambesi Rivers', *JRGS*, 41: 100–131; *PRGS*, 15, 1870–71: 147–158.

Baker, Samuel. 1863. 'Journey to Abyssinia in 1862', *JRGS*, 33: 237–241; *PRGS*, 7, 1862–63: 21, 46–48, 78–80.

    1866. 'Account of the discovery of the second great lake of the Nile, Albert Nyanza', *JRGS*, 36: 1–18; *PRGS*, 10, 1865–66: 6–24.

    1867. 'On the relations of the Abyssinian tributaries of the Nile and the Equatorial lakes to the inundations of Egypt', *RBA 1866, TS*: 102–104.

    1868. 'Address of the President of the Geography and Ethnology Section', *RBA 1867, TS*: 104–111.

Barton, Alfred. 1862. Notes on the Yang-tsze-kiang', *JRGS*, 32: 26–41; *PRGS*, 6, 1861–62: 85–95.

Basalla, George. 1967. 'The spread of Western science', *Science*, 156 (5 May 1967): 611–621.

Bauermann, Hilary. 1860. 'On the geology of the south-eastern part of Vancouver Island', *QJGS*, 16: 198–202.

    1869. 'Notes on a geological reconnaissance made in Arabia Petraea in the spring of 1868', *QJGS*, 25: 17–37.

Bayliss, Robert A. 1978. 'The travels of Joseph Beete Jukes, F.R.S.', *Notes and Records of the Royal Society of London*, 32: 201–212.

Beasley, W. G. 1951. *Great Britain and the Opening of Japan*. London.

Beaver, S. H. 1982. 'Geography in the British Association for the Advancement of Science', *The Geographical Journal*, 148: 173–181.

Beaver, William C. 1976. 'The development of the Intelligence Division and its role in aspects of imperial policy-making, 1854–1901'. D. Phil. thesis. Oxford University.

Bedingfeld, Norman. 1859–60. 'On the Congo', *PRGS*, 4: 66–72.

Beechey, F. W. 1856. 'Address to the Royal Geographical Society', *JRGS*, 26: clxxi–ccxxxiv.

Bell, Gerda. 1976. *Ernst Dieffenbach: Rebel and Humanist*. Palmerston North.

Bell, Herbert. 1936. *Lord Palmerston*. 2 vols. London.

Berman, Morris. 1975. '"Hegemony" and the amateur traditon in British science', *Journal of Social History*, 8: 30–50.

1977. *Social Change and Scientific Organization: the Royal Institution, 1799–1844*. London.

Bernstein, Henry T. 1960. *Steamboats on the Ganges: An Exploration in the History of India's Modernization through Science and Technology*. Bombay.

Bickmore, Albert S. 1868. 'Sketch of a journey through the interior of China, from Canton to Hankow', *JRGS*, 38: 50–68; *PRGS*, 12,1867–68: 51–54.

Blackney, William. 1860. 'Ascent of the Yang-tze-kiang', *JRGS*, 30: 93–100.

Blaikie, W. G. 1910. *The Life of David Livingstone*. London.

Blainey, Geoffrey. 1969. *The Rush That Never Ended: a History of Australian Mining* (reprint ed.). Melbourne.

1970. 'A theory of mineral discovery: Australia in the nineteenth century', *Economic History Review*, 2nd ser., 23: 298–313.

Blanc, Henry. 1869. 'From Metemma to Damot, along the western shores of the Tana Sea', *JRGS*, 39: 36–50; *PRGS*, 13, 1868–69: 39–51.

Blanford, William. 1869. 'On the geology of a portion of Abyssinia', *QJGS*, 25: 401–406.

1870a. 'Notes on a journey in northern Abyssinia', *RBA 1869,TS*: 159–160.

1870b. *Observations on the Geology and Zoology of Abyssinia, Made during the Progress of the British Expedition to that Country in 1867–8*. London.

1878. *Scientific Results of the Second Yarkand Expedition: Geology*. Calcutta.

Blanpied, William. 1973. 'Notes for a study on the early scientific work of the Asiatic Society of Bengal', *Japanese Studies in the History of Science*, 12: 121–144.

'Boundaries between British Guiana and Venezuela'. 26 Nov. 1857. *Foreign Office Confidential Print*.

Bowen, George F. 1889. *Thirty Years of Colonial Government; a Selection from the Despatches and Letters of the Right Hon. Sir George Bowen*, ed. S. Lane-Poole. 2 vols. London.

Branagan, D. F. 1985. 'Phillip Parker King: colonial anchor man', in A. Wheeler and J. Price (eds.), *From Linnaeus to Darwin*, pp. 179–193. London.

1986. 'Strzelecki's geological map of southeastern Australia; an eclectic synthesis', *Historical Records of Australian Science*, 6: 375–392.

Brannigan, Augustine. 1981. *The Social Basis of Scientific Discoveries*. Cambridge.

Bridges, R. C. 1963a. 'The British exploration of East Africa, 1788–1885, with specific reference to the activities of the Royal Geographical Society'. Ph.D. thesis. University of London.

1963b. 'The R.G.S. and the African Exploration Fund, 1876–80', *Geographical Journal*, 129: 25–35.

1973. 'Europeans and East Africans in the Age of Exploration', *Geographical Journal*, 139: 220–232.

Brockway, Lucile. 1979. *Science and Colonial Expansion: the Role of the British Royal Botanical Gardens*. London.

Brooke, James. 1838. 'Proposed exploring expedition to the Asiatic Archipelago', *JRGS*, 8: 443–448.

    1852. 'On the geography of the northern portion of Borneo', *RBA 1851, TS*:89–90.

Brown, R. 1868–69. 'On the physical geography of the Queen Charlotte Islands', *PRGS*, 13: 381–392.

Browne, C. A. (ed.). 1871. *Letters and Extracts from the Addresses and Occasional Writings of J. Beete Jukes*. London.

Browne, Janet. 1983. *The Secular Ark: Studies in the History of Biogeography*. New Haven and London.

Brunner, Thomas. 1850. 'Journal of an expedition to explore the interior of the middle island of New Zealand', *JRGS*, 20: 344–378.

Buckland, William. 1840. 'Address to the Geological Society', *PGS*, 3: 210–267.

    1841. 'Address delivered on the anniversary', *PGS*, 3: 469–540.

Bullock, Charles. 1864. 'Notes to accompany some fossils from Japan', *QJGS*, 20:44–45.

Burman, Jose. 1968. 'Andrew Geddes Bain', in W. J. de Kock (ed.), *Dictionary of South African Biography*, i: 35–38. Cape Town.

Burnett, Robert. 1936. 'The life and work of Sir James Hector, with special reference to the Hector Collection'. M.A. thesis. University of Otago.

Burton, Peggy. 1965. *The New Zealand Geological Survey, 1865–1965*. Wellington.

Cage, R.A, (ed.) 1985. *The Scots Abroad: Labour, Capital, Enterprise, 1750–1914*. London.

Cain, P. J., and A. G. Hopkins. 1986–87. 'Gentlemanly capitalism and British expansion overseas I: the old colonial system, 1688–1850; II: new imperialism, 1850–1945', *Economic History Review*, 2nd ser., 39: 501–525, and 40: 1–26.

Calvert, John. 1853a. *The Gold Rocks of Great Britain and Ireland, and a General Outline of the Gold Regions of the World*. London.

    1853b. 'Mineralogy of Australia', *The Mining Journal*, 23: 580–581.

Campbell, George. 1866–67. 'On the geography and climate of India, in reference to the best site for a capital', *PRGS*, 11: 54–77.

Cannon, Susan Faye. 1978. *Science in Culture: the Early Victorian Period*. New York.

Carr, D. J., and S. G. M. Carr (eds.). 1981. *Plants and Man in Australia*. London.

Casada, James A. 1974. 'James A. Grant and the Royal Geographical Society', *Geographical Journal*, 140: 245–253.

Cawood, John. 1979. 'The magnetic crusade: science and politics in early Victorian Britain', *Isis*, 70: 493–518.

Chesney, Francis R. 1850. *The Expedition for the Survey of the Euphrates and Tigris, Carried On by Order of the British Government, in the Years 1835, 1836, and 1837*. 2 vols. London.

    1853. 'Observations on the Euphrates line of communication with India', *RBA 1852,TS*: 104–110.

Chesney, Francis R. and William Ainsworth. 1837. 'A general statement of the labours and proceedings of the expedition to the Euphrates, under the command of Colonel Chesney, R.A., F.R.S.', *JRGS*, 7: 411–439.

Chikhachëv, Pëtre. 1849. 'Notice of researches in Asia Minor', *QJGS*, 5: 360–362.

    1870. 'On Central Asia', *RBA 1869, TS*: 168–172.

Chorley, R. J., A. J. Dunn and R. P. Beckinsale. 1964. *The History of the Study of Landforms or the Development of Geomorphology. vol.1*. London.

Clarendon, Earl of. 1846. 'The Royal College of Chemistry', *London Medical Gazette*, new ser., 2: 1009–1011.

Clarke, William B. 1852. 'On the discovery of gold in Australia', *QJGS*, 8:131–134.
1860. *Researches in the Southern Gold Fields of New South Wales*. London.
1867–68. 'The auriferous and other metalliferous districts of northern Queensland', *PRGS*, 12: 138–144.

Cobbold, R. H. 1856. 'On the occurrence of coal near the city of E-U in China', *QJGS*,12: 358–359.

Collingwood, C. 1868. 'On some sources of coal in the eastern hemisphere, namely Formosa, Labuan, Siberia, and Japan', *QJGS*, 24: 98–102.

Collins, William F. 1918. *Mineral Enterprise in China*. London.

Collinson, John. 1867–68. 'Explorations in Central America, accompanied by surveys and levels from Lake Nicaragua to the Atlantic Ocean', *PRGS*, 12: 25–48.

Conybeare, William, D. 1833. 'Report on the progress, actual state, and ulterior prospects of geological science', *RBA 1831–32*: 365–414.

[Conybeare, William D.]. 1846. Review of R.I. Murchison *et al.*, *The Geology of Russia in Europe, Quarterly Review*, 77: 348–380.

Cooper, T. T. 1869–70. 'Travels in western China and eastern Thibet', *PRGS*, 14: 335–346, 353–356.
1870–71. 'On the Chinese province of Yunnan and its borders', *PRGS*, 15: 163–174.

'Copies of correspondence between the Colonial Office and the Hudson's Bay Company since 1st January 1862, relative to a road and telegraph from Canada to British Columbia, and the transfer of the property and rights of the Hudson's Bay Company to other parties'. 1863. *Parliamentary Papers*, XXXVIII (438.): 81–102.

'Copies of the reports of 1846, 1848, and 1850, of committees of the Legislative Councils of New South Wales, New Zealand, and the Mauritius, on the subject of steam communication with Europe'. 1851. *Parliamentary Papers*, XXXV (349.): 123–222.

'Copy of Dr. M'Clelland's report on the coal fields of India'. 1863. *Parliamentary Papers*, LXV (372.): 289–473.

'Copy or extracts of any correspondence between the Colonial Office and the authorities in Canada and British Columbia, on the subject of the proposed telegraphic communication between Canada and the Pacific'. 1864. *Parliamentary Papers*, XLI (402.): 115–132.

'Correspondence on the subject of the discovery of metals in Namaqualand'. 1854. *Cape Colony Parliamentary Report:* A.39.

'Correspondence on the subject of the geological survey of the colony'. 1868. *Select Documents Presented to the Legislative Council, Natal, 1866–74:* Document No. 17.

'Correspondence respecting the demarcation of the frontier between Turkey and Persia'. 1865. *Parliamentary Papers*, LVII (3504.): 829–847.

Cotton, Arthur. 1867. 'On a communication between India and China by the line of the Burhampooter and Yang-tsze', *JRGS*,37: 231–239; *PRGS*, 11, 1866–67: 255–259.

Crawfurd, John. 1853. 'A sketch of the geography of Borneo', *JRGS*, 23: 69–86.

Crespigny, C. de. 1855–56. 'Proposed exploration of Borneo', *PRGS*, 1: 205–209. 1857–58. 'Notes on Borneo', *PRGS*, 2: 342–350.

Crosby, Alfred W. 1986. *Ecological Imperialism: the Biological Expansion of Europe, 900–1900*. Cambridge.

Crosse, Mrs Andrew [Cornelia]. 1892. *Red-Letter Days of My Life*. 2 vols. London.

Cullen, Edward. 1851. 'Gold mines of the Isthmus of Darien, emigration to New Granada, and canalization of the Isthmus of Darien', *RBA 1850, TS*: 79.

Cumpston, J. H. L. 1972. *Augustus Gregory and the Inland Sea*. Canberra.

Daintree, Richard. 1872. 'Notes on the geology of the colony of Queensland', *QJGS*, 28: 271–360.

Darragh, Thomas A. 1987. 'The Geological Survey of Victoria under Alfred Selwyn, 1852–1868', *Historical Records of Australian Science*, 7: 1–25.

Davies, Gordon L. 1969. *The Earth in Decay: a History of British Geomorphology, 1578–1878*. London.

Davis, John F. 1856–57. 'Memoir on the neighbourhood of Canton and Hong Kong, and the east coast of China', *PRGS*, 1: 330–341.
    1858. 'On China, in more immediate reference to pending operations in that quarter', *RBA 1857, TS*: 129.
    1858–59. 'View of the great valley of the Yang-tse-keang before and since its occupation by rebels', *PRGS*, 3: 164–171.

Davison, Simpson. 1861. *The Discovery and Geognosy of Gold Deposits in Australia* (2nd edition).

De la Beche, Henry, and Lyon Playfair. 1848. 'First report on coals suited to the steam navy', *Memoirs of the Geological Survey of Great Britain, and of the Museum of Practical Geology*, 2 (pt.2): 539–630.

Denison, William. 1870. *Varieties of a Vice-Regal Life*. 2 vols. London.

Derry, Duncan. 1980. *A Concise World Atlas of Geology and Mineral Deposits*. London.

Desmond, Adrian. 1985. 'The making of institutional zoology in London, 1822–1836', *History of Science*, 23:153–185, 223–250.

'Despatch and report on the subject of the consumption and supply of coal in Natal'. 1867. *Select· Documents Presented to the Legislative Council, Natal, 1866–74:* Document No. 35.

Dieffenbach, Ernst. 1841. 'An account of the Chatham Islands', *JRGS*, 11: 195–215.
    1843. *Travels in New Zealand, with Contributions to the Geology, Geography, Botany and Natural History of that Country*. 2 vols. London.
    1846. 'On the geology of New Zealand', *RBA 1845, TS*: 50.

Dionne, Russell, and Roy MacLeod. 1979. 'Science and policy in British India, 1858–1914: perspectives on a persisting belief', *Proceedings of the Sixth European Conference on Modern South Asian Studies, Colloques Internationaux du C.N.R.S.* Paris.

Drew, Fredrick. 1875. *The Jummoo and Kashmir Territories: A Geographical Account*. London.

Duckworth, H. 1862. 'New commercial route to China', *RBA 1861, TS*: 194–195.

Duthie, John L. 1983. 'Sir Henry Creswicke Rawlinson and the art of Great Gamesmanship', *Journal of Imperial and Commonwealth History*, 11: 253–274.

Dutton, Geoffrey. 1977. *Edward John Eyre: the Hero as Murderer* (reprint ed.). Harmondsworth, Middlesex.

Earl, George Windsor. 1845. 'On the physical structure and arrangement of the islands of the Eastern Archipelago' *JRGS*, 15: 358–365.

Elie de Beaumont, J. B. 1831. 'Researches on some of the revolutions which have taken place on the surface of the globe', *Philosophical Magazine*, new ser., 10: 241–264.

Elster, John. 1983. *Explaining Technical Change: a Case Study in the Philosophy of Science.* Cambridge.

Emery, F. V. 1984. 'Geography and imperialism: the role of Sir Bartle Frere (1815–84)', *Geographical Journal*, 150: 342–350.

Erskine, Vincent. 1869. 'Journey of exploration to the mouth of the River Limpopo', *JRGS*, 39: 233–276; *PRGS*, 13, 1868–69: 320–342.

Eyles, V. A. 1971. 'Roderick Murchison, geologist and promoter of science', *Nature*, 234: 387–389.

Farber, Paul L. 1982. *The Emergence of Ornithology as a Scientific Discipline: 1760–1850.* Dordrecht.

Fermor, Lewis. 1951. 'Geological Survey of India', *Nature*, 167: 10–12.

Fieldhouse, David. 1981. *Colonialism 1870–1945: an Introduction.* London.

Fitton, William H. 1856–57. 'On the structure of north-western Australia', *PRGS*, 1: 501–503.

Fitzpatrick, Kathleen. 1949. *Sir John Franklin in Tasmania, 1837–43.* Melbourne.

Fitzroy, Robert. 1850. 'Considerations on the Great Isthmus of Central America', *JRGS*, 20: 161–189.

  1853. 'Further considerations on the Great Isthmus of Central America', *JRGS*, 23: 171–190.

Fleming, Andrew. 1853a. 'On the Salt Range of the Punjaub', *QJGS*, 9: 189–200.

  1853b. 'On the geology of part of the Sooliman Range', *QJGS*, 9: 346–349.

Fleming, C. A. 1987. 'Some nineteenth century trans-Tasman influences in geology', *Australian Journal of Earth Sciences*, 34: 261–277.

Flett, J. S. 1937. *The First One Hundred Years of the Geological Survey of Great Britain.* London.

Flint, John. 1974. *Cecil Rhodes.* London.

Foote, George. 1954. 'Science and its functions in early nineteenth century England', *Osiris*, 11: 438–454.

Forbes, Charles. 1855. 'On the geology of New Zealand; with notes on its Carboniferous deposits', *QJGS*, 11: 521–530.

Forbes, David. 1861a. 'On the geology of Bolivia and southern Peru', *QJGS*, 17: 7–62.

  1861b. *Correspondence with Lord John Russell, and Memoranda Relating to the Appointment of a Representative of H.M. Govt. in Bolivia.* London.

  1866. 'On the existence of gold-bearing eruptive rocks in South America which have made their appearance at two very distinct geological periods', *RBA 1865, TS*: 52–53.

Forbes Edward. 1852. 'On the discovery by Dr. Overweg of Devonian rocks in North Africa', *RBA 1851, TS*: 58

[Forbes Edward]. 1854. Review of R. I. Murchison, Siluria, *Literary Gazette*, no. 1953: 581–582.

Forster, E. M. 1941. 'The new disorder', *Horizon*, 4: 379–380.

Forsyth, Thomas Douglas. 1868–69. 'On the transit of tea from North-west India to Eastern Turkestan', *PRGS*, 13: 198–200.

  1870. 'Trade routes between Northern India and Central Asia', *RBA 1869, TS*: 161–162.

1871. 'Letter on Eastern Turkestan', *RBA 1870, TS*: 169–170.

Foster, C. Le Neve. 1869. 'On the existence of Sir Walter Raleigh's El Dorado', *QJGS*, 25: 336–343.

Fox, Cyril. 1947. 'The Geological Survey of India, 1846 to 1947', *Nature*, 160: 889–891.

Frere, Henry Bartle. 1870. 'Address of the President of the Geography Section', *RBA 1869, TS*: 152–153.

'Further papers relative to the recent discovery of gold in Australia'. 1852–53. *Parliamentary Papers*, LXIV (1684.): 465–700.

'Further papers relative to the exploration by the expedition under Captain Palliser'. 1860. *Parliamentary Papers*, XLIV (2732.): 427–507.

Gallagher, John, and Ronald Robinson. 1953. 'The imperialism of free trade', *Economic History Review*, 2nd ser., 6: 1–15.

Galton, Francis. 1852. 'Recent expedition into the interior of South-western Africa', *JRGS*, 22: 140–163.

Galvin, Robert James. 1959. 'Palmerston's policy towards east and west Africa, 1830–1865'. Ph.D. thesis. Cambridge University.

Geikie, Archibald. 1875. *The Life of Sir Roderick I. Murchison, based on his Journals and Letters with Notices of his Scientific Contemporaries and a Sketch of the Rise and Growth of Palaeozoic Geology in Britain*. 2 vols. London.

'Geological Survey'. 1859–60. *Victoria. Votes and Proceedings of the Legislative Assembly*, i: 1209–1216.

Gilbert, Edmund, and Andrew Goudie. 1971. 'Sir Roderick Impey Murchison, BART, KCB, 1792–1871', *Geographical Journal*, 137: 505–511.

Gisborne, Lionel, and Mr Ford. 1853. 'Recent survey for a ship canal through the Isthmus of Central America', *RBA 1852, TS*: 110.

Gleason, John H. 1950. *The Genesis of Russophobia in Great Britain: a Study in the Interaction of Policy and Opinion*. Harvard, Mass.

Godwin-Austen, H. H. 1864a. 'Geological notes on part of the North-western Himalayas', *QJGS*, 20: 383–388.

1864b. 'On the glaciers of the Mustakh Range', *JRGS*, 34: 19–56; *PRGS*, 8, 1863–64: 34–42.

1867. 'Notes on the Pangong Lake district of Ladakh, from a journal made during a survey in 1863', *JRGS*, 37: 343–363; *PRGS*, 11, 1866–67: 32–33; also in *RBA 1866, TS*: 100.

Geotzmann, William. 1966. *Exploration and Empire: the Explorer and the Scientist in the Winning of the American West*. New York.

Goldsmid, Frederick J. 1863. 'Diary of proceedings of the mission into Mekran for political and survey purposes', *JRGS*, 33: 181–214; *PRGS*, 7, 1862–63: 91–95.

Gordon, Lieut. 1849. 'Observation on coal in the northeast part of the island of Formosa', *JRGS*, 19: 22–25.

Gough, Barry M. 1971. *The Royal Navy and the Northwest Coast of North America, 1810–1914: a Study of British Maritime Ascendancy*. Vancouver.

Gould, Charles. 1862. 'Results of the Geological Survey of Tasmania', *RBA 1861, TS*: 112–113.

Gowing, Margaret, 1978. 'Science, technology and education: England in 1870',. *Oxford Review of Education*, 4: 3–17.

Graham, Gerald S. 1967. *Great Britain in the Indian Ocean: a Study of Maritime Enterprise, 1810–1850*. Oxford.

1978. *The China Station: War and Diplomacy, 1830–1860.* Oxford.

Grainger, Elena. 1982. *The Remarkable Reverend Clarke: the Life and Times of the Father of Australian Geology.* Melbourne.

Grant, C. Mitchell. 1863. 'Route from Pekin to St. Petersburg via Mongolia', *JRGS*, 33: 166–177; *PRGS*, 7, 1862–63: 27–35.

Grant, William. 1857. 'Description of Vancouver Island', *JRGS*, 27: 268–320; *PRGS*, 1, 1856–57: 487–490.

1861. 'Remarks on Vancouver Island, principally concerning townsites and native population', *JRGS*, 31: 208–212.

Greenberg, Michael. 1951. *British Trade and the Opening of China, 1800–1860.* Oxford.

Greene, Mott T. 1982. *Geology in the Nineteenth Century: Changing Views of a Changing World.* Ithica.

Gregory, A. C. 1855–56. 'Progress of the North Australian Expedition', *PRGS*, 1: 32–35.

1858. 'Journal of the North Australian Exploring Expedition', *JRGS*, 28: 1–137; *PRGS*, 1, 1856–57: 490–501.

Gregory, A. C., and F. T. Gregory, 1884. *Journals of Australian Explorations, 1846–1858.* Brisbane.

Gregory, F. T. 1858–59. 'Exploration of the Murchison, Lyons, and Gascoyne Rivers in Western Australia', *PRGS*, 3: 34–54.

1861. 'On the geology of a part of Western Australia', *QJGS*, 17: 475–483.

1862. 'Expedition to the north-west coast of Australia', *JRGS*, 32: 372–429; *PRGS*, 6, 1861–62: 54–59.

Groot, Cornelius de. 1863. 'Notes on the mineralogy and geology of Borneo and the adjacent islands', *QJGS*, 19: 513–517.

Guntau, Martin. 1978. 'The emergence of geology as a scientific discipline', *History of Science*, 16: 280–290.

Gutzlaff, Karl. 1848. 'Geography of the Cochin-Chinese empire', *JRGS*, 18: 84–143.

Haast, Heinrich F. von, 1948. *The Life and Times of Sir Julius von Haast.* Wellington.

Haast, Julius von. 1864. 'Notes on the mountains and glaciers of the Canterbury Province, New Zealand', *JRGS*, 34: 87–96; *PRGS*, 8, 1863–64: 56–58.

1865. 'Notes on the causes which have led to the excavation of deep lake-basins in hard rocks in the Southern Alps', *QJGS*, 21: 130–132.

1870. 'Notes to accompany the topographical map of the Southern Alps, in the Province of Canterbury, New Zealand', *JRGS*, 40: 433–441.

Hall, D. H. 1976. *History of the Earth Sciences During the Scientific and Industrial Revolutions.* Oxford.

Hall, Marie Boas. 1984. *All Scientists Now: the Royal Society in the Nineteenth Century.* Cambridge.

Hall, Philip B. 1987. 'Robert Swinhoe (1836–1877): a Victorian naturalist in treaty port China', *Geographical Journal*, 153: 37–47.

Hamilton, William J. 1842. *Researches in Asia Minor, Pontus, and Armenia.* London.

Hardy, Phillip Dixon (ed.). 1835. *Proceedings of the Fifth Meeting of the British Association for the Advancement of Science, Held in Dublin.* Dublin.

Hargraves, Edward, 1863–64. 'Report on the non-auriferous character of Western Australia', *PRGS*, 8: 32–34.

Harrington, Bernard, 1883. *The Life of Sir William E. Logan.* London.

Hattersley, A. F. 1968. 'Peter Cormac Sutherland', in W. J. de Kock (ed.), *Dictionary of South African Biography*, i: 281–282. Cape Town.

Havard, W. L. 1941. 'Sir Paul Edmund de Strzelecki', *Journal and Proceedings of the Royal Australian Historical Society*, 26: 20–97.

Hayward, George W. 1870. 'Journey from Leh to Yarkand and Kashgar, and exploration of the sources of the Yarkand River', *JRGS*, 40: 33–166; *PRGS*, 14, 1869–70: 41–74.

1871. 'Letters from Mr. G. W. Hayward on his explorations in Gilgit and Yassin', *JRGS*, 41: 1–46; *PRGS*, 15, 1870–71: 10–22.

Headrick, Daniel. 1981. *The Tools of Imperialism: Technology and European Imperialism in the Nineteenth Century*. Oxford.

Heaphy, Charles. 1854. 'On the Coromandel gold-diggings, New Zealand', *QJGS*, 10: 322–324.

Hector, James. 1861. 'On the geology of the country between Lake Superior and the Pacific Ocean visited by the government exploring expedition under the command of Captain J. Palliser (1857–60)', *QJGS*, 17: 387–445.

1864. 'Expedition to the west coast of Otago, New Zealand; with an account of the discovery of a low pass from Martin's Bay to Lake Wakatipu', *JRGS*, 34: 96–111; *PRGS*, 8, 1863–64: 46–50.

1865a. 'On the geology of Otago, New Zealand', *QJGS*, 21: 124–128; also in *RBA 1864, TS*: 54–55.

1865b. 'Letter to Murchison (origin of rock basins in New Zealand)', *The Geological Magazine*, 2: 377–378.

1879. *Handbook of New Zealand*. Wellington.

1886. *Outlines of New Zealand Geology*. Wellington.

Helly, Dorothy. 1969. ' "Informed" opinion on tropical Africa in Great Britain, 1860–1890', *African Affairs*, 68: 195–217.

Henderson, Thomas, and S. A. Paynter. 1850. 'Reports on coal formations in the Straits of Magellan', *JRGS*, 20: 151–153.

Heney, Helen M. E. 1961. *In a Dark Glass: the Story of Paul Edmund Strzelecki*. London.

Henwood, William J. 1841. 'Notes to accompany a series of specimens from Chaleur Bay and the River Ristigouche in New Brunswick', *PGS*, 3: 454–456.

Hicks, E. 1853. 'On certain ancient mines., *RBA 1852, TS*: 110–112.

Hippius, M. 1865. 'On Russian trade with Bokhara', *RBA 1864, TS*: 145.

Hoare, Michael. 1969. 'All things are queer and opposite – scientific societies in Tasmania in the 1840s', *Isis*, 60: 198–209.

1974. 'Science and scientific associations in eastern Australia, 1820–1890'. Ph.D. thesis. Australian National University.

Hochstetter, Ferdinand von. 1864. *Geology of New Zealand: Contributions to the Geology of the Provinces of Auckland and Nelson*. Vienna (1st edition: new edition translated and edited by C. A. Fleming, Wellington, 1959).

Hodgson, Christopher P. 1860–61. 'Account of four excursions in the Japanese island of Yesso', *PRGS*, 5: 113–118.

Holmes, W. H., and W. H. Campbell. 1857–58. 'Report of an expedition to explore a route by the rivers Waini, Barama, and Cuyuni, to the gold fields of Caratal, and thence by Upata to the Orinoco', *PRGS*, 2: 154–157.

Home, Roderick W. 1988. *Australian Science in the Making*. Sydney.

Hooykaas, R. 1970. *Catastrophism in Geology, its Scientific Character in Relation to Actualism and Uniformitarianism*. Amsterdam and London.

Howard, Dora. 1931–32. 'The English activities on the north coast of Australia in the first half of the nineteenth century', *Proceedings of the Royal Geographical Society, Australasia, South Australia Branch*, 33: 21–194.

Hutton, Frederick W. 1869. 'Description of Nga Tutura, an extinct volcano in New Zealand', *QJGS*, 25: 13–15.

Huxley, Leonard, 1918. *The Life and Letters of Sir Joseph Dalton Hooker*. 2 vols. London.

Inkster, Ian, and J. B. Morrell (eds.). 1983. *Metropolis and Province: Science and British Culture, 1780–1850*. London.

Isbister, A. K. 1855. 'On the geology of the Hudson's Bay Territories, and of portions of the Arctic and North-western regions of America', *QJGS*, 11: 497–520.

Jackson, Santiago. 1858. 'On routes from Lima to the navigable branches of the Amazon, with notes on eastern Peru as a field for colonization', *RBA 1857, TS*: 145–146.

Jameson, William. 1861. 'Journey from Quito to Cayembe', *JRGS*, 31: 184–189.

Jeal, Tim. 1974. *Livingstone*. 2nd ed., New York.

Jervis, James. 1944. 'Rev. W. B. Clarke, "the Father of Australian Geology" ', *Royal Australian Historical Society, Journal and Proceedings*, 30: 345–458.

Jochmus, Baron August. 1854. 'Notes on a journey into the Balkan, or Mount Haemus, in 1847', *JRGS*, 24: 36–85.

Johns, R. K. (ed.). 1976. *History and Role of Government Geological Surveys in Australia*. Adelaide.

Joll, James. 1977. *Gramsci*. London.

Jordanova, L. G. and Roy Porter (eds.). 1979. *Images of the Earth: Essays in the History of the Environmental Sciences*. Chalfont St Giles.

'The journals, detailed reports, and observations relative to the exploration, by Captain Palliser, . . . of British North America, . . . during the years 1857, 1858, 1859, and 1860'. 1863. *Parliamentary Papers*, XXXIX(3164.): 441–769.

Jukes, Joseph Beete. 1842. *Excursions in Newfoundland*. 2 vols. London.
    1847. *Narrative of the Voyage of H.M.S. Fly*. 2 vols. London.
    1850. *Sketch of the Physical Structure of Australia*. London.

Jukes, Joseph Beete, *et al.* 1852. *Lectures on Gold*. London.

Kelly, F. M. 1856. 'On the connection between the Atlantic and Pacific Oceans, via the Atrato and Truando Rivers', *JRGS*, 26: 174–182; *PRGS*, 1, 1855–56: 63–74.

Kelly, William. 1861–62. 'British Columbia', *PRGS*, 6: 107–111.

Kenwood, A. G., and A. L. Lougheed. 1971. *The Growth of the International Economy, 1820–1960*. London.

Khanikov, Nicholas. 1870. 'On the latitude of Samarcand', *RBA 1869, TS*: 164.

Kiernan, V. G. 1953–55. 'Kashgar and the politics of Central Asia 1868–78', *Cambridge Historical Journal*, 11: 317–342.

Kingsmill, T. W. 1869. 'Notes on the geology of China, with more especial reference to the provinces of the Lower Yangse', *QJGS*, 25: 119–138.

Kumar, Deepak. 1985. 'Science policy of the Raj (1857–1905)'. Ph.D. thesis. University of Delhi.

Lane-Poole, Stanley (ed.). 1885. *The Life of the Late General F. R. Chesney*. London.

1888. *The Life of the Rt. Hon. Stratford Canning, Viscount Stratford de Redcliffe*. 2 vols. London.

Larwood, H. J. C. 1958. 'Science in India before 1850', *British Journal of Educational Studies*, 7: 36–49.

Latour, Bruno. 1985. 'Les "Vues" de L'esprit', *Culture Technique*, 14: 4–29.

Laudan, Rachel. 1987. *From Mineralogy to Geology: the Foundations of a Science, 1650–1830*. Chicago.

Lay, George Tradascent. 1840. 'On part of Borneo proper', *PGS*, 3: 290–291.

Layard, Austen Henry. 1903. *Autobiography and Letters*. 2 vols. London.

Lindsay, William Lauder. 1863a. 'On the geology of the goldfields of Otago, N.Z.', *RBA 1862, TS*: 77–80.

　　1863b. 'On the geology of the gold-fields of Auckland, N.Z.', *RBA 1862, TS*: 80–82.

Lingwood, Peter F. 1985. 'Admiral Sir Edward Belcher (1790–1877): Natural History Catalyst or Catastrophe?' in A. Wheeler and J. Price (eds.), *From Linnaeus to Darwin*, pp. 195–203. London.

Livingstone, David. 1851. 'Second visit to the South African lake, Ngami', *JRGS*, 21: 18–24.

　　1855–57. 'Explorations into the interior of Africa', *JRGS*, 25: 218–237; 26: 78–84; 27: 349–387.

　　1857a. 'Return journey across southern Africa', *RBA 1856, TS*: 113–114.

　　1857b. *Missionary Travels and Researches in South Africa*. London.

　　1859–60a. 'Latest accounts of the Central African Expedition', *PRGS*, 4: 19–29.

　　1959–60b. 'On Lakes Nyinyesi, or Nyassa and Shirwa, in Eastern Africa', *PRGS*, 4: 87–90.

Livingstone, David, and Charles Livingstone. 1865. *A Narrative of an Expedition to the Zambesi and its Tributaries, and of the Discovery of the Lakes Shirwa and Nyassa, 1858–1864*. London.

Livingstone, David N. 1984. 'The history of science and the history of geography: interactions and implications', *History of Science*, 22: 271–302.

Lloyd, Christopher. 1970. *Mr. Barrow of the Admiralty: a Life of Sir John Barrow, 1754–1848*. London.

Lloyd, J. A. 1853. 'The mines of Copiapo', *JRGS*, 23: 196–212; also in *RBA 1852, TS*: 53.

　　1854. 'Report of a journey across the Andes, between Cochabamba and Chimoré, to the westward of the traders' route, with remarks on the proposed communication between Bolivia and the Atlantic via the Amazon', *JRGS*, 24: 259–265.

Lochhead, Elspeth. 1980. 'The emergence of academic geography in Britain in its historical context'. Ph.D. thesis. University of California at Berkeley.

Lockhart, William. 1858. 'The Yang-tse-keang and the Hwang-Ho, or Yellow River', *JRGS*, 28: 288–297; *PRGS*, 2, 1857–58: 201–209.

Loftus, William Kennet. 1855. 'On the geology of portions of the Turko-Persian frontier, and of the districts adjoining', *QJGS*, 11: 247–344.

　　1856. 'Notes of a journey from Baghdad to Busrah, with descriptions of several Chaldaean remains', *JRGS*, 26: 131–153.

Logan, William E. 1842. 'On the coal-fields of Pennsylvania and Nova Scotia', *PGS*, 3: 707–712.

　　1863. *The Geology of Canada*. London.

Lugg, H. C. 1949. *Historical Natal and Zululand.* Pietermaritzburg.

Lyell, Charles. 1843. 'On the coal-formation of Nova Scotia, and on the age and relative position of the gypsum and accompanying marine limestor.es', *PGS*, 4: 184–186.

Lynch, Henry B. 1839. 'Notes on a part of the River Tigris, between Baghdad and Samarrah', *JRGS*, 9: 471–476.

McCartney, Paul J. 1977. *Henry De la Beche: Observations on an Observer.* Cardiff.

MacDonagh, Oliver. 1961. *A Pattern of Government Growth 1800–60: the Passenger Acts and their Enforcement.* London.

MacDonald, John. 1857. 'Proceedings of the expedition for the exploration of the Rewa River and its tributaries, in Na Viti Levu, Fiji Islands', *JRGS*, 37: 232–268.

MacDonnell, Richard. 1864. *Australia: What It Is, and What It May Be.* Dublin.

MacKay, David. 1985. *In the Wake of Cook: Exploration, Science, and Empire, 1780–1801.* London and New York.

McKay, D. V. 1943. 'Colonialism in the French geographical movement', *Geographical Review*, 33: 214–232.

McKerrow, James. 1864. 'Reconnaissance survey of the Lake Districts of Otago and Southland, New Zealand', *JRGS*, 34: 56–82; *PRGS*, 8, 1863–64: 46–50.

McLeod, J. Lyons. 1853. 'On the proposed expedition to ascend the Niger to its source', *RBA 1852, TS*: 112.

MacLeod, Roy. 1975. 'Scientific advice for British India: imperial perceptions and administrative goals, 1898–1923', *Modern Asian Studies*, 9: 343–384.

  1982. 'On visiting the "moving metropolis": reflections on the architecture of imperial science', *Historical Records of Australian Science*, 5: 1–15.

  1983. 'Whigs and savants: reflections on the reform movement in the Royal Society, 1830–48', in Ian Inkster and J. B. Morrell (eds.), *Metropolis and Province*, pp. 55–90. London.

MacLeod, Roy, and Peter Collins (eds.). 1981. *The Parliament of Science: the British Association for the Advancement of Science, 1831–1981.* Northwood.

Mann, Robert. 1869a. 'On the coal-field of Natal', *RBA 1868, TS*: 73.

  1869b. 'The gold-fields of South Africa', *RBA 1868, TS*: 137–138.

Mantell, Gideon. 1848. 'On the fossil remains of birds collected in various parts of New Zealand by Mr. Walter Mantell, of Wellington', *QJGS*, 4: 225–241.

Manten, A. A. 1968. 'Geo-scientific aspects of the discovery, exploration, and development of New Zealand', *Earth Science Review/Atlas*, 4 (4): A228–A252.

Markham, Clements R. 1861. 'The province of Caravaya, in southern Peru', *JRGS*, 31: 190–203; *PRGS*, 5, 1960–61: 224–225.

  1868a. 'The Portuguese expeditions to Abyssinia in the 15th, 16th, and 17th centuries', *JRGS*, 38: 1–12; *PRGS*, 12, 1867–68: 11–19.

  1868b. 'Geographical results of the Abyssinian Expedition, No. 1, No. 2', *JRGS*, 38: 12–25; *PRGS*, 12, 1867–68: 113–119.

  1868c. 'Geographical results of the Abyssinian Expedition, No. 3, No. 4', *JRGS*, 38: 25–49; *PRGS*, 12, 1867–68: 298–301.

  1869. 'On the physical geography of the portion of Abyssinia traversed by the English Expeditionary Force', *RBA 1868, TS*: 138–140.

Markham, J. 1870. 'Notes on a journey through Shantung', *JRGS*, 40: 207–228; *PRGS*, 14, 1869–70: 137–144.

Marshall-Cornwall, James. 1976. *History of the Geographical Club*. Brentford.

May, Daniel T. 1860. 'Journey into the Yóruba and Núpe Countries in 1858', *JRGS*, 30: 212–233; also in *RBA 1860, TS*: 170.

Mayne, R. C., H. S. Palmer, and M. Begbie. 1861. 'British Columbia. Journeys in the districts bordering on the Fraser, Thompson, and Harrison Rivers', *JRGS*, 31: 213–248; *PRGS*, 4, 1859–60: 35–37.

'Memorandum on Reform in Turkey'. 5 March 1857. *Foreign Office Confidential Print*.

Merivale, Herman. 1863. 'On the utility of colonization', *RBA 1862, TS*: 161–162.

Michie, Alexander. 1859–60. 'Notes of a cruise in the Gulf of Pe-che-li and Leo-tung in 1859', *PRGS*, 4: 58–62.

Middleton, Dorothy. 1986. 'The early history of the Hakluyt Society, 1847–1923', *Geographical Journal*, 152: 217–224.

Mill, Hugh Robert. 1930. *The Record of the Royal Geographical Society, 1830–1930*. London.

Mitford, A. B. 1868. 'Memorandum on the coal-mines of Iwani, island of Jesso, Japan', *QJGS*, 24: 511.

Monteith, William. 1857. 'Notes on the routes from Bushire to Shiraz', *JRGS*, 27: 108–119; *PRGS*, 1, 1856–57: 279–280.

Montgomerie, T. G. 1869. 'Report of the Trans-Himalayan explorations during 1867', *JRGS*, 39: 146–187; *PRGS*, 13, 1868–69: 183–198.

1869–70. 'Report of the Trans-Himalayan explorations made during 1868', *PRGS*, 14: 207–214.

1871. 'Report of "The Mirza's" exploration from Caubul to Kashgar', *JRGS*, 41: 132–193; *PRGS*, 15, 1870–71: 181–204.

Moore, D. T. 1982. 'Geological collectors and collections of the India Museum, London, 1801–79', *Archives of Natural History*, 10 (3): 399–427.

Morgan, E. Delmar. 1869–70. 'Progress of Russian explorations in Turkestan', *PRGS*, 14: 229–234.

Morrell, J. B. 1971. 'Individualism and the structure of British science in 1830', *Historical Studies in the Physical Sciences*, 3: 183–204.

1983. 'Economic and ornamental geology: the Geological and Polytechnic Society of the West Riding of Yorkshire, 1837–53', in Ian Inkster and J. B. Morrell (eds.), *Metropolis and Province*, pp. 231–256. London.

Morrell, J. B., and Arnold Thackray. 1981. *Gentlemen of Science: Early Years of the British Association for the Advancement of Science*. Oxford.

Morrell, W. P. 1940. *The Gold Rushes*. London.

Motley, James. 1853. 'On the geology of Labuan', *QJGS*, 9: 54–57.

Moyal, Ann. See also Mozley. (ed.), 1976. *Scientists in Nineteenth Century Australia*. Sydney.

Moyle, H. J., and C. B. Hillier. 1857. 'Notice of the occurrence of metalliferous ores and of coal in Siam', *QJGS*, 13: 188.

Mozley, Ann. 1965. 'The foundation of the Geological Survey of New South Wales', *Royal Society of New South Wales Proceedings*, 98 (2): 91–100.

Mumford, Lewis. 1944. *The Condition of Man*. London.

Mundy, Godfrey C. 1852. *Our Antipodes*. 3 vols. London.

Murchison, Kenneth. 1940. *Family Notes and Reminiscences*. Rushden.

Murchison, Roderick I. 1829a. 'A geological sketch of the north-western extremity of

Sussex, and the adjoining parts of Hants and Surrey', *TGS*, second ser., 2: 97–107.

1829b. 'On the coal-field of Brora, in Sutherlandshire, and some other stratified deposits in the north of Scotland', *TGS*, second ser., 2: 293–326.

1832. 'Address to the Geological Society', *PGS*, 1: 362–386.

1833. 'Address to the Geological Society', *PGS*, 1: 438–464.

1839a. 'Address', *RBA 1838*: xxxi–xliv.

1839b. *The Silurian System, Founded on a Series of Geological Researches in the Counties of Salop, Hereford, Radnor, Montgomery, Caermarthen, Brecon, Pembroke, Monmouth, Gloucester, Worcester, and Stafford: with Descriptions of the Coal-fields and Overlying Formations.* 2 vols. London.

1840–41. 'Tours in the Russian Provinces', *Quarterly Review*, 67: 344–375.

1842. 'Anniversary address of the president', *PGS*, 3: 637–687.

1843a. 'Anniversary address of the president', *PGS*, 4: 65–151.

1843b. 'A few observations on the Ural Mountains, to accompany a new map of a southern portion of that chain', *JRGS*, 13: 269–278.

1844. 'Address to the Royal Geographical Society', *JRGS*, 14: xlv–cxxviii. (Hereafter cited as 'Address to the RGS . . . ')

1845a. 'Address to the RGS', *JRGS*, 15: xli–cxi.

1845b. 'On Russia and the Ural Mountains', discourse at the Royal Institution reported in *Athenaeum Journal*, no. 920, 15 June 1845: 591–592.

1846. 'A brief review of the classification of the sedimentary rocks of Cornwall', *Transactions of the Royal Geological Society of Cornwall*, 6: 317–326.

1847. 'President's address', *RBA 1846*: xxvii–xliii.

1850a. 'Siberia and California', *Quarterly Review*, 87: 395–434.

1850b. 'On the distribution of gold in the crust and on the surface of the earth', discourse at the Royal Institution reported in *Athenaeum Journal*, no. 1167, 9 March 1850: 265–269.

1850c. 'On the distribution of gold ore in the crust and on the surface of the earth', *RBA 1849, TS*: 60–63.

1852a. 'Address to the RGS', *JRGS*, 22: lxii–cxxvi.

1852b. 'On the anticipation of the discovery of gold in Australia with a general view of the conditions under which the metal is distributed', *QJGS*, 8: 134–136.

1853. 'Address to the RGS', *JRGS*, 23: lxii–cxxxviii.

1854. *Siluria. The History of the Oldest Known Rocks Containing Organic Remains, with a Brief Sketch of the Distribution of Gold Over the Earth.* London.

1856. *Annual Report of the Director-General of the Geological Survey of the United Kingdom, the Museum of Practical Geology, and the School of Science Applied to Mining and the Arts, for the Year 1855.* London. (Hereafter cited as *Annual Report*, (date)).

1857a. *Annual Report, 1856.* London.

1857b. 'Address to the RGS', *JRGS*, 27: xciv–cxcviii.

1858a. *Annual Report, 1857.* London.

1858b. 'Address to the RGS', *JRGS*, 28: cxxiii–ccxviii.

1859a. *Annual Report, 1858.* London.

1859b. 'Address to the RGS', *JRGS*, 29: cii–ccxxiv.

1859c. 'On the commercial and agricultural value of certain phosphatic rocks of the Anguilla Isles, in the Leeward Islands', *Journal of the Royal Agricultural Society*, 20: 31–32.

1860. *Annual Report, 1859*. London.

1861a. *Annual Report, 1860*. London.

1861b. 'Address to the RGS', *JRGS*, 31: cxi–clxxxvi.

1861c. 'Address of the President of the Geography and Ethnology Section', *RBA 1860, TS*: 148–153.

1862a. *Annual Report, 1861*. London.

1862b. 'Address of the President of the Geology Section', *RBA 1861, TS*: 95–108.

1863a. *Annual Report, 1862*. London.

1863b. 'Address to the RGS', *JRGS*, 33: cxiii–cxcii.

1864a. *Annual Report, 1863*. London.

1864b. 'Address to the RGS', *JRGS*, 34: cix–cxciii.

1864c. 'Address of the President of the Geography and Ethnology Section', *RBA 1863, TS*: 126–133.

1864d. 'On the antiquity of the physical geography of inner Africa', *JRGS*, 34: 201–205; *PRGS*, 8: 1863–64: 151–154.

1865a. *Annual Report, 1864*. London.

1865b.'Address to the RGS', *JRGS*, 35: cviii–clxxxvii.

1865c. 'Address of the President of the Geography Section', *RBA 1864, TS*: 130–136.

1865d. 'Notes on communicating the notes and map of Dr. J. Haast upon the glaciers and rock-basins of New Zealand', *QRGS*, 21: 129–130.

1866a. *Annual Report, 1865*. London.

1866b. 'Address to the RGS', *JRGS*, 36: cxviii–cxcvii.

1866c. 'Address of the President of the Geology Section', *RBA 1865, TS*: 41–48.

1867a. *Annual Report, 1866*. London.

1867b, 'Address to the RGS', *JRGS*, 37: cxv–clix.

1867c. *Siluria. A History of the Oldest Rocks in the British Isles and Other Countries; with Sketches of the Origin and Distribution of Native Gold, the General Succession of Geological Formations, and Changes of the Earth's Surface* (4th edition). London.

1868a. *Annual Report, 1867*. London.

1868b. 'Address to the RGS', *JRGS*, 38: cxxxiii–cxcviii.

1869a. *Annual Report, 1868*. London.

1869b. 'Address to the RGS', *JRGS*, 39: cxxxv–cxciv.

1870a. *Annual Report, 1869*. London.

1870b. 'Address to the RGS', *JRGS*, 40: cxxxiii–clxxviii.

1871a. *Annual Report, 1870*. London.

1871b. 'Address to the RGS', *JRGS*, 41: cxlvi–ccx.

1871c. 'Address of the President of the Geography Section', *RBA 1870, TS*: 158–166.

1872. *Annual Report, 1871*. London.

Murchison, Roderick I., and Edouard de Verneuil. 1841. 'On the geological structure of the northern and central regions of Russia in Europe', *PGS*, 3: 398–408.

Murchison, Roderick I., Edouard de Verneuil, and Alexander von Keyserling. 1845. *The Geology of Russia in Europe and the Ural Mountains*. vol 1. London. *Géologie de la Russie d'Europe et des Montagnes d'Oural*. vol 2. Paris.

Neale, Edward St John. 1866. 'On the discovery of new gold-deposits in the district of Esmeraldas, Ecuador', *QJGS*, 22: 543–544.

Newland, Elizabeth. 1983. 'Sir Roderick Murchison and Australia: a case study of

British influence on Australian geological science'. M.A. thesis. University of New South Wales.

Nzemeke, Alexander. 1982. *British Imperialism and African Response: the Niger Valley, 1851–1905: a Case Study of Afro-British Contacts in West Africa*. Paderborn.

O'Conner, J. G., and A. J. Meadows. 1976. 'Specialization and professionalization in British geology', *Social Studies of Science*, 6: 77–89.

Odernheimer, Friedrich. 1855. 'On the geology of part of the Peel River district in Australia', *QJGS*, 11: 399–402.

Oldroyd, David R. 1967. 'Geology in New Zealand prior to 1900'. M.Sc. thesis. University of London.

1972. 'Nineteenth-century controversies concerning the Mesozoic/Tertiary boundary in New Zealand', *Annals of Science*, 29: 39–57.

1973. 'Haast's glacial theories and the opinions of his European contemporaries', *Journal of the Royal Society of New Zealand*, 3: 5–14.

Oliphant, Laurence. 1860. 'Notes of a voyage up the Yang-tse-keang from Wusung to Hankow', *JRGS*, 30: 75–93; *PRGS*, 3, 1858–59: 162–164.

1863. 'A visit to the island of Tsusima', *JRGS*, 33: 178–181; *PRGS*, 7, 1862–63: 61–64.

1865. 'On the Bayones River, Isthmus of Panama', *JRGS*, 35: 142–147; *PRGS*, 9, 1864–65: 276–280.

Orange, A. D. 1972. 'The origins of the British Association for the Advancement of Science', *British Journal for the History of Science*, 5: 152–176.

Osborn, Sherard. 1858–59. 'Notes, geographical and commercial, made during the passage of *H.M.S. Furious*, in 1858, from Shanghai to the Gulf of Pecheli and back', *PRGS*, 3: 55–87.

Osten Sacken, Baron P. R. 1870. 'Expedition to the trans-Naryn country in 1867', *JRGS*, 40: 250–268; *PRGS*, 14, 1869–70: 221–229.

Page, Leroy E. 1976. 'The rivalry between Charles Lyell and Roderick Murchison', *British Journal for the History of Science*, 9: 156–165.

Palgrave, William Gifford. 1864. 'Observations made in central, eastern, and southern Arabia during a journey through that country in 1862 and 1863', *JRGS*, 34: 111–154; *PRGS*, 8, 1863–64: 63–82.

1869. 'On the north-east Turkish frontier and its tribes', *RBA 1869, TS*: 140–141.

Palliser, John and James Hector. 1858–59. 'Discovery by Capt. John Palliser, FRGS, and Dr. Hector, of practicable passes through the Rocky Mountains within the British possessions' *PRGS*, 3: 122–127.

Palliser, John, James Hector, and Mr Sullivan. 1860. 'Latest explorations in British North America', *JRGS*, 30: 268–314; *PRGS*, 4, 1859–60: 228–234.

Palmer, H. S. 1864. 'Remarks upon the geography and natural capabilities of British Columbia, and the condition of its principal gold-fields', *JRGS*, 34: 171–195; *PRGS*, 8, 1863–64: 87–94.

Parkes, Harry. 1856. 'Notes on the Hindu-Chinese nations and Siamese rivers, with an account of Sir John Bowring's mission to Siam', *RBA 1855. TS*: 149–150; *JRGS*, 26: 1856: 71–78.

Pelly, Lewis. 1865. 'A visit to the Wahabee capital, central Arabia', *JRGS*, 35: 169–190; *PRGS*, 9, 1864–65: 293–296.

Pentland, Joseph Barclay. 1827. *Report on Bolivia, 1827* (ed.) J. Valerie Fifer, Royal Historical Society, Camden Fourth Series, 13, 1974: 169–267.

Petherick, John. 1859–60. 'Journey up the White Nile to the Equator, and travels in the interior of Africa, in the years 1857–8', *PRGS*, 4: 39–44.

Pim, Bedford. 1861–62. 'Proposed transit route across Central America, from a new harbour in Nicaragua', *PRGS*, 6: 75–79.

    1868. 'On the mining district of Chontales, Nicaragua', *RBA 1867, TS*: 127–128.

Platt, D. C. M. 1968. *Finance, Trade, and Politics in British Foreign Policy, 1815–1914*. Oxford.

Poole, Henry. 1856. 'Report of a journey in Palestine', *JRGS*, 26: 55–70.

Porter, Roy. 1973. 'The Industrial Revolution and the rise of the science of geology', in M. Teich and R. M. Young (eds.), *Changing Perspectives in the History of Science*, pp. 320–343. London.

    1978. 'Gentlemen and geology: the emergence of a scientific career, 1660–1920', *Historical Journal*, 21 (4): 809–836.

    1982. 'The Natural Science Tripos and the "Cambridge school of geology", 1850–1914', *History of Universities*, 2: 193–216.

Postans, Captain. 1844. 'Routes through Kach'hí Gandará. And an account of the Belúchi' and other tribes in upper Sind'h and Kach'hí', *JRGS*, 14: 193–219.

Pratt, Mary Louise. 1985. 'Scratches on the face of the country; or, what Mr. Barrow saw in the land of the bushmen', *Critical Inquiry*, 12: 119–143.

Prevost, J. C. 1854. 'Official report of the proceedings of the exploring party sent to cross the Isthmus of Darien', *JRGS*, 24: 249–256.

Pyenson, Lewis. 1984. 'Astronomy and imperialism: J. A. C. Oudemans, the topography of the East Indies, and the rise of the Utrecht Observatory, 1850–1900', *Historia Scientiarum*, 26: 39–81.

    1985. *Cultural Imperialism and Exact Sciences: German Expansion Overseas, 1900–1930*. New York.

Ranger, Terence. 1976. 'From humanism to the science of man: colonialism in Africa and the understanding of alien societies', *Transactions of the Royal Historical Society*, 5th ser., 26: 115–141.

Rao, M. A. 1975. *Indian Railways*. New Delhi.

Rawlinson, Henry C. 1856–57. 'Observations on the geography of Southern Persia, with reference to the pending military operations', *PRGS*, 1: 280–299.

    1865. 'The Russians in Central Asia', *Quarterly Review*, 118: 529–581.

    1866. 'Notes on the Russian frontiers in Central Asia', *RBA 1865, TS*: 128.

    1866–67. 'On the recent journey of Mr. W. H. Johnson from Leh, in Ladakh, to Ilchi in Chinese Tartary', *PRGS*, 11: 6–14.

    1868–69. 'On trade routes between Turkestan and India', *PRGS*, 13: 10–23.

    1871. 'Notes on the site of the terrestrial paradise, and early traditions regarding the Oxus River', *RBA 1870, TS*: 172–174.

Reeks, Margaret. 1920. *Register of the Associates and Old Students of the Royal School of Mines, and History of the Royal School of Mines*. London.

Reese, Trevor. 1968. *The History of the Royal Commonwealth Society, 1868–1968*. London.

'Report from the Select Committee on steam navigation to India', 1834. *Parliamentary Papers*, XIV(478.): 389–827.

Richards, G. H. 1869. 'Address of the President of the Geography and Ethnology Section', *RBA 1868, TS*: 121–130.

Ritchie, G. S. 1967. *The Admiralty Chart: British Naval Hydrography in the Nineteenth Century*. London.

Roberts, G. K. 1976. 'The establishment of the Royal College of Chemistry: an investigation of the social context of early-Victorian chemistry', *Historical Studies in the Physical Sciences*, 7: 437–485.

Robinson, Hercules. 1862. 'A letter relating to the journey of Major Sarel, Captain Blakiston, Dr. Barton, and another, who are endeavouring to pass from China to the north of India', *RBA 1861, TS*: 196–197.

Robinson, Ronald, and John Gallagher, with Alice Denny. 1968. *Africa and the Victorians: the Climax of Imperialism* (reprint ed.). Garden City, New York.

Rogers, A. W. 1937. 'The pioneers in South African geology and their works', *Transactions of the Geological Society of South Africa*, annex to 39: 1–139.

Royal Geographical Society. 1854. *Hints to Travellers*. London.

Royle, J. Forbes. 1846. 'On the tin mines of Tenasserim Province', *PGS*, 4: 165–167.

Rubidge, R. N. 1855. 'On the occurrence of gold in the trap dykes intersecting the dicynodon strata of South Africa', *QJGS*, 11: 1–7.

Rudwick, Martin J. S. 1963. 'The foundation of the Geological Society of London: its scheme for co-operative research and its struggle for independence', *British Journal for the History of Science*, 1: 325–355.

1972. *The Meaning of Fossils: Episodes in the History of Palaeontology*. London.

1974a. 'Poulett Scrope on the volcanoes of Auvergne: Lyellian time and political economy', *British Journal for the History of Science*, 7: 205–242.

1974b. 'Roderick Impey Murchison', in Charles C. Gillespie (ed.), *Dictionary of Scientific Biography*, 9, pp. 582–585. New York.

1976. 'The emergence of a visual language for geological science, 1760–1840', *History of Science*, 14: 149–195.

1982. 'Charles Darwin in London: the integration of public and private science', *Isis*, 73: 186–206.

1985. *The Great Devonian Controversy: the Shaping of Scientific Knowledge among Gentlemanly Specialists*. Chicago and London.

Rupke, Nicholaas A. 1983. *The Great Chain of History: William Buckland and the English School of Geology, 1814–49*. Oxford.

Said, Edward. 1979. *Orientalism* (reprint ed.). New York.

St John, Spencer. 1862. 'On the north-west coast of Borneo', *JRGS*, 32: 217–234; *PRGS*, 7, 1861–62: 83–84.

Sandison, D. 1855. 'Notice of the occurrence of coal near the Gulf of Nicomedia', *QJGS*, 11: 476.

Sarel, Henry A. 1862. 'Notes on the Yang-tsze-kiang, from Hank-kow to Ping-shan', *JRGS*, 32: 1–25; *PRGS*, 6, 1861–62: 2–4.

Saunders, Trelawny. 1853. *The Asiatic Mediterranean, and its Australian Port: the Settlement of Port Flinders, and the Province of Albert, in the Gulf of Carpentaria, Practically Proposed* (preliminary edition). London.

1870. 'The Himalayas and Central Asia', *RBA 1869, TS*: 167.

Sawkins, James G. 1869. 'Reports on the Geology of Jamaica', *Memoirs of the Geological Survey of the United Kingdom*. London.

Sawkins, James G. and Charles B. Brown. 1875. 'Reports on the physical, descriptive, and economic geology of British Guiana', *Memoirs of the Geological Survey of the United Kingdom*. London.

Schomburgk, Robert. 1845a. 'Remarks on the geology of British Guiana', *QJGS*, 1: 298–300.

1845b. 'Journal of an expedition from Pirara to the upper Corentyne, and from thence to Demarara', *JRGS*, 15: 1–103.

1846. 'On the Lake Parima, the El Dorado of Sir Walter Raleigh, and the geography of Guiana', *RBA 1845, TS*: 50–51.

1861. 'Boat excursion from Bangkok to Pecha-buri', and 'General report on the trade of Siam', *JRGS*, 31: 302–320; *PRGS*, 4, 1859–60: 211–218.

Seaver, George. 1957. *David Livingstone: His Life and Letters*. London.

'Second progress report of the commissioners appointed to inquire into the mining resources of the colony', *Victoria. Votes and Proceedings of the Legislative Assembly*, 1856–57, IV: 1463–1474.

Secord, James A. 1982. 'King of Siluria: Roderick Murchison and the imperial theme in nineteenth-century British geology', *Victorian Studies*, 25: 413–442.

1986a. *Controversy in Victorian Geology: the Cambrian–Silurian Dispute*. Princeton.

1986b . 'The Geological Survey of Great Britain as a research school, 1839–1855', *History of Science*, 24: 223–275.

Sedgwick, Adam. 1835. [Address delivered 14 Aug. 1835] in P. D. Hardy (ed.), *Proceedings of the Fifth Meeting of the British Association*, p. 118. Dublin.

Seeman, Berthold, 1862. 'Remarks on a government mission to the Fiji Islands', *JRGS*, 32: 51–62; *PRGS*, 6, 1861–62: 96–102.

Shapin, Steven. 1982. 'History of science and its sociological reconstructions', *History of Science*, 20: 157–211.

Sharpe, Daniel and John W. Salter. 1856. 'Description of Palaeozoic fossils from South Africa', *TGS*, 2nd ser., 7(4): 203–225.

Shaw, A. G. L. 1966. *Convicts and Colonies: a Study of Penal Transportation from Great Britain and Ireland to Australia and Other Parts of the British Empire*. London.

Shaw, Robert. 1869–70. 'A visit to Yarkand and Kashgar', *PRGS*, 14: 124–137.

Silver, Lynnette Ramsay. 1986. *A Fool's Gold? William Tipple Smith's Challenge to the Hargraves Myth*. Milton, Queensland.

Sladon, E. 1871. 'Burma: exploration via the Irrawaddy and Bhamo to south-western China', *JRGS*, 41: 257–281; *PRGS*, 15, 1870–71: 343–364.

Smith, Vincent. 1981. *The Oxford History of India* (4th ed.). Delhi.

Smyth, Warington W., Trenham Reeks, and F. W. Rudler. 1864. *A Catalogue of the Mineral Collections in the Museum of Practical Geology*. London.

Spry, Irene M. 1963. *The Palliser Expedition: An Account of John Palliser's British North American Expedition, 1857–1860*. Toronto.

(ed.), 1968. *The Papers of the Palliser Expedition, 1857–1860*. Toronto.

Sprye, Richard. 1860–61. 'Communication with the south-west provinces of China from Rangoon in British Pegu', *PRGS*, 5: 45–47, 50–54.

Stafford, Robert A. 1984. 'Geological surveys, mineral discoveries, and British expansion, 1835–71', *Journal of Imperial and Commonwealth History*, 12: 5–32.

1988a. 'The long arm of London: Sir Roderick Murchison and imperial science in Australia', in R. W. Home (ed.), *Australian Science in the Making*, pp. 69–101. Cambridge.

1988b. 'Roderick Murchison and the structure of Africa: a geological prediction and its consequences for British expansion', *Annals of Science*, 45: 1–40.

1988c. 'Preventing the "Curse of California": advice for English emigrants to the Australian gold-fields', *Historical Records of Australian Science*, 7(3): 215–230.

(forthcoming). 'Annexing the landscapes of the past: British imperial geology in the nineteenth century', in John Mackenzie (ed.), *Imperialism and the Natural World*. Manchester.

Stanger, William. 1843. 'On the geology of some points on the west coast of Africa, and of the banks of the river Niger', *PGS*, 4: 190–193.

Stoddart, D. R. 1980. 'The R.G.S. and the "new geography": changing aims and roles in nineteenth-century science', *The Geographical Journal*, 146: 190–202.

Stokes, Eric. 1959. *The English Utilitarians and India*. Oxford.

Stokes, John Lort. 1851. 'Survey of the southern part of the Middle Island of New Zealand', *JRGS*, 21: 25–35.

1856. 'On steam communication between England, Australia, and the Cape of Good Hope', *JRGS*, 26: 183–188; *PRGS*, 1, 1855–56: 79–82.

Strachey, Henry. 1853. 'Physical geography of western Tibet', *JRGS*, 23: 1–69.

Strachey, Richard. 1851a. 'On the physical geography of the provinces of Kumaon and Garhwal in the Himalaya Mountains, and of the adjoining parts of Tibet', *JRGS*, 21: 57–85; also in *RBA 1851, TS*: 92–94.

1851b. 'On the geology of part of the Himalayan Mountains and Tibet', *QJGS*, 7: 292–310; also in *RBA 1851, TS*: 69–70.

Strangford, Viscountess (ed.). 1869. *A Selection From the Writings of Viscount Strangford on Political, Geographical, and Social Subjects*. 2 vols. London.

Strickland, Hugh, Jr and William J. Hamilton. 1842. 'On the geology of the western part of Asia Minor', *TGS*, 2nd ser., 6: 1–39.

Strzelecki, Paul Edmund de. 1845. *Physical Description of New South Wales and Van Diemen's Land*. London.

Stuart, John MacDouall. 1861. 'Journal of an expedition across the centre of Australia, from Spencer Gulf on the south to Latitude 18″ 47′ on the north', *JRGS*, 31: 65–145; *PRGS*, 5, 1860–61: 55–60, 104–106.

Sutherland, K. 1858. 'Observations on Vancouver Island', *RBA 1857. TS*; 153–154.

Sutherland, Peter C. 1853. 'On the geological and glacial phenomena of the coasts of Davis' Strait and Baffin's Bay', *QJGS*, 9: 296–312.

1855a. 'Remarks on a series of three-hourly meteorological observations made during a passage from London to Algoa Bay from July to October, 1853', *JRGS*, 25: 256–260.

1855b. 'Notes on the geology of Natal, South Africa', *QJGS*, 11: 465–468.

1868. *The Geology of Natal*. London.

1869. 'Note on the auriferous rocks of south-eastern Africa', *QJGS*, 25: 169–171.

1870. 'Notes on an ancient boulder clay of Natal', *QJGS*, 26: 514–517.

Swinhoe, Robert. 1864. 'Notes on the island of Formosa', *JRGS*, 34: 6–18; *PRGS*, 8, 1863–64: 23–28.

1870. 'Special mission up the Yang-tsze-kiang', *JRGS*, 40: 268–285; *PRGS*, 14, 1869–70: 235–243.

Synge, Millington. 1852. 'Proposal for a rapid communication with the Pacific and the east, via British North America', *JRGS*, 22: 174–200.

1862–63. 'Rupert Land, the colony and its limits', *PRGS*, 7: 71–76.

Tabler, Edward C. (ed.). 1963. *The Zambesi Papers of Richard Thornton*. 2 vols. London.

Tarling, Nicholas. 1982. *The Burthen, the Risk, and the Glory, a Biography of Sir James Brooke*. Oxford.

Teich, M., and R. M. Young (eds.). 1973. *Changing Perspectives in the History of Science: Essays in Honour of Joseph Needham*. London.

Thackray, John C. 1972. 'Essential source material of Roderick Murchison', *Journal of the Society for the Bibliography of Natural History*, 6: 162–170.

　　1978. 'R. I. Murchison's *Silurian System* (1839)', *Journal of the Society for the Bibliography of Natural History*, 7: 61–73.

　　1979. 'R. I. Murchison's *Geology of Russia* (1845)', *Journal of the Society for the Bibliography of Natural History*, 8: 421–433.

　　1981. 'R. I. Murchison's *Siluria* (1854 and later)', *Archives of Natural History*, 10:37–43.

Thomson, J. 1858. 'Extracts from a journal kept during the performance of a reconnaissance survey of the southern districts of the province of Otago, New Zealand', *JRGS*, 28: 298–332; *PRGS*, 2, 1857–58: 354–357.

Thornton, A. P. 1953–55. 'Afghanistan in Anglo-Russian diplomacy, 1869-73', *Cambridge Historical Journal*, 11: 204–218.

　　1954. 'British policy in Persia, 1858–90, Pt. I', *English Historical Review*, 69: 554–579.

　　1956. 'The reopening of the Central Asian Question, 1864–9', *History*, 41:122–136.

Thornton, Edward. 1867. 'Report on the existence of a large coal-field in the province of St. Catherine's, Brazil', *QJGS*, 23: 368–387.

Thornton, Richard. 1859. 'On the coal found by Dr. Livingstone at Tete, on the Zambesi, South Africa', *QJGS*, 15: 556.

　　1862.'On the geology of Zanzibar', *QJGS*, 18: 447–449.

　　1864. 'Notes on the Zambesi and Shiré', *JRGS*, 34: 196–199.

　　1865. 'Notes on a journey to Kilima-ndjaro, made in company of the Baron von der Decken', *JRGS*, 35: 15–21; *PRGS*, 9, 1864–65: 15–16.

Vallance, Thomas G. 1975. 'Presidential Address: origins of Australian geology', *Proceedings of the Linnean Society of New South Wales*, 100: 13–43.

　　1981. 'The fuss about coal: troubled relations between palaeobotany and geology', in D. J. and S. G. M. Carr (eds.), *Plants and Man in Australia*, pp. 136–176. London.

Vámbéry, Armin. 1863–64. 'Sketch of a journey through Central Asia to Khiva, Bokhara, and Samarcand', *PRGS*, 3: 267–274.

Vecchi, Vittorio de. 1978. 'Science and government in nineteenth-century Canada.' Ph.D. thesis. University of Toronto.

Vicary, N. 1847. 'Notes on the geological structure of parts of Sinde', *QJGS*, 3: 334–349.

　　1851. 'On the geology of the upper Punjab and Peshaur', *QJGS*, 7: 38–46.

Vogel, Eduard. 1855. 'Mission to Central Africa', *JRGS*, 25: 237–245.

Wall, George P. 1860. 'On the geology of a part of Venezuela and Trinidad', *QJGS*, 16: 460–470.

Wall, George P. and James G. Sawkins. 1860. 'Report on the geology of Trinidad', *Memoirs of the Geological Survey of the United Kingdom*. London.

Wallace, Alfred Russel. 1855–56. 'Notes of a journey up the Sadong River, in north-west Borneo', *PRGS*, 1: 193–205, 206–209.

1905. *My Life: A Record of Events and Opinions*. 2 vols. London.

Wallis, J. P. R. (ed.). 1956. *The Zambesi Expedition of David Livingstone, 1858–1863*. 2 vols. London.

1976. *Thomas Baines: His Life and Explorations in South Africa, Rhodesia and Australia, 1820–1875* (reprint ed.). Cape Town and Rotterdam.

Ward, J. 1834. 'On the geology of Pulo Pinang & the neighbouring islands', *PGS*, 1: 392.

Waugh, Andrew Scott, *et al*. 1869. 'Report of a committee appointed for the purpose of waiting upon the Secretary of State for India to represent the desirability of an exploration being made of the district between the Brahamputra, the upper Irawadi, and the Yang-tse-kiang, with a view to a route being established between the navigable parts of these rivers', *RBA 1868, TS*: 430–431.

Weindling, Paul J. 1979. 'Geological controversy and its historiography: the prehistory of the Geological Society of London', in L. J. Jordanova and Roy Porter (eds.), *Images of the Earth*, pp. 248–271. Chalfont St Giles.

1983. 'The British Mineralogical Society: a case study in science and social improvement', in Ian Inkster and Jack Morrell (eds.), *Metropolis and Province*, pp. 120–150. London.

Wheeler, Alwyne, and James H. Price (eds.). 1985. *From Linnaeus to Darwin: Commentaries on the History of Biology and Geology*. London.

Wheelwright, William. 1861. 'Proposed railway route across the Andes, from Caldera in Chile to Rosario on the Parana, via Cordová', *JRGS*, 31: 155–162; *PRGS*, 4, 1859–60: 45–50.

1867.'On the discovery of coal on the eastern slope of the Andes', *QJGS*, 23: 197.

Wileman, David. 1971. 'Fyodor Petrovich Litke: a review', *The Geographical Journal*, 137: 75–76.

Wilkinson, J. Fenwick. 1868–69. 'Journey through the gold country of South Africa', *PRGS*, 13: 134–137.

Williams, Donovan. 1962. 'C. R. Markham and the introduction of the cinchona tree into British India, 1861', *Geographical Journal*, 78: 431–442.

Williamson, Alexander. 1869. 'Notes on Manchuria', *JRGS*, 39: 1–36; *PRGS*, 13, 1868–69: 26–38.

Wilson, H. E. 1985. *Down to Earth: One Hundred and Fifty Years of the British Geological Survey*. Edinburgh.

Wilson, James. 1858. 'Notes on the physical geography of northwest Australia', *JRGS*, 28: 137–153; *PRGS*, 2, 1857–58: 210–217.

Worboys, Michael. 1980. 'Science and British colonial imperialism, 1895–1940'. D.Phil. thesis. University of Sussex.

Wyld, James. 1853. *Notes on the Distribution of Gold Throughout the World* (3rd ed.: 1st ed., 1852). London.

Young, Allen. 1864–65. 'On Korea', *PRGS*, 9: 296–300.

Yule, Henry. 1872. 'Address by the President of the Geography Section', *RBA 1871, TS*: 162–174.

Zaslow, Morris. 1975. *Reading the Rocks, the Story of the Geological Survey of Canada*. Toronto.

INDEX

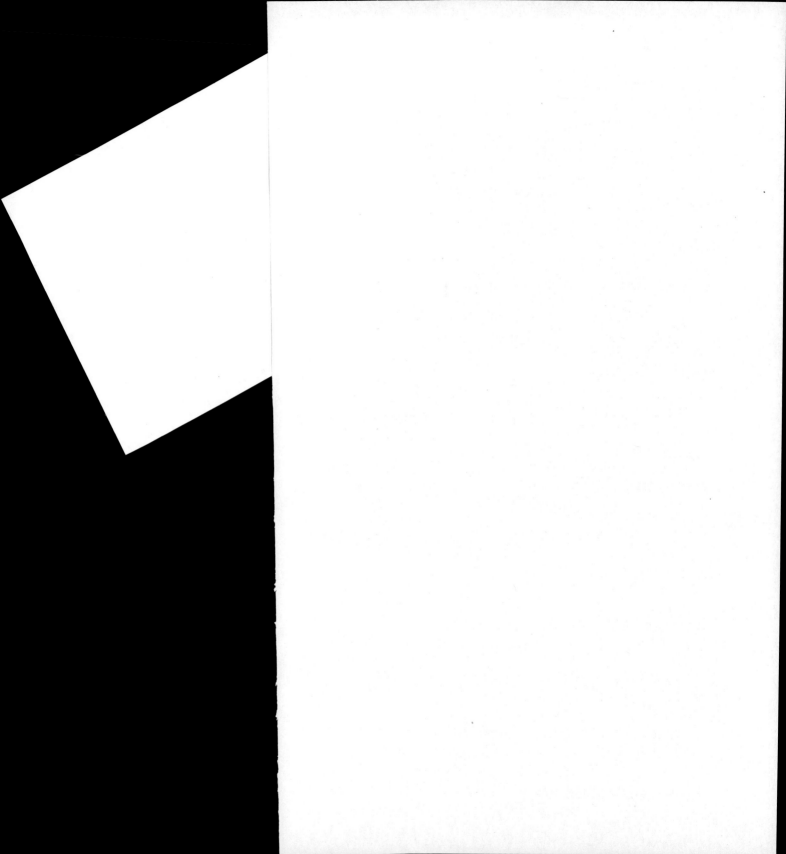

DATE DUE

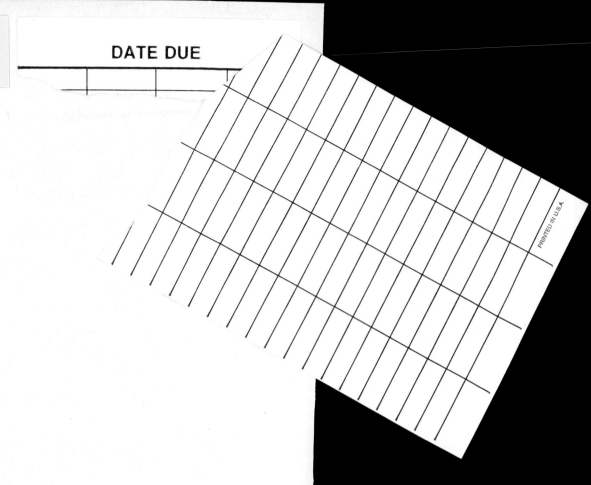

PRINTED IN U.S.A.